PHYSICAL ORIGIN OF HOMOCHIRALITY IN LIFE

PHYSICAL ORIGIN OF HOMOCHIRALITY IN LIFE

Santa Monica, California   February 1995

EDITOR
David B. Cline
University of California, Los Angeles

AIP CONFERENCE
PROCEEDINGS 379

American Institute of Physics   **Woodbury, New York**

L.C. Catalog Card No. 96-86631
ISBN 1-56396-507-0
DOE CONF- 9502169

Printed in the United States of America

CONTENTS

I. BRIEF HISTORICAL INTRODUCTION

II. HOMOCHIRALITY AND LIFE

III. MODELS OF PHYSICAL CHIRAL SYMMETRY BREAKING

PREFACE

For many years there have been several issues associated with the homochiral structure of biomolecules, first observed by L. Pasteur in 1848:

a) Is a homochiral structure necessary for life as we know it?
b) Did homochirality precede the formation of life (homochiral prebiotic medium hopothesis)?
c) Is there any reasonable physical mechanism that could have produced the large chiral symmetry breaking in the prebiotic medium or in the observed homochiral structure?
d) Is the homochiral structure an accident that occurred in the biological systems and was later amplified?
e) Can the homochirality be used as a signature for existing or previous living systems in the solar system or other parts of our galaxy?
f) Are there any experiments that can be carried out now to clarify the origin of homochirality?

These and many other questions were the basis for organizing the first symposium on this subject in Santa Monica, California. Some of the leaders of this field attended the meeting, made presentations, and joined the discussions as well. This note is not meant to be a summary of the meeting, but rather a collection of comments on the discussion at the meeting to accompany this book of the proceedings.

The meeting started with an overview of the importance of homochirality in biomolecular life by Bonner (Stanford) and Goldanskii (Moscow). Their conclusions were that homochirality is almost certainly required for the formation of the complex self-reproducing biomolecular structures (i.e., DNL) required for life. They both stated that the prebiotic medium should have been homochiral; although, they differed on how and even where this may have happened. Mann (U. Penn.) and Cline (UCLA) discussed the possible physical processes that could have helped produce chiral symmetry breaking. Radioactive beta particle interactions to the influence of weak neutral current all "violate parity" but are also very small effects. There was a brief discussion of the role that nearby supernovas in the ISM before the solar system formed could have played. It was agreed that some powerful amplification mechanism must have intervened to achieve any large effect.

On Feburary 16, the first session was "Biomolecular Aspects of Homochirality." There were four very interesting presentations: Goldanskii (Moscow), Gilat (Technion), Avetisov (Moscow), and Miller (UCSD). Goldanskii discussed the Frank process, a process invented by the British chemist Frank that can be used to amplify physical or spontaneous chiral symmetry breaking in so far as no definite chemical reaction has been proposed as an example of the Frank process. Gol'danskii showed that formaldehyde could be assembled in a Frank-type process. Gilat indicated some other types of chiral symmetry breaking processes that should be studied in the laboratory. Avetisov discussed the very interesting question of assembling a complex structure like DNA out of homochiral monomers. He indicated that even with a homochiral prebiotic medium, a "complexity threshold" would be reached that is hard to understand passing through. Part of the final talk in this session by Miller dealt with the famous Miller-Urey experiment that showed many years ago how amino acids can be formed using an electrical discharge in a methane-water, etc. gas medium. In his talk Miller indicated that homochirality is not needed to give birth to biomolecules; however, it is a very great aid to biology and life. Miller also pointed out that the organic materials produced in the famous Miller-Urey experiment were tested and found to be racemic!

The next part of the meeting, "Physical Origin of Homochirality—Symmetry Breaking," took the better part of February 16 and the first part of the morning of February 17 and caused the most controversy. There were three themes to this part:

a) The weak interaction which is the only universal chiral symmetry breaking system–MacDermott (Oxford U.) and Hegstrom (Wake Forest U.).

b) Small effects might be amplified by auto catalytic Frank-type processes–Kondepudi (Wake Forest U.), Gol'danskii, and MacDermott.

c) There may be more useful mechanisms to break the symmetry either by a biochemical process or if the organic molecules are formed in the ISM and subjected to polarized light from neutron stars–Eschenmoser (Zurich), Nielsen (Copenhagen), and Bonner.

Let us elaborate a little on some of the newer results at the meeting. MacDermott has carried out some very interesting calculations of the energy difference on different chirality biomolecular configurations, due to the weak neutral current, and has found that some systems could have $10^{-14}/kT$ energy difference—orders of magnitude larger than all previous estimates. Kondepudi once again emphasized the importance of the Frank process, the auto catalytic process, and has even tested one of the ideas by studying salt crystals where the symmetry is broken by stirring the liquid! Eschenmoser showed how p-RNA could have symmetry breaking effects. Finally, Bonner elaborated on the "abiotic" possibilities of the homochirality and indicated that the time available on the early Earth for such a process to take place was less than 300 million years (and could be even less). He indicated that a more viable possibility could be formation of homochiral molecules in the ISM and proposed a specific mechanism due to the absorption of polarized UV light from one or many neutron stars in the ISM. These ideas and others led to a healthy debate at the meeting. One issue was associated with so-called false chirality due to electric and magnetic fields, discussed by Barron (U. Glasgow), who's talk was presented by MacDermott; similar ideas were discussed by Gilat.

The third session was devoted to "Astrophysical and Planetary Aspects of Homochirality and Origin of Life on Earth." The presentations by Bada (Scripps/UCSD) and Greenberg (Leiden) were very interesting. Bada first showed that biomolecular systems (i.e., human teeth) progressively recemize as they age and indicated that the age of the Swiss Iceman is being determined this way. Then Bada pointed out that there is no evidence for non-racemized organic molecules in micrometeorites in the polar ice or at the K-T edge in sediments. This is a serious constraint on theories where the organic molecules were brought in by comets, etc. Of course, it may not be directly relevant to what happened four billion years ago. The highlights of Greenberg's talk, titled "Photochemical Production of Non Chiral and Chiral Organics in the Interstellar Dust: Laboratory, Observation and Theory," were: organic materials make up approximately 10^{-4} of the mass in the galaxy (as dust); the Halleys Comet carries 10% of the biomass on earth; and organisms in dust undergo complex chemical reactions with the help of UV light (this has been studied in the laboratory where amino acids and other biochemicals are produced). Greenberg's main theme was that homochiral biomolecules could be produced from the polarized light from neutron stars over a long period of evolution of the large gas clouds in the galaxy. Greenberg went on to describe an experiment carried out with polarized UV light where selective destruction of D or L molecules was observed, thus confirming the Bonner hypothesis who's claim is that comets could have brought in both homochiral organisms and water to the early earth approximately 4 billion years ago. (There was also a discussion that a supernova in the ISM could provide a chiral impulse by the anti neutrino and ^{26}Al production).

The final section of the meeting was "Future Perspectives and Experiments." It is clear that some experimental results are needed if this is to be a viable field of study. Khriplovich (Novosibirsk) discussed the possible study of high frequency radiation on homochiral substances; a careful study could yield the energy differences due to the weak neutral current. An open discussion on future experiments was held, with many participants providing a suggestion for an experiment. One interesting possibility suggested using the new high intensity polarized electron guns for linear colliders to study the effects of polarized electrons on homochiral materials. This suggestion and other proposed experiments will be summarized in the proceedings.

All in all, it seemed to be a lively and productive meeting, and there was talk of holding a second meeting in three years or so, most likely in Santa Monica, California.

I wish to thank J. George, M. Wieden, J. Kolonko, and D. Sanders for their help with this meeting and these proceedings.

David Cline

ACKNOWLEDGEMENTS

Acknowledgement is made to the following for permission to reprint copyrighted material:

"Extraterrestrial amino acids and terrestrial life" by C. F. Chyba.
Reprinted with permission from *Nature*, vol **348**, pp. 113-114.
Copyright © 1990 Macmillian Magazines Ltd.

"Carbon isotope composition of individual amino acids in the
Murchison meteorite" by M. H. Engel. Reprinted with permission
from *Nature*, vol **348**, pp. 47-49.
Copyright © 1990 Macmillian Magazines Ltd.

"β Decay and the origins of biological chirality: theoretical results" by R. A. Hegstrom. Reprinted with permission from *Nature*,
vol **297**, pp. 643-647.
Copyright © 1982 Macmillian Magazines Ltd.

"Weak neutral currents and the origin of biomolecular chirality"
by D. K. Kondepudi. Reprinted with permission from *Nature*, vol
314, pp. 438-441.
Copyright © 1985 Macmillian Magazines Ltd.

"Selection of handedness in prebiotic chemical processes" by D.
K. Kondepudi. Reprinted from AIP Conference Proceedings, vol
300, edited by A. K. Mann and D. B. Cline, pp. 491-498.
Copyright © 1994 American Institute of Physics.

Attendees List

Avetisov, Vladik A.: N.N. Semenov Institute of Chemical Physics, Russian Academy of Sciences, Ulitza Kossygina 4, 117977 Moscow, Russia, fax: 7-095 938-2156, phone: 7-095 137-6708

Bada, Jeffrey: University of California San Diego, Scripps Institution of Oceanography, La Jolla, CA 92093, USA, e-mail: jbada@ucsd.edu

Bonner, William A.: Stanford University, Department of Chemistry, Stanford, CA 94305, USA

Chakravarti, Soumya: California State University Pomona, Physics Department, 3801 West Temple Avenue, Pomona, CA 91768, USA, e-mail: schakravarti@csupomona.edu, fax: (909) 869 4396, phone: (909) 869 4006

Cheng, Cheng-Wei: University of California Los Angeles, Dept. of Physics & Astronomy, Box 951547, Los Angeles, CA 90095-1547, USA, fax: (310) 206-1091

Cline, David: University of California Los Angeles, Dept. of Physics & Astronomy, Box 951547, Los Angeles, CA 90095-1547, USA, e-mail: dcline@physics.ucla.edu, fax: (310) 206-1091, phone: (310) 825-1673

Cohen, Jon: Science, 1148 Hymettus Avenue, Leucadia, CA 92024, USA, e-mail: cohenj@cerf.net

Cooper, George: NASA/Ames Research Center, Exobiology, MS-239-4, Moffett Field, CA 94035, USA, e-mail: george_cooper@qmgate.arc.nasa.gov, fax: (415) 604-1088, phone: (415) 604-5968

Crick, Francis: The Salk Institute, Molecular Genetics Laboratory, 10010 North Torrey Pines Road, La Jolla, CA 92037, USA, phone: (619) 453-4100 ext 261

Eschenmoser, Albert: Laboratorium für Organishe Chemie, ETH-Zentrum Universitätsstrasse 16, CH-8092, Zurich, Switzerland, e-mail: kraeutler@czheth5a.bitnet@cearn.cern.ch, fax: 01 63 2 1043, phone: 01 63 2 2893

Gelmini, Graciela: University of California Los Angeles, Dept. of Physics & Astronomy, Box 951547, Los Angeles, CA 90095-1547, USA, e-mail: gelmini@physics.ucla.edu, phone: (310) 825-4293

Gilat, Gideon: TECHNION, Israel Institute of Technology, Department of Physics, Haifa 32000 Haifa, Israel, e-mail: phr02gg@technion.technion.ac.il, fax: 00972-4-221514, phone: 00972-4-330644

Goldanskii, Vitallii: N.N. Smenov Institute of Chemical Physics, Russian Academy of Sciences, Ulitsa Kosygina 4, 117 334 Moscow, Russia

Greenberg, J. Mayo: Huggens Laboratory, Neils Bohr weg 2, P.O. Box 9504, 2300 RA Leiden, Netherlands, e-mail: mayo@rulhl1.leidenuniv.nl

Hegstrom, Roger A.: Wake Forest University, Department of Chemistry, Box 7507, Winston-Salem, NC 27109, USA, e-mail: hegstrom@wfu.edu

Huebner, Walter: NASA Headquarters, Code SL, Wahsington DC 20546, USA, e-mail: walter@swri.space.swri.edu, phone: (202) 453-1413

Khriplovich, Iosif: Institute of Nuclear Physics, CEI-630090 Novosibirsk, Russia, e-mail: khriplovich@inp.nsk.su, phone: 7 38 32 352 163

Kolonko, Jim: University of California Los Angeles, Dept. of Physics & Astronomy, Box 951547, Los Angeles, CA 90095-1547, USA, e-mail: kolonko@physics.ucla.edu, fax: (310) 206-1091, phone: (310) 206-4548

Kondepudi, Dilip K.: Wake Forest University, Department Chemistry, Box 7486, Winston-Salem, NC 27109, USA, e-mail: dilip@wfu.edu, fax: (910) 759-4656, phone: (910) 759-5131 or (910) 759-4541

Laraneta, Melinda: University of California Los Angeles, Dept. of Physics & Astronomy, Box 951547, Los Angeles, CA 90095-1547, USA, e-mail: laraneta@physics.ucla.edu, fax: (310) 206-1091, phone: (310) 206-4549

Lippmann, David: Southwest Texas State University, Department of Chemistry, 601 University Drive, San Marcos, TX 78666, USA, e-mail: lippmann@order.ph.utexas.edu, fax: (512) 245-2374, phone: (512) 245-3119

Lu, Kau U.: California State University Long Beach, Department of Mathematics, Long Beach, CA 90840, USA, e-mail: lunotwen@csulb.edu, fax: (310) 985-8227, phone: (310) 985-9731

MacDermott, Alexandra J.: University of Cambridge, Department of Chemistry, Cambridge CB2 1EW, Great Britian, e-mail: ajm58@cus.cam.ac.uk, fax: +44 (1223) 336-362, phone: +44 (1223) 336-351

Mann, Alfred K.: University of Pennsylvania, Department of Physics, 209 South 33rd Street, Philadelphia, PA 19104-6396, USA, e-mail: edu%"oboyle@penndrls.upenn.edu", phone: (215) 898-8155

Miller, Stanley: University of California, San Diego, Department of Chemistry, 9500 Gilman Dr., La Jolla, CA. 92093-0317, USA, fax: (619) 534-6128, phone: (619) 534-3365

Mohan, Ram: Max Planc Institut für Biophysikalische Chemie, Nikolansberg, Am Faßberg 11, 37077 Göttingen, Germany, e-mail: ram@ictp.trieste.it or ram@karl.dnet.gmdg.de, fax: 49 551 201 467, phone: fax at weisman inst., israel 11-12 972 8-344 112

Nielsen, Peter E.: University of Copenhagen, Dept of Med. Biochem & Geneitcs, Blegdamsvej 3, DK-2200 Copenhagen N, Denmark, fax: 31 39 60 42, phone: 35 32 77 62

Rajpoot, Subhash: California State University Long Beach, Physics Department, Long Beach, CA 90840, USA, e-mail: rajpoots@beach1.csulb.edu

Sanders, David: University of California Los Angeles, Dept. of Physics & Astronomy, Box 951547, Los Angeles, CA 90095-1547, USA, e-mail: sanders@physics.ucla.edu, fax: (310) 206 1091, phone: (310) 825-1214

Saxon, Prof. David: University of California Los Angeles, Dept. of Physics & Astronomy, Box 951547, Los Angeles, CA 90095-1547, USA, e-mail: saxon@physics.ucla.edu

The Origin of Homochirality in Life

February 15-17, 1995 — Holiday Inn BayView Plaza Hotel, Santa Monica, California

Agenda

Wednesday, February 15:

13:00 - 15:00	*Registration*
16:00 - 18:00	Introductory Historical Overview Lectures Chair: D. Cline (UCLA)
16:00 - 16:30	W. Bonner (Stanford University)
	The Quest for Chirality
16:35 - 17:15	V. Goldanskii (N.N. Semenov Institute of Chemical Physics)
	Chirality, Tunneling, and the Prebiotic Evolution
17:20 - 18:00	A.K. Mann (Univ. of Pennsylvania) & D. Cline
	Comments on the Physical Sources of Chiral Symmetry Breaking
18:00 - 19:30	*Welcoming Reception*

Thursday, February 16:

08:00 - 08:30	*Continental Breakfast & Registration*
08:30 - 11:30	Biomolecular Aspects of Homochirality Chair: A.K Mann (U. Penn)
08:30 - 09:05	V. Goldanskii (N.N. Semenov Institute of Chemical Physics)
	Formaldehyde as a Hypothetical Primer of "Biohomochirality"
09:10 - 09:45	G. Gilat (Technion)
	Chiral Interaction, Biomolecular Evolution and Homochirality
09:45 - 10:20	V. Avetisov (N.N. Semenov Institute of Chemical Physics)
	Origin of Homochirality: What, How and When?
10:25 - 10:40	*Coffee Break*
10:40 - 11:10	S. Miller (UCSD) Chair: R. Hegstrom (Wake Forest)
	Homochirality is Not Required for the Origin of Life
11:15 - 19:05	Physical Origin of Homochirality
	Symmetry Breaking
11:15 - 11:45	W. Bonner (Stanford University)
	The Cosmic Connection
11:50 - 12:20	I. Khriplovich (Novosibirsk)
	Weak Interaction Influence on Phase Transition in Helical Ferromagnets
12:25 - 13:30	*Lunch*
13:30 - 14:30	*Break*
14:30 - 15:05	A. MacDermott (Oxford) Chair: V. Avetisov (Moscow)
	The Weak Force and SETH, the Search for Extraterrestrial Homochirality
15:10 - 15:40	D. Kondepudi (Wake Forest)
	Amplification and Propagation of Chiral Asymmetry
15:45 - 16:00	*Coffee Break* Chair: G. Gilat (Technion)
16:00 - 16:40	A. Eschenmoser (ETH Zurich)
	Base-pairing to quasi-Racemates in p-RNA

16:45 - 17:20   P. Nielsen (University of Copenhagen)
    Induced Chirality in PNA Double Helices

17:25 - 18:05   J. Bada (UCSD/Scripps)
    Amino Acid Stereochemistry in Cosmogeochemical Samples and the
    Origin of Amino Acid Homochirality on Earth

19:00 - 19:30   *Reception*

19:30 - 20:30   *Dinner*

Friday, February 17:

08:30 - 09:00   *Continental Breakfast*

09:00 - 10:15   <u>Physical Origin of Homochirality</u>
    <u>Symmetry Breaking</u>   Chair: D. Kondepudi (Wake Forest)

09:00 - 09:40   R. Hegstrom (Wake Forest)
    Electroweak Interactions and the Origins of Biomolecular Homochirality

09:45 - 10:30   A. MacDermott (presenting talk for L. Barron of U. Glasgow)
    False Chirality, CP Violation and the Breakdown of Microscopic Reversibility

10:35 - 10:50   *Coffee Break*

10:50 - 11:10   C.-W. Cheng (UCLA)
    Electronic Simulation of Chiral Symmetry Breaking

11:10 - 13:00   <u>Astrophysical and Planetary Aspects of</u>
    <u>Homochirality - Origin of Life on Earth</u>
    Chair: A. Eschenmoser (Zurich)

11:10 - 12:00   J.M. Greenberg (U. Leiden)
    Photochemical Production of Non-chiral and Chiral Organics in
    Interstellar Dust: Laboratory, Observation, and Theory

12:05 - 12:30   R. Mohan (Max Planck Institute)
    Neutral Weak Current Chemistry

12:35 - 13:00   D. Lippman (U. Texas)
    Symmetry Breaking by Autocatalysis

13:00 - 14:00   *Lunch*

14:00 - 16:00   <u>Future Perspectives and Experiments</u>   Chair: D. Saxon (UCLA)

14:00 - 14:40   I. Khriplovich (Novosibirsk)
    High-frequency Asymptotics of Optical Activity and P-odd Energy
    Difference of Chiral Radicals Dependent on Nuclear Spin

14:40 - 15:00   K. U. Lu (CSU Long Beach)
    Molecular Structures Derived from Deterministic Theory of Atomic Structure

15:05 - 16:00   Discussion of Experiments that Could Clarify the Origin of
    Homochirality

ADJOURN

I. BRIEF HISTORICAL INTRODUCTION

Extraterrestrial amino acids and terrestrial life

Christopher F. Chyba

Since the Swedish chemist Baron Jöns Jacob Berzelius first analysed the Alais meteorite for organic molecules[1] in 1834, attempts to forge a link between extra-terrestrial organic materials and terres-trial life have remained alluring, but often deceptive. Two distinct investigations are demanded in exploring such links: accur-ately characterizing the organics present in extraterrestrial bodies; and evaluating possible mechanisms for delivering these molecules intact to Earth. New studies eported in this and last week's issues[2,3] hold the promise of important advances in both endeavours. The results of Engel *et al.*[2], if upheld, would suggest that the characteristic 'handedness' of biochemis-try on Earth may ultimately have been determined by an asymmetry already existing in extraterrestrial amino acids. The paper on page 157 of this issue[3] by Zahnle and Grinspoon presents one mechanism, the gentle collection of organic-rich dust evolved from an evapor-ating comet, whereby amino acids might in fact reach the Earth without being destroyed. A similar suggestion has been made previously in the prebiotic context[4], but Zahnle and Grinspoon directly con-front the most important datum currently available to test such hypotheses: the apparently extraterrestrial amino acids discovered[5] in abundance in sediments at Stevns Klint, Denmark, marking the Cretaceous–Tertiary (K/T) boundary, the stratum associated with the major extinc-tion 65 million years ago.

Murchison meteorite

The first identifications of amino acids in meteorites failed to survive doubts concerning terrestrial contamination, and were made in the midst of an acrimonious controversy over claimed observations of microfossils and biogenic organic mole-cules in meteorites[6]. In 1970, Kvenvolden *et al.*[7] published the first of a series of studies which unequivocally demonstrated that amino acids in the Murchison meteo-rite, which had fallen the previous year,

were extraterrestrial and probably abio-genic. The evidence was threefold: Murchison's amino acids were nearly racemic (equal) mixtures of the D and L enantiomers (see the figure), whereas terrestrial organisms use almost exclusively the L enantiomer; extractable Murchison

The L and D enantiomers (from the Greek *enantios*: oppo-site) of the amino acid alanine. The two molecules are mirror images; no rotation through three dimensions can super-impose them. The enantiomers rotate the plane of polariza-tion of light in opposite directions; unequal mixtures are therefore optically active. A mixture consisting of equal numbers of each is said to be racemic. Terrestrial life uses almost exclusively the L-amino-acid enantiomers.

organic compounds had $\delta^{13}C$ values (which measure the carbon isotope ratio $^{13}C/^{12}C$) that were greater than those in any naturally occurring terrestrial organics; and Murchison contained non-biological amino acids — those not among the 26 or so commonly found in terrestrial organisms. At least 74 amino acids have now been identified in Murchison extract[8].

There is no difficulty in the non-destruc-tive delivery of organic molecules to Earth in objects as small as the Murchison frag-ments. The bulk of the mass in potential terrestrial impactors, however, is concen-trated in much larger bodies. For extra-terrestrial material to have had a substan-tial influence on prebiotic chemistry, it would seem that this reservoir must somehow be tapped. Yet even objects as small as 100 metres in radius cannot be sufficiently decelerated by the Earth's present atmosphere for organic inclusions to survive the ensuing violent impact[4].

This is one reason why the reported[5] high abundance of the presumably extra-

terrestrial amino acids α-amino-iso-butyric acid (AIB) and isovaline at the K/T boundary is so remarkable. (AIB and isovaline are extremely rare in the biosphere, but common in meteorites; moreover, the K/T isovaline is racemic.) A cosmochemical interpretation of the iridium (Ir) abundance at the boundary implies that it was created by an impactor with a diameter of 10 km; such an object would hit the Earth with a kinetic energy equivalent to 10^8 megatons of explosive, enough to incinerate even the hardiest organics[9]. Yet the AIB/Ir ratio at Stevns Klint is substantially higher than for Murchison. More puzzling still, the amino acids were found tens of centimetres above and below the K/T boundary, but not in the boundary clay itself.

Assuming that these data are not anomalous, how are they to be explained? Zahnle and Grinspoon take the sedimen-tary record at face value, and suggest that Earth swept up the amino acids "gently and non-destructively" in cometary dust over a period of around 10^5 years. Organic molecules in interplanetary dust particles (IDPs) reach the Earth today[3]; a sufficiently large comet trapped in the inner Solar System could have provided the high IDP flux needed to deposit the amino acids around the K/T boundary. The boun-dary clay itself would then record the impact of a fragment of the parent comet, an impact which amino acids did not survive.

Zahnle and Grinspoon cite a report by D. Carlisle of amino acids found within, rather than around, the boundary clay at a Canadian K/T site. These results apparently corroborate the existence of K/T amino acids; it will be important to know if Stevns Klint AIB/Ir ratios are typical or anomalously high. Only the depositional history of the Canadian site, combined with detailed knowledge of the amino acid distribution, will show whether these results are consis-tent with the comet-dust model.

A difficulty that has received too little attention in all IDP organic-delivery schemes is the question of whether IDP atmospheric deceleration is really gentle enough for amino acids to survive. Models predict cometary IDPs to be typically heated to around 900 K for 5–15 seconds during atmospheric entry[10]. A naive appli-cation of thermal decomposition rates for amino acids[9] suggests these molecules would survive around 800 K for 1 s. But the sole data available are for decomposi-tion in solution, and these rates are of dubious relevance to degradation of amino acids in a mineral matrix. Labora-tory rate parameters for a variety of matrices are badly needed to resolve this

Reprinted with permission from *Nature*, vol. 348, pp. 113-114, 8 November 1990. © 1990 Macmillan Magazines Ltd.

question. In any case, the hardier organics[9], after some thermal processing, will certainly survive, and one is free to speculate[3] that surviving amino acid precursors may then be synthesized into the observed products.

One apparent problem with the comet-dust model may be less severe than has been suggested[3]. Data collected by R. Rocchia et al., cited by Officer and Drake[11], show that Ir values at Stevns Klint are correlated with the clay fraction; expressed on a carbonate-free basis, Ir concentrations stay within a factor of two of the peak clay abundance out to 30–40 cm above and below the boundary, and remain above background levels out to around 50 cm. These distances correspond to 10^4–10^5-year depositional times, and suggest that the putative comet dust was delivering Ir as well as amino acids throughout its accretion. (However, high AIB/Ir levels in the dust still appear to be required.) I emphasize that such long-term cometary depositional histories rebut the argument[11] that claimed 10^4–10^5-year timescales for K/T Ir deposition necessarily favours explanations of the K/T catastrophe in terms of extended terrestrial volcanism.

In any case, altogether different explanations for the K/T amino acids are possible. Extensive experimental data show that amino acids can be synthesized by quenched shock heating of reducing gases[12]. It is commonly asserted that such recombination could not have occurred for vaporized impactor material in an O_2-rich background atmosphere[3,5]. In fact, production efficiency remains high even in background air[12], so post-impact shock synthesis is not yet ruled out[9]. A final possibility is organic delivery following catastrophic impactor fragmentation. A 10-km diameter bolide would be too large to be affected by the atmosphere, but if several smaller objects or fragments were responsible for the K/T geochemical signatures, each might have airburst in the atmosphere before hitting the ground. A carbonaceous chondrite did exactly this over Revelstoke, Canada, in 1965; photomicrographs of the millimetre-sized fragments recovered reveal unheated interiors[13] within which organics should have survived[9].

Significant sources

Our preliminary quantitative work[14] suggests that IDPs[4], airbursts and, possibly, quench syntheses, could have been significant sources of prebiotic organics on early Earth. In the first two cases, the report last week by Engel et al.[2] of an excess of L-type alanine in Murchison is potentially of great importance. No convincing explanation yet exists for why terrestrial life uses L, not D, amino acids. Abiotic explanations must either account for L/D ratios which differ from unity by more than statistical fluctuations, or provide an amplification mechanism that proceeds faster than natural racemization. Neither programme has yet been successful[15], and the possibility that the predominance of L-forms is simply an accident of early biology remains a leading alternative. Finding a meteoritic preference for L-amino acids would 'explain' terrestrial biology's choice by pushing the problem out into the cosmos — or it might imply yet-undiscovered abiotic processes that could also have operated on the early Earth.

If the data of Engel et al. are valid, there is a third possibility that must be forthrightly addressed. Non-racemic mixtures of amino acids have long been suggested as indicators of biologically derived molecules[16]. Entire papers at 'exobiology' meetings have been devoted to instrumentation designed to detect optical activity. Does not scientific honesty then require us to consider the new results as prima facie evidence of extraterrestrial life?

Scientific humility answers no. A simple failure to find an abiotic mechanism cannot in itself require an appeal to biology. (Indeed, A. Salam and J. Strathdee (personal communication) have just proposed a low-temperature quantum amplification of the inequal ground-state energies — around 10^{-19} electron volts, due to parity-violating weak nuclear interactions — of L and D amino acid enantiomers, which they argue could create non-racemic mixtures.) Results from the Viking biology packages on Mars are powerful reminders that prejudices about supposedly unambiguous attributes of terrestrial biology may crumble in the face of unanticipated extraterrestrial chemistry. In any case, the complete structural diversity of the Murchison amino acids argues for non-catalytic, thermodynamically-controlled (that is, abiotic) synthesis[6].

Any detection of non-racemic amino acid mixtures must exclude the possibility of contamination due to L-rich terrestrial microorganisms. This concern was raised over an earlier report of non-racemic Murchison amino acids; it is especially worrying that only protein amino acids have been claimed to be non-racemic, and that it is the L enantiomer of alanine that is found in excess[17]. Both these results are, of course, consistent with contamination. The key result of Engel et al.[2] is that the L and D enantiomers of alanine in Murchison have extraterrestrial $\delta^{13}C$ values of +27 and +30 parts per thousand (‰), respectively. A putative terrestrial alanine contaminant, they show, would need an isotopic composition of +10 ‰ to arrive at the observed L-alanine bulk value, and the authors argue that no such terrestrial sources are apparent. (Terrestrial inorganic carbon has $\delta^{13}C$ 0‰; microorganisms typically fractionate carbon isotopes to lighter values, producing organics with $\delta^{13}C$ as low as −40‰.)

Unfortunately, at least one plausible such contaminant does exist. It has recently been shown[18] that a wide variety of common bacteria are able to obtain their carbon and energy needs from abiotically produced 'tholins', laboratory materials that provide good analogues to meteoritic and cometary organics[9]. This raises the possibility of contamination of Murchison by terrestrial bacteria that consume ^{13}C-rich organics, then produce $\delta^{13}C$-rich amino acids. Methane from Murchison, for example, has $\delta^{13}C$ = +9.2‰, and certain other organics have higher values[8]. Moreover, at least one bacterium, Methanobacterium thermoautotrophicum, fractionates (admittedly under special circumstances) carbon to heavier isotopes[19], by as much as +13 ‰. A bacterium capable of such fractionation while metabolizing, say, the Murchison benzene–methanol extract (which includes[8] nonvolatile aromatics and the higher alkanes, and has $\delta^{13}C$ = +5 ‰), could produce the required effect.

Unanticipated chemistry and uncertain contamination have been the bane of 'exobiology'. Occam's razor, the nineteenth century name given to various versions of William of Ockham's fourteenth century maxim, remains a useful warning. The actual formulation was "pluralitas non est ponenda sine necessitate": plurality is not to be posited without necessity. This was meant as a methodological, not an ontological, admonition[20]. Understood in this way, it puts the pursuit of familiar explanations before that of more encompassing ones: the former must be excluded before we may embrace the latter. □

Christopher F. Chyba is in the Laboratory for Planetary Studies, Cornell University, Ithaca, New York 14853, USA.

1. Berzelius, J.J. Ann. phys. Chem. 33, 113–148 (1834).
2. Engel, M.H., Macko, S.A. & Silfer, J.A. Nature 348, 47–49 (1990).
3. Zahnle, K. & Grinspoon, D. Nature 348, 157–160 (1990).
4. Anders, E. Nature 342, 255–257 (1989).
5. Zhao, M. & Bada, J.L. Nature 339, 463–465 (1989).
6. Urey, H.C. Science 151, 157–166 (1966).
7. Kvenvolden, K. et al. Nature 228, 923–926 (1970).
8. Cronin, J.R., Pizzarello, S. & Cruikshank, D.P. in Meteorites and the Early Solar System (eds Kerridge, J.F. & Matthews, M.S.) 819–857 (University of Arizona Press, 1988).
9. Chyba, C.F., Thomas, P.J., Brookshaw, L. & Sagan, C. Science 249, 336–373 (1990).
10. Flynn, G.J. Icarus 77, 287–310 (1989).
11. Officer, C.B. & Drake, C.L. Science 227, 1161–1167 (1985).
12. Barak, I. & Bar-Nun, A. Origin Life 6, 483–506 (1975).
13. Folinsbee, R.E., Douglas, J.A.V. & Maxwell, J.A. Geochim. Cosmochim. Acta 31, 1625–1635 (1967).
14. Chyba, C. & Sagan, C. Bull. Am. astr. Soc. 22, 1097 (1990).
15. Bonner, W.A., Blair, N.E. & Dirbas, F.M. Origin Life 11, 119–134 (1981).
16. Lederberg, J. Nature 207, 9–13 (1965).
17. Bada, J.L. et al. Nature 301, 494–497 (1983).
18. Stoker, C.R. et al. Icarus 85, 241–256 (1990).
19. Fuchs, G., Thauer, R., Ziegler, H. & Stichler, W. Arch. Microbiol. 120, 135–139 (1979).
20. Thorburn, W.M. Mind 27, 345–353 (1918).

4

Carbon isotope composition of individual amino acids in the Murchison meteorite

M. H. Engel*, S. A. Macko† & J. A. Silfer*

* School of Geology and Geophysics, 100 E Boyd Street, University of Oklahoma, Norman, Oklahoma 73019, USA
† Department of Environmental Sciences, University of Virginia, Charlottesville, Virginia 22903, USA

A SIGNIFICANT portion of prebiotic organic matter on the early Earth may have been introduced by carbonaceous asteroids and comets[1]. The distribution and stable-isotope composition of individual organic compounds in carbonaceous meteorites, which are thought to be derived from asteroidal parent bodies, may therefore provide important information concerning mechanistic pathways for prebiotic synthesis[2] and the composition of organic matter on Earth before living systems developed[3]. Previous studies[11,12] have shown that meteorite amino acids are enriched in ^{13}C relative to their terrestrial counterparts, but individual species were not distinguished. Here we report the ^{13}C contents of individual amino acids in the Murchison meteorite. The amino acids are enriched in ^{13}C, indicating an extraterrestrial origin. Alanine is not racemic, and the ^{13}C enrichment of its D- and L-enantiomers implies that the excess of the L-enantiomer is indigenous rather than terrestrial contamination, suggesting that optically active materials were present in the early Solar System before life began.

Hypotheses concerning extraterrestrial prebiotic synthesis of organic compounds are often based on the composition of organic matter in carbonaceous meteorites[4,5]. Attention continues to be focused on the organic-rich Murchison meteorite (Type CM, ~2% total organic carbon), which was collected shortly after impact in Australia in 1969. Compared to terrestrial materials, the overall distribution and relative abundance of amino acids in Murchison meteorite stones is exotic, as evidenced by high concentrations of non-protein amino acids that are uncommon in living systems (such as α-aminoisobutyric acid and isovaline) and the apparent absence or extremely low concentration of many of the amino acids common in proteins (such as serine, phenylalanine, tyrosine, lysine, histidine and arginine)[4,6].

Several amino acids that are common in living systems also occur in fairly high abundance in the Murchison meteorite (glycine, alanine, glutamic acid, aspartic acid and proline, for example)[4,7]. The expectation has been that these amino acids, when found in meteorites, will be racemic because laboratory simulations of prebiotic synthesis result in the formation of racemic amino acids[8]. Our previous finding[6] that some of these amino acids are not racemic (L-enantiomer predominating) led to speculation on whether a percentage of the L-enantiomers of these amino acids reflects terrestrial contamination or whether diagenetic reactions occurring in space explain this anomaly. So far, no one has found a terrestrial contaminant that could rapidly and selectively introduce into a meteorite stone small quantities of some of the protein amino acids (such as glycine, alanine, aspartic acid and glutamic acid), but not others (such as phenylalanine, serine, tyrosine and lysine)[7].

Pillinger[9] has suggested that the stable-isotope composition of the D- and L-enantiomers of the individual amino acids might indicate whether non-racemic amino acids were indigenous to the meteorite or a consequence of contamination. This idea is based on the observation that terrestrial organic matter is depleted in ^{13}C (ref. 10) and that bulk extracts of the amino acids in carbonaceous meteorites are enriched in ^{13}C (refs 11, 12). A significant amount of an L-amino-acid contaminant in a meteorite stone should therefore result in a marked stable-isotope depletion for this component. Because the concentrations of amino acids in the Murchison meteorite are of the order of nanomoles per gram, previously available methods[13–15] would require the destruction of large quantities of sample to isolate individual enantiomers for stable-isotope analysis. The recent development of a method for amino acid analysis that combines gas chromatography and isotope-ratio mass spectrometry (GC-IRMS), however, allows direct measurement of the stable-carbon-isotope composition of individual enantiomers in a water extract of a Murchison stone.

We isolated samples weighing 50 mg and 3.8 g from the interior of a 52.3-g Murchison stone (R. A. Langheinrich meteorite collection) that had a largely intact fusion crust. Portions of the 50-mg sample were acidified (6 N HCl), dried and then prepared for stable carbon and nitrogen isotope analysis by the Dumas sealed-tube method[17]. Analyses of the resultant gases were performed with a VG PRISM isotope-ratio mass spectrometer. Other portions of the bulk, untreated sample of the meteorite were analysed directly for stable-carbon-isotope composition and elemental abundance (C, N) using a VG Isogas Carlo Erba combustion system, and for stable-isotope (C, O) composition of the carbonate component using the VG PRISM.

The 3.8-g sample was ground to a fine powder and placed in a Pyrex tube. Deionized, distilled H_2O (20 ml) was added to the tube, which was sealed under N_2 and heated for 8 h in an oven at 100 °C. After heating, the water extract and a procedural blank were desalted and prepared for amino acid derivatization as previously reported[6]. Unlike previous studies[6,12], we did not hydrolyse the water extract with 6 N HCl before desalting. Although hydrolysis increases the recovery of amino acids (presumably by the alteration of labile precursor[18]), we decided to try to obtain stable-isotope compositions for the water-extractable amino acids alone, as the addition of amino acids formed by hydrolysis might complicate interpretation of the data. We determined the abundances of amino acids in a portion of the water extract using high-performance liquid chromatography (HPLC), as previously reported[19]. The remainder of the water-extractable amino acid fraction was derivatized to (N, O)-TFA-isopropyl esters[16,20] for GC and GC-IRMS analysis using reagents (trifluoroacetic anhydride, isopropanol) of known stable-carbon isotope composition.

The VG Isochrom GC-IRMS system consists of a Hewlett-Packard 5890 GC coupled to a combustion furnace/water trap, which in turn is interfaced to a VGSIRA isotope-ratio mass spectrometer[21]. We used a Chirasil-Val 50 m × 0.25 mm (inner diameter) fused-silica capillary column capable of resolving (N, O)-TFA-isopropyl esters of D and L amino acids[20].

Unlike GC-IRMS analyses of hydrocarbons[22,23], the analysis of amino acids is complicated by the addition of carbon in the required derivatization. Previous $\delta^{13}C$ analyses of standard amino acid derivatives by GC-IRMS and conventional static-combustion IRMS were in good agreement[16]. For example, the $\delta^{13}C$ values for N-TFA-isopropyl esters of D-alanine and L-alanine determined by conventional static-combustion IRMS were -29.9 ± 0.1 and -28.0 ± 0.1‰, respectively. The $\delta^{13}C$ obtained for the same derivatives of D- and L-alanine by the GC-IRMS method were -30.1 ± 0.1 and -27.5 ± 0.2‰, respectively. In general, the error for replicate derivatization and analysis of amino acid standards by either method is better than 0.2‰. The $\delta^{13}C$ values for amino acids are corrected for the carbon introduced during derivatization[16].

The C and N contents determined for the non-acidified sample of the Murchison stone were 2.2% and 0.27%, respectively. The $\delta^{13}C$ value for this sample was -9.9‰ (relative to the PDB standard). These values are in agreement with previously reported values for bulk samples of the Murchison meteorite[24]. The $\delta^{13}C$ and $\delta^{15}N$ values for the acidified bulk sample of this Murchison stone are -12.3‰ and $+34.7$‰ relative to PDB and atmospheric N_2, respectively. Stable carbon and oxygen isotope values for the carbonate component of this stone were $+59.9$‰ (PDB) and $+37.8$‰ (standard mean ocean water), respectively. The $\delta^{13}C$ and $\delta^{18}O$ values for the carbonate are enriched in ^{13}C and ^{18}O relative to previous reports[25].

The concentrations of amino acids in the non-hydrolysed water extract of this stone (Table 1) were in good agreement

TABLE 1 $\delta^{13}C$ values

Amino acid	Concentration (nmol g^{-1})	$\delta^{13}C$* (‰)
α-aminoisobutyric acid	16.0	+5
L-glutamic acid	4.6†	+6
Isovaline	7.5‡	+17
Glycine	28.1	+22
D-alanine	†	+30
L-alanine	12.9†	+27

* The $\delta^{13}C$ values are corrected for carbon introduced during derivatization[16]. $\delta^{13}C(‰) = [(^{13}C/^{12}C)_{sample}/(^{13}C/^{12}C)_{PDB} - 1] \times 10^3$.

† Concentrations reported for L-glutamic acid and L-alanine represent the total contribution of both enantiomers.

‡ This value includes a minor contribution of valine that co-eluted with isovaline during HPLC analysis.

Reprinted with permission from *Nature*, vol 348, pp. 47–49, 1 November 1990. © 1990 Macmillan Magazines Ltd.

with a previous analysis of this fraction in another Murchison stone[26]. Several common protein amino acids were not detected in this sample (phenylalanine, tyrosine, lysine) or were present in trace amounts (serine, threonine, leucine, isoleucine). Again, this is consistent with a previous study of the non-hydrolysed water extract of a Murchison stone[26]. GC/mass-selective detector analyses confirmed which amino acids were present in this water extract.

The carbon isotope compositions of the major amino acid constituents of the water extract of this Murchison stone are enriched in ^{13}C relative to terrestrial biogenic materials (Table 1), and clearly indicate an extraterrestrial origin. The stable-isotope values for the individual amino acids cover a range of 25‰ and seem to decrease with increasing carbon number, similar to a trend reported by Yuen et al.[2] for hydrocarbons and monocarboxylic acids in the Murchison meteorite. To confirm and interpret this trend with respect to possible kinetic effects during synthesis, however, we must acquire a larger data set.

The D/L values for amino acids in the water extract of this interior stone sample were not racemic and were similar to the D/L values that we previously reported for the hydrolysed water extract of another Murchison stone[6]. Here we find that the D/L value for alanine is 0.85 ± 0.03. Assuming that the D-enantiomer of alanine is entirely extraterrestrial in origin, the data indicate that the excess L-enantiomer is not from a terrestrial source (Table 1). The L-alanine is 3.0‰ lighter than the D-alanine. If the relative amounts of L-alanine and D-alanine were originally identical, the excess L-alanine required to produce a D/L value of 0.85 would have to have an isotopic composition of ~+10‰ to account for the observed value of +27‰ for L-alanine in the water extract. We are not aware of natural terrestrial sources for alanine that are so enriched.

Here and in our previous work alanine was the most racemized amino acid in the water extract. The D/L values for the other amino acids were lower (for example, D/L glutamic acid = 0.54). The $\delta^{13}C$ value for L-glutamic acid is +6‰. Although we cannot, at present, measure the $\delta^{13}C$ value of D-glutamic acid, we are modifying the GC-IRMS system to allow analysis of the enantiomers (such as D-glutamic acid) that are present in lower abundances.

Our finding of non-racemic amino acids cannot be adequately explained by terrestrial contamination, and may indicate that laboratory simulations of prebiotic syntheses resulting in racemic amino acids are not entirely appropriate models for the chemical reactions that actually took place during formation of the Solar System. Alternatively, extraterrestrial diagenetic reactions after synthesis may have altered the D/L values. The recent finding[27] that the Murchison meteorite contains a significant level of ^{14}C formed by cosmic-ray-induced nuclear reactions indicates that organic matter (such as amino acids) in the Murchison meteorite may have been susceptible to extraterrestrial diagenetic reactions, at least since the time of its existence as a smaller body (~1 million years) before impact. Although the significance of such reactions and the reactions resulting from β-decay of ^{14}C over the extended residence time in space are unknown, it has been hypothesized that β-decay may have resulted in the preferential destruction of D amino acids on Earth (and thus enrichment in L amino acids) before the origin of life[28].

Further refinement of the GC-IRMS method will eventually permit stable-isotope analyses to be performed on other pairs of amino acid enantiomers occurring in lower concentrations in extracts of the Murchison meteorite. Clearly, stable-isotope geochemistry at the molecular level is helping us to understand extraterrestrial, prebiotic organic reactions at about the time of formation of our Solar System. □

Received 4 May; accepted 12 September 1990.

1. Chyba, C. F., Thomas, P. J., Brookshaw, L. & Sagan, C. Science 249, 366–373 (1990).
2. Yuen, G., Blair, N., Des Marais, D. J. & Chang, S. Nature 307, 252–254 (1984).
3. National Research Council The Search for Life's Origins (National Academy Press, Washington, DC, 1990).
4. Cronin, J. R., Pizzarello, S. & Cruikshank, D. P. in Meteorites and the Early Solar System (eds Kerridge, J. F. & Matthews, M. S.) 819–857 (University of Arizona Press, 1988).
5. Shock, E. L. & Schulte, M. D. Geochim. cosmochim. Acta (in the press).
6. Engel, M. H. & Nagy, B. Nature 296, 837–840 (1982).
7. Engel, M. H. & Nagy, B. Nature 301, 496–497 (1983).
8. Khare, B. N. et al. Icarus 68, 176–184 (1986).
9. Pillinger, C. T. Nature 296, 802 (1982).
10. Schidlowski, M. Nature 333, 313–318 (1988).
11. Chang, S., Mack, R. & Lennon, K. Lunar planet. Sci. 9, 157–158 (1978).
12. Epstein, S., Krishnamurthy, R. V., Cronin, J. R., Pizzarello, S. & Yuen, G. U. Nature 326, 477–479 (1987).
13. Engel, M. H. & Macko, S. A. Analyt. Chem. 56, 2598–2600 (1984).
14. Engel, M. H. & Macko, S. A. Nature 323, 531–533 (1986).
15. Serban, A., Engel, M. H. & Macko, S. A. Org. Geochem. 13, 1123–1129 (1988).
16. Silfer, J. A., Engel, M. H., Macko, S. A. & Jumeau, E. J. Analyt. Chem. (submitted).
17. Macko, S. A. thesis, Univ. Texas (Austin) (1981).
18. Cronin, J. R. & Moore, C. B. Science 172, 1327–1329 (1971).
19. Hare, P. E., St John, P. A. & Engel, M. H. in Chemistry and Biochemistry of the Amino Acids (ed. Barrett, G. C.) 415–425 (Chapman & Hall, London, 1985).
20. Engel, M. H. & Hare, P. E. in Chemistry and Biochemistry of the Amino Acids (ed. Barrett, G. C.) 462–479 (Chapman & Hall, London, 1985).
21. Freedman, P. A., Gillyon, E. C. P. & Jumeau, E. J. Am. Lab. 114–119 (June 1988).
22. Freeman, K. H., Hayes, J. M., Trendel, J-M. & Albrecht, P. Nature 343, 254–256 (1990).
23. Kennicutt, M. C. II & Brooks, J. M. Org. Geochem. 15, 193–197 (1990).
24. Kerridge, J. F. Geochim. cosmochim. Acta 49, 1707–1714 (1985).
25. Grady, M. M., Wright, I. P., Swart, P. K. & Pillinger, C. T. Geochim. cosmochim. Acta 52, 2855–2866 (1988).
26. Cronin, J. R. Origins of Life 7, 337–342 (1976).
27. Jull, A. J. T., Donahue, D. J. & Linick, T. W. Geochim. cosmochim. Acta 53, 2095–2100 (1989).
28. Noyes, H. P., Bonner, W. A. & Tomlin, J. A. Origins of Life 8, 21–23 (1977).

ACKNOWLEDGEMENTS. We thank E. C. P. Gillyon and N. Crossley for the use of the instrumentation at VG Isogas. We also thank J. Jumeau for her help with the GC-IRMS analyses and P. Harrigan-Ostrom for her help with the GC/MSD analyses. S.A.M. thanks C. Hillaire-Marcel for providing office and lab space during his sabbatical leave. This work was supported in part by the NSF.

β Decay and the origins of biological chirality: theoretical results

Roger A. Hegstrom

Department of Chemistry, Wake Forest University, Winston-Salem, North Carolina 27109, USA

A dynamical mechanism is found whereby a dissymmetric molecule and its mirror image are ionized at different rates by longitudinally polarized electrons such as produced by nuclear β decay. An enhancement is predicted for molecules containing heavy atoms. Order-of-magnitude estimates indicate that the asymmetric effect of this mechanism may be detectable by current experiments on positronium formation.

THE postulated existence of a causal relationship between the parity nonconserving aspect of the weak interaction and the observed dissymmetry of the molecules present in living organisms[1-3] seems reasonable as the weak interaction is universal and always acts in the same chiral sense. Hence if an effective mechanism exists involving the weak interaction in molecular evolution, this inherent universal chirality in nature could have influenced the selection of exclusively D sugars for RNA and DNA and L amino acids for proteins.

Two candidates for such a mechanism have been proposed, each of which involves a different aspect of the weak interaction. One produces an energy difference between a chiral molecule and its mirror image[4]; but the calculated difference[5] seems too small to have been effective in molecular evolution. The other, which seems the most popular theory at present, postulates the asymmetric radiolysis of racemic mixtures of prebiotic chiral molecules by the longitudinally polarized electrons produced in nuclear β decay[6]. A precise test of some aspects of this mechanism is the object of new experiments to measure an asymmetry in the rate of triplet positronium (Ps) formation in chiral molecules discussed in the accompanying article[7] (where references to previous work may be found).

The present article gives a theoretical basis for, and estimates the magnitudes of, the cross-section asymmetries for both Ps formation and radiolysis. The relevance of the results to the question of the origin of chirality in living organisms is discussed extensively in the preceding article[7].

Ps formation in optically active molecules

The treatment of the asymmetry in the cross-section for positronium formation given here and that for radiolysis which follows are based on nonrelativistic scattering theory. The notation of Taylor[ ] is followed except that atomic units are used here ($h/2\pi = e = m_e = 1$, where h is Planck's constant, e the electron charge, and m_e the electron mass).

Consider a positron with momentum $\vec{p}$ and helicity $\lambda = \vec{\sigma} \cdot \hat{p} = \pm 1$, where $\hat{p} = \vec{p}/p$ and $p = |\vec{p}|$, incident on an optically active molecule, the nuclei of which are taken to be fixed in the laboratory frame of reference (Fig. 1). This initial channel is labelled α, and α' labels the final channel which consists of the ionized molecule and a Ps atom with momentum $\vec{P}'$. Spin is quantized along $\vec{p}$ and hence the spin projection quantum number of the positron is related simply to the helicity by $m_\lambda = \frac{1}{2}\lambda$. The cross-section depends on the helicity λ, the chirality of the target molecule (L or D), and on the spin state S of the Ps, and is given by

$$\sigma_S^\lambda(L) = (2\pi)^4 \frac{P'}{p} \int |t_S^\lambda(L)|_{\lambda v}^2 \, d\Omega' \qquad (1)$$

(with a corresponding expression for the D isomer) where $P' = |\vec{P}'|$, where $d\Omega'$ denotes an integration over the solid angles corresponding to directions of $\vec{P}'$, and where Av denotes an average over all orientations of the molecule (random orientations are assumed). The quantity t_S^λ, which is related to the on-shell T-matrix, is defined below.

In a notation consistent with that of the preceding paper[7], the asymmetry in the cross-section is defined as

$$H_{Ps}(L) \equiv \frac{\sigma_S^+(L) - \sigma_S^-(L)}{\sigma_S^+(L) + \sigma_S^-(L)} = \frac{\sigma_S^+(L) - \sigma_S^+(D)}{\sigma_S^+(L) + \sigma_S^+(D)} \qquad (2)$$

(with a corresponding equation for the D isomer) where $\sigma_S^+(L)$ and $\sigma_S^-(L)$ are obtained from equation (1) with $\lambda = +1$ and $\lambda = -1$ respectively. The second equation in equation (2) is a consequence of parity conservation for the electromagnetic interaction [equations (6, 7)], from which the general relation-

ship $\sigma_S^\lambda(L) = \sigma_S^{-\lambda}(D)$ follows. For the case of a positron beam of helicity $h(e^-)$ and with $h \equiv |h(e^+)| \le 1$, it can be shown that, for a given L or D isomer (from now on the isomer is not denoted explicitly), the asymmetry in the cross-section obtained by reversing the beam helicity is

$$A_{Ps} \equiv \frac{\sigma_S^{+h} - \sigma_S^{-h}}{\sigma_S^{+h} + \sigma_S^{-h}} = hH_{Ps} \qquad (3)$$

The dynamics which produce the cross-section asymmetry are now considered. For simplicity, a single-electron treatment of the molecular bound state is given here and illustrated in Fig. 1. The positron 'a', is incident on a molecule consisting of electron 'b' bound to a core 'c'. The core approximates the fixed nuclear framework and the other electrons in the molecule and provides an effective potential for b. The spin of b, like that of a, is quantized along $\hat{p}$ with its direction given by $s \equiv 2m_s = \pm 1$, where m_s is the spin projection quantum number. The electron bound state is denoted ψ_s.

The cross-section is given by equation (1), where, for a spin-unpolarized target molecule

$$|t_S^\lambda|^2 = \frac{1}{2} \sum_{m_\lambda} \sum_{M_S} |t_{SM_S}^{m_\lambda m_s}|^2 \qquad (4)$$

where the second summation is taken over the spin projection quantum number M_S corresponding to the spin S. The on-shell T-matrix is given by

$$t_{SM_S}^{m_\lambda m_s} = \langle \vec{P}' \phi_{Ps} S M_S - |V^\circ| \hat{p} m_\lambda \psi_s \rangle \qquad (5)$$

where ϕ_{Ps} denotes the spatial part of the Ps wave function, V° denotes the potential energy giving the interaction between the incident positron and the molecule, and the minus sign denotes the prior form[8] of the T-matrix.

It can be shown on the basis of equations (1)–(5) that H_{Ps} vanishes unless (1) the molecule is dissymmetric and (2) spin-dependent interactions are involved. According to equation (5) there are three possible ways for the effects of spin-dependent forces to occur: (1) in the interaction potential V°, (2) in the positronium wave function, and (3) in the molecular wave function ψ_s. An investigation of each of these possibilities indicates that for molecules containing atoms as large as carbon, or larger, the dominant contribution to the asymmetry H_{Ps} is expected to come from the perturbation of the electron bound state ψ_s by the spin–orbit interaction of the bound electron moving in the electric field of the molecular core. In what follows, only this bound-state spin–orbit interaction is kept and all other spin-dependent terms in the hamiltonian are ignored. The resulting description is called the bound helical electron model, because, as will be seen later the theory predicts the existence of a helicity density for the electrons bound in an optically active molecule. The qualitative aspects of this model resemble the "helical electron gas" model of Hrasko and Garay[9].

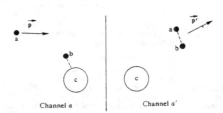

Channel a   Channel a'

Fig. 1 The formation of positronium.

Reprinted with permission from *Nature*, vol. 297, pp. 643-647, 24 June 1982. © 1992 Macmillan Magazines Ltd.

The dynamics of the bound helical electron model are expressed in the total Hamiltonian $H = H^\alpha + V^\alpha$ for the system, where the channel α Hamiltonian is given by

$$H^\alpha = T_a + T_b + V_{bc}^{COUL} + V_{bc}^{SO} \qquad (6)$$

and the interaction V^α by

$$V^\alpha = V_{ab}^{COUL} + V_{ac}^{COUL} \qquad (7)$$

where the subscripts on the kinetic energy operators T and potential energy operators V denote the particles a, b and core c, and where 'COUL' and 'SO' denote the Coulomb and spin-orbit interactions, respectively. Now the asymptotic stationary state $|\hat{p}m_s,\psi_s\rangle$ appearing in equation (5) is an eigenfunction of H^α with eigenvalue E, the total energy of the system, so that when the spin-orbit energy is small compared with E first-order perturbation theory may be used to obtain the result

$$|\psi_s\rangle = \left(\phi_0 + s\sum_n \varepsilon_n^0 \phi_n\right)|m_s\rangle + \sum_n \varepsilon_n^* \phi_n|\bar{m}_s\rangle \qquad (8)$$

where $\phi_0|m_s\rangle$ is the unperturbed molecular bound state, $\phi_n|m_s\rangle$ are excited states, $\bar{m}_s = -m_s$, and

$$V_{bc}^{SO} = \vec{\Lambda}_{bc} \cdot \vec{\sigma}_b \equiv \tfrac{1}{2}\alpha^2 \vec{E}_{bc} x \vec{p}_b \cdot \vec{\sigma}_b \qquad (9a)$$

$$\vec{\varepsilon}_n = (E_0 - E_n)^{-1}\langle\phi_n|\vec{\Lambda}_{bc}|\phi_0\rangle \qquad (9b)$$

$$\varepsilon_n^0 = \vec{\varepsilon}_n \cdot \hat{e}_3 = \vec{\varepsilon}_n \cdot \hat{p} \qquad (9c)$$

$$\varepsilon_n^\pm = \vec{\varepsilon}_n \cdot \hat{e}_\pm = \vec{\varepsilon}_n \cdot (\hat{e}_1 \pm i\hat{e}_2) \qquad (9d)$$

Here $\alpha \sim 137^{-1}$ is the fine structure constant, $\vec{E}_{bc}$ is the electric field produced by the core c at the location of electron b, $\vec{p}_b$ is the momentum of b, $\vec{\sigma}_b$ is the Pauli spin matrix, and $\hat{e}_k$ are unit vectors with $\hat{e}_3 = \hat{p}$.

The asymmetry of the positronium formation cross-section is now considered within the bound helical electron model. It can be shown that the 'spin-flip' term in the perturbed molecular wave function, that is, the last term of equation (8), does not contribute to the cross-section to first-order in the spin-orbit interaction and hence can be ignored. The contributing part of equation (8) is now abbreviated $\phi_s|m_s\rangle$, where

$$\phi_s = \phi_0 + s\sum_n \varepsilon_n^0 \phi_n \qquad (10)$$

Substitution of $\phi_s|m_s\rangle$ for $|\psi_s\rangle$ in equation (5) then gives

$$t_{SM_S}^{m_s m_s} = V^s(SM_S|m_s m_s) \qquad (11)$$

where

$$V^s = \langle \vec{P}'\phi_{p'} - |V^\alpha|\hat{p}\phi_s\rangle \qquad (12)$$

The spin-dependent factors appearing in equation (11) are easily evaluated to obtain the contributing T-matrix elements in terms of V^s, and then equations (1)–(4) are used to obtain the cross-section asymmetry H_{Ps} or A_{Ps} to first order in the spin-orbit interaction. For the triplet state the result is

$$H_{Ps} = \frac{1}{6}\frac{\int\{|V^+|_{Av}^2 - |V^-|_{Av}^2\}\,d\Omega'}{\int|V^0|_{Av}^2\,d\Omega'} \qquad (13)$$

where $V^\pm$ are defined by equation (12) and where V^0 is obtained by replacing ϕ_s by ϕ_0 on the right-hand side of equation (12). The result for the singlet state is obtained by replacing the factor $1/6$ in equation (13) by $-1/2$.

The precise calculation of H_{Ps} is an extremely difficult collision problem. Its order of magnitude can be estimated simply, however, using equations (9)–(13), from which it follows that H_{Ps} is first order in the spin-orbit coefficient ε_n. On the basis of dimensional arguments H_{Ps} would then be expected to be of order $|\varepsilon_n| \sim (\alpha Z)^2 = (Z/137)^2$ where Z is the electric charge on the heaviest atom or atoms in any asymmetric environment in the molecule[4]. Following the treatment of the parity nonconserving energy difference[4], H_{Ps} is expressed

$$H_{Ps} = \eta_{Ps}(\alpha Z)^2 \qquad (14)$$

where the molecular asymmetry factor η_{Ps} is included to take into account the effect of the molecular environment on the magnitude of H_{Ps}. An approximate calculation of H_{Ps} (R.H. and R. Stewart, unpublished data; see also ref. 5) for the simplest dissymmetric molecules for which a wave function has been published, namely twisted ethylene, suggests $\eta_{Ps} \sim 10^{-2}$ to 10^{-3}, which may be typical, although values outside this range cannot be ruled out until more extensive calculations have been performed.

At present the Michigan experiments[7] have examined three amino acids, two of which give the result $|H_{Ps}| < 10^{-2}$ in agreement with the present theoretical predictions (taking $\eta_{Ps} = 10^{-2}$ to 10^{-3} gives $H_{Ps} \sim 10^{-5}$–10^{-6} for amino acids). The third amino acid, leucine, gives the preliminary result $H_{Ps} = (4 \pm 2) \times 10^{-2}$ in apparent disagreement with theory. Although it is conceivable that the present theory could obtain such a large value of H_{Ps} in atypical cases, it is premature to attempt to make a more detailed analysis for the case of leucine at present due to the difficulty of even crude calculations involving amino acid wave functions and also due to the preliminary nature of the experimental result.

Finally, in view of the treatment of radiolysis to be given next, it should be noted that the asymmetry in the total inclusive cross-section for Ps formation, obtained by summing equation (1) over $S = 0$ and 1, vanishes identically due to the cancellation of the singlet and triplet contributions. A vanishing result is also obtained for ionization without Ps formation, which is closely related to the process of ionization by polarized electrons treated next.

Electron impact ionization of optically active molecules (radiolysis)

The theoretical treatment of radiolysis is similar to that just given for Ps formation. The most important difference is that, because the electrons are identical, the two-electron state of incident plus target electron is antisymmetric with respect to the interchange of these particles, and as a consequence both direct and exchange terms appear. In fact it is the existence of these exchange terms which leads to a nonvanishing asymmetry in the total cross-section, as is shown below, in contrast to the case of ionization by positrons discussed above.

The scattering process is illustrated in Fig. 2. In the initial channel α the incident electron 1 has helicity $\lambda = \pm 1$ and momentum $\vec{p}$ and the target electron 2 is bound to the core c. In the two final channels α' and $\tilde{\alpha}'$, which differ by an exchange of the two electrons, the momenta are designated $\vec{p}_a$ and $\vec{p}_b$. The spins are again quantized along $\hat{p}$ with initial spin projection quantum numbers m_λ, m_s and final spin quantum numbers S, M_S describing singlet and triplet states. The total ionization cross-section, inclusive of electrons in singlet and triplet final states, is given (for either the D or L isomer) by

$$\sigma^\lambda = \tfrac{1}{2}(2\pi)^4\frac{1}{p}\int_0^{E_p-I} dE_b p_a p_b \int d\Omega_a\,d\Omega_b |\hat{t}|_{Av}^2 \qquad (15)$$

where $E_p = p^2/2$ is the energy of the incident electron, I is the molecular ionization energy, $E_b = p_b^2/2$, and where

$$|\hat{t}|^2 = \tfrac{1}{2}\sum_{m_s}\sum_S\sum_{M_S} |\hat{t}_{SM_S}^{m_\lambda m_s}|^2. \qquad (16)$$

The on-shell T-matrix elements are given by

$$\hat{t}_{SM_S}^{m_\lambda m_s} = t_{SM_S}^{m_\lambda m_s}(\text{dir}) + (-)^S t_{SM_S}^{m_\lambda m_s}(\text{exch}) \qquad (17)$$

with the direct and exchange terms given by

$$t_{SM_S}^{m_\lambda m_s}(\text{dir}) = \langle\vec{p}_a\vec{p}_b SM_S - |V^\alpha|\hat{p}m_\lambda\psi_s\rangle \qquad (18a)$$

$$t_{SM_S}^{m_\lambda m_s}(\text{exch}) = \langle\vec{p}_b\vec{p}_a SM_S - |V^\alpha|\hat{p}m_\lambda\psi_s\rangle \qquad (18b)$$

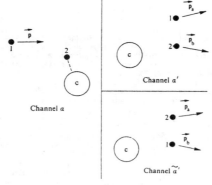

Fig. 2 Ionization by electron impact.

Equations (18a, b) are analogous to equation (5) and ψ_s is again given by equation (8). The same procedure which led from equation (5) to equation (13) is again followed here using equations (15)–(18) and gives the following result for the asymmetry in the radiolysis cross-section, correct to first order in the spin–orbit interaction:

$$H_R = \frac{\sigma^+ - \sigma^-}{\sigma^+ + \sigma^-}$$

$$= -\frac{1}{2}\frac{\int_0^{E_p-I} dE_b\, p_a p_b \int d\Omega_a\, d\Omega_b\, Re\{V_d^{-*}V_e^+ - V_d^{-*}V_e^-\}_{Av}}{\int_0^{E_p-I} dE_b\, p_a p_b \int d\Omega_a\, d\Omega_b\, \{|V_d^0|^2 + |V_e^0|^2 - Re\,V_d^{0*}V_e^0\}_{Av}} \quad (19)$$

where

$$V_d^* = \langle \vec{p}_a \vec{p}_b - |V^a| \vec{p}\phi_s\rangle \quad (20a)$$

$$V_e^* = \langle \vec{p}_b \vec{p}_a - |V^a| \vec{p}\phi_s\rangle \quad (20b)$$

are direct and exchange integrals with ϕ_s again given by equation (10). The integrals V_d^0 and V_e^0 are obtained by replacing ϕ_s with ϕ_0 in equations (20a, b). If the incident electron state is a mixture of pure helicity states with a net helicity $h(e^-)$, then the asymmetry in the cross-section is

$$A_R = \frac{\sigma^{+h} - \sigma^{-h}}{\sigma^{+h} + \sigma^{-h}} = h H_R \quad (21)$$

where $h = |h(e^-)|$. A negative value of H_R or A_R for a given L or D isomer implies that the isomer is preferentially ionized by natural β electrons, for which $h(e^-)$ is negative.

Although the precise calculation of H_R is difficult, it is possible to make rough estimates of its order of magnitude as was done for H_{Pn}. It is useful to distinguish two cases. For the case of incident energies E_p near the ionization threshold, the final momenta $\vec{p}_a$, $\vec{p}_b$ both approach zero, so it follows from equations (20a, b) that $V_e^* \to V_d^*$ (note that only scattering into the singlet state contributes in this limit), and then from equation (19) that

$$H_R \xrightarrow[E_p \to I]{} -\frac{1}{2}\frac{|V_d^*|_{Av}^2 - |V_d^-|_{Av}^2}{|V_d^0|_{Av}^2} \quad (22)$$

which, on the basis of dimensional arguments similar to those which led to equation (14), can be shown to have the same order of magnitude as H_{Pn}. Hence H_R is expressed

$$H_R = \eta_R (\alpha Z)^2 \quad (E_p \to I) \quad (23)$$

where the molecular asymmetry factor for radiolysis η_R is expected to have the same order of magnitude as η_{Pn}. For the case of incident energies near 100 keV, which is typical of β-decay electrons, use of the nonrelativistic Bethe approximation[10] in equation (19) gives the result

$$H_R = \frac{\eta_R'(\alpha Z)^2}{(2E_p)\ln(2E_p)} \quad (1 \ll E_p \ll \alpha^{-2}) \quad (24)$$

where η_R' is formally of the order of one but is expected to be less than one for the same reasons given in connection with equations (14) and (23). The range for E_p corresponds to the range of validity for the nonrelativistic Bethe approximation[10]; for more precise calculations relativistic effects should be included. For $E_p = 100$ keV $\approx 3,700$ AU equation (24) gives $H_R = 10^{-5}\eta_R'(\alpha Z)^2$ which is roughly four orders of magnitude smaller than the threshold value. This large reduction is due primarily to the decreasing magnitudes of the exchange integrals V_e^* with increasing energy. Taking η_R' to be of the order 10^{-2}–10^{-3} in equation (24) gives an estimated asymmetry $H_R \sim 10^{-10}$–10^{-11} for the radiolysis of amino acids ($Z = 6$) with 100-keV electrons and a magnitude roughly 100 times larger for a molecule containing a dominant heavy atom with Z near 100.

Recently Zel'dovich and Saakyan[11] have calculated a contribution to H_R which is not included in equation (19) but rather is due to the direct spin-dependent interaction between the incident and target electrons, and which may be expressed

$$H_R \text{ (ref. 10)} \approx \alpha^2 \frac{v}{c}\frac{\sum_i Im(\vec{d}_n \cdot \vec{m}_n^*)}{\sum_i |\vec{d}_n|^2} \quad (25)$$

where v/c is the ratio of the speed of the incident electron to the speed of light, and where $\vec{d}_n$ and $\vec{m}_n$ are the electric and magnetic transition moments, respectively, between the initial and final molecular states. The ratio of sums appearing in equation (25) may be estimated from spectral data on optically active molecules[12]. For strong absorption bands, which are

expected to dominate the cross-section, this ratio is typically of the order of 10^{-6}, which is much less than the estimated value 10^{-2}–10^{-3} based on dimensional arguments[11]. Equation (25) then gives the result H_R (ref. 11) $\approx 10^{-11}$ for 100 keV electrons, which is comparable with the result for the bound helical electron model at the same energy and with $Z \approx 6$. Hence the contribution to H_R from equation (19) dominates at low energies and high Z, but the contributions from both equations (19) and (25) are important at high energies.

More recently, Mann and Primakoff[13] have estimated H_R to be relatively large (of the order of $\alpha^2 \approx 10^{-4}$) due to the interaction between the spin of the incident beta electron and the orbital angular momentum of the target electron. This estimate is in direct disagreement with that from ref. 12. My preliminary calculation gives the result that the contribution to H_R from this spin–other-orbit interaction, although formally of the order of α^2, vanishes in the first Born approximation when the average over molecular orientations is taken.

I conclude this section with the following remarks: (1) β Electrons are emitted with a spectrum of energies ranging from zero up to a maximum which is typically hundreds of keV. In view of the present results, the low energy end of this spectrum may be more important for asymmetric radiolysis than previously thought, even though the magnitude of the helicity is less at low energies. (2) Experiments which attempt to detect this asymmetry are expected to find a larger effect for low-energy polarized electrons, particulary with dissymmetric target molecules containing heavy atoms.

Electron helicity in bound states

For a one-electron state ψ_s which is characterized by the spin direction $s \equiv 2m_s$, the helicity density is defined in coordinate space by the equation

$$h_s(\vec{x}) = Re\{\psi_s^*(\vec{x})\vec{\sigma} \cdot \hat{p}_e \psi_s(\vec{x})\} \quad (26a)$$

or in momentum space by

$$h_s(\vec{p}_e) = Re\{\psi_s^*(\vec{p}_e)\vec{\sigma} \cdot \hat{p}_e \psi_s(\vec{p}_e)\} \quad (26b)$$

where $\vec{\sigma}$ is the Pauli spin operator, $\vec{p}_e$ is the electron momentum, and $\hat{p}_e = \vec{p}_e/|\vec{p}_e|$. The two-component spin functions $\psi_s(\vec{x})$ and $\psi_s(\vec{p}_e)$ are Fourier transforms of each other. For a one-electron atom or molecule in a pure spin state, the helicity density can be relatively large, of the order of several atomic units.

The helicity density may also be defined in a general way for a many-electron system, but here it is sufficiently general to consider the simple case of a two-electron singlet state

$$\Psi(1, 2) = 2^{-1/2}\{\psi_+(1)\psi_-(2) - \psi_-(1)\psi_+(2)\} \quad (27)$$

for which the helicity density is given by

$$h(\vec{p}_e) = \sum_{s=\pm 1} Re\{\psi_s^*(\vec{p}_e)\vec{\sigma} \cdot \hat{p}_e \psi_s(\vec{p}_e)\} \quad (28a)$$

$$= \sum_{s=\pm 1} h_s(\vec{p}_e) \quad (28b)$$

In this case it can be easily shown that, in the absence of spin-orbit coupling, the helicity density vanishes due to cancellation of equal and opposite contributions from the two spin states. If spin-orbit coupling corrections to the wave function $\Psi(1, 2)$ are now included by perturbing the spin orbitals ψ_s according to equation (8), then equation (28) gives

$$h(\vec{p}_e) = 4Re \sum_n \vec{\varepsilon}_n \cdot \hat{p}_e \phi_0^*(\vec{p}_e)\phi_n(\vec{p}_e) \quad (29)$$

where $\vec{\varepsilon}_n$ is defined in equation (9b). The same result (divided by two) is obtained for a spin-unpolarized one-electron state. The following comments regarding equation (29) can be made: (1) h is formally of the order of $|\varepsilon_n| \sim (\alpha Z)^2$ atomic units; (2) h vanishes for an atom or for a molecule with a centre of inversion symmetry; (3) h is in general non-zero for a molecule with a fixed orientation and without a centre of inversion; (4) if an average over orientations is taken for an ensemble of randomly oriented molecules, the resulting average helicity density h_{Av} is independent of the direction of $\vec{p}_e$ and is non-zero only in the case of a dissymmetric molecule; (5) the expectation value $\langle \vec{\sigma} \cdot \hat{p}_e\rangle$, which is obtained from the right-hand side of equation (29) or equation (28a) by replacing $\hat{p}_e$ with $\vec{p}_e$ and integrating over all momentum space, is identically zero to first order in the spin–orbit interaction. Proofs of these statements will be given elsewhere. The helicity density is an observable property of atoms and molecules which apparently has not been considered previously.

The asymmetry in the cross-sections for triplet Ps formation in dissymmetric molecules may be calculated from the helicity density by means of an approximate equation derived from equation (13):

$$H_{Ps} \approx \frac{1}{6} \frac{\int d\Omega' |F_{Ps}(\bar{Q})|^2 \hat{\mathbf{p}} \cdot \hat{\mathbf{q}} h_{Av}(q)}{\int d\Omega' |F_{Ps}(\bar{Q})|^2 |\phi_0(\bar{q})|^2_{Av}} \tag{30}$$

where $F_{Ps}(\bar{Q})$ is the Fourier transform of the product of the Ps wave function and an effective potential energy for the interaction between the incident and target particles, $\hat{\mathbf{q}} = \bar{\mathbf{P}}' - \bar{\mathbf{p}}$, $\bar{\mathbf{Q}} = \frac{1}{2}(\bar{\mathbf{q}} - \bar{\mathbf{p}})$, and where the other quantities have been defined previously. A similar approximate relation is obtained for the radiolysis asymmetry:

$$H_R \approx -\frac{1}{2} \frac{\int_0^{(E_e - \Gamma)/2} dE_b p_e p_b \int d\Omega_e \, d\Omega_b F_R(\bar{Q}_e) F_R(\bar{Q}_b) \hat{\mathbf{p}} \cdot \hat{\mathbf{q}} h_{Av}(q)}{\int_0^{(E_e - \Gamma)/2} dE_b p_e p_b \int d\Omega_e \, d\Omega_b [F_R^2(\bar{Q}_e) + F_R^2(\bar{Q}_b) - F_R(\bar{Q}_e) F_R(\bar{Q}_b)] |\phi_0(\bar{q})|^2_{Av}} \tag{31}$$

where F_R is defined similarly to F_{Ps}, and where $\bar{Q}_e = \bar{\mathbf{p}}_e - \bar{\mathbf{p}}$, $\bar{Q}_b = \bar{\mathbf{p}}_b - \bar{\mathbf{p}}$, and $\bar{\mathbf{q}} = \bar{\mathbf{p}}_e + \bar{\mathbf{p}}_b - \bar{\mathbf{p}}$. The effect of F_{Ps} or F_R is to weight the integrands in equations (30) and (31) heavily towards small values of $\bar{Q}$, $\bar{Q}_e$, and $\bar{Q}_b$ to the extent allowed by energy and momentum conservation. The primary advantage of writing H_{Ps} and H_R in terms of the helicity density is a simple unified semiquantitative understanding of the mechanism producing these asymmetries which is reminiscent of the qualitative "helical electron gas" model of Garay and Hrasko[9].

Equations (30) and (31) indicate that the signs of H_{Ps} and H_R are determined by the function $\hat{\mathbf{p}} \cdot \hat{\mathbf{q}} h_{Av}(q)$, since all of the other quantities appearing in these equations have a positive sign. In view of the expected difficulty of accurately calculating the sign of h_{Av} theoretically, an important question is whether the sign of H_R at high energies can be predicted from a knowledge of the sign of H_{Ps} at low energies, as the latter may be determined experimentally by the methods of ref. 7. As yet the dependence of h_{Av} on the bound-electron momentum q is not understood sufficiently to answer this question, but if it can eventually be answered in the affirmative, the positronium formation asymmetry measurements may then be able to provide the answer to the important question of the relative susceptibility of D compared with L isomers to ionization by the polarized electrons from nuclear β decay.

I thank G. W. Ford, D. W. Gidley, A. Rich, P. G. H. Sandars, J. Van House, and P. W. Zitzewitz for hospitality and for helpful and stimulating discussions.

Received 30 December 1981; accepted 7 April 1982.

1. Kizel, V. A. Soviet Phys. Usp. 23, 277–295 (1980).
2. Norden, B., J. molec. Evolut. 11, 313–332 (1978).
3. Keszthelyi, L. Origins of Life 8, 299–340 (1977); Origins of Life 11, nos. 1/2 (March/June 1981).
4. Rein, D. W. J. molec. Evolut. 4, 15–22 (1974).
5. Hegstrom, R. A., Rein, D. W. & Sandars, P. G. H. J. chem. Phys. 73, 2329–2341 (1980).
6. Vester, F., Ulbricht, T. L. V. & Krauss, H. Naturwissenschaften. 46, 68 (1959).
7. Gidley, D. W., Rich, A., Van House, J. C. & Zitzewitz, P. W. Nature 297, 639–643 (1982).
8. Taylor, J. R. Scattering Theory (Wiley, New York, 1972).
9. Garay, A. S. & Hrasko, P. J. molec. Evolut. 6, 77–89 (1975).
10. Inokuti, M. Rev. Mod. Phys. 43, 297–347 (1971).
11. Zel'dovich, Ya. B. & Saakyan, D. B. Soviet Phys. JETP 51, 1118–1120 (1980).
12. Condon, E. U. Rev. Mod. Phys. 9, 432–457 (1937).
13. Mann, A. K. & Primakoff, H. Origins of Life 11, 255–265 (1981).

Reprinted from Nature, Vol. 314, No. 6010, pp. 438-441, 4 April 1985
© Macmillan Journals Ltd., 1985

Weak neutral currents and the origin of biomolecular chirality

D. K. Kondepudi & G. W. Nelson

Center for Statistical Mechanics and Thermodynamics,
The University of Texas, Austin, Texas 78712, USA

It has long been known that Earth's biochemistry is overwhelmingly dissymmetric or chiral[1-4]. In model chemical systems[5-7] that spontaneously evolve to a state dominated by either the L or the D enantiomer, parity violation in β-decay and that attributable to weak neutral currents (WNC) in molecules[8,9] is thought to be too small to have any significant influence on the emergent chirality[10,11]. Other conceivable systematic chiral influences are generally even weaker[12-14]. We show here that there is a simple and extremely sensitive mechanism by which a minute but systematic chiral interaction, no stronger than the WNC interaction in amino acids, can, over a period of ~15,000 yr, determine which enantiomer will dominate. Such a mechanism is especially interesting when considering the origins of terrestrial biochemistry, particularly in view of the work by Mason and Tranter[15], who found that it is the terrestrially dominant L amino acids that are favoured by the WNC interaction.

The process occurs in a randomly fluctuating environment. Chemical systems that, in conditions that are thermodynamically far from equilibrium, can evolve spontaneously to a chirally asymmetric state—that is 'break chiral symmetry'—exhibit a universal behaviour that is a consequence of the symmetry properties of the system[14,16-18]. The amplitude α of the chiral dissymmetry, with the inclusion of fluctuations, obeys the stochastic (Langevin) equation

$$\frac{d\alpha}{dt} = -A\alpha^3 + B(\lambda - \lambda_c)\alpha + Cg + C'\eta f_2(t) + \varepsilon_1^{1/2} f_1(t) \quad (1)$$

in which A, B and C are constants that depend on the kinetics[14,16]; $g \equiv (\Delta E/kT)$, where k is the Boltzmann constant and T the temperature, is the factor by which the Arrhenius reaction rate constants for the L and D enantiomers differ because of a small difference, ΔE, in their reaction barrier energies caused by an extremely weak chiral interaction[14]—such as WNC. In addition to such systematic effects, there is the fluctuating chiral influence from the environment, such as that attributable to circularly polarized ultraviolet light, represented by $C'\eta f_2(t)$, where $f_2(t)$ is assumed to be a normalized gaussian white noise. We estimate the numerical value of $C'\eta$ from the best known data on circularly polarized light. The intrinsic thermodynamic fluctuations[19,20] are represented by $\varepsilon_1^{1/2} f_1(t)$; again $f_1(t)$ is assumed to be a normalized gaussian white noise; $\varepsilon_1 = (Q/VN_A)$, where Q can be calculated from the chemical kinetics; V is the volume over which the concentration of the reactants may be assumed homogeneous and N_A is the Avogadro number. The most general stochastic equation must include fluctuations in A and B but, as they have no significant effect on the mechanism[21,22], we ignore them.

As an example and for later numerical considerations, we shall consider the following model scheme of reactions studied in detail in ref. 14

$$S + T \underset{k_{-1}}{\overset{k_1}{\rightleftharpoons}} X_{L(D)} \quad (i)$$

$$S + T + X_{L(D)} \underset{k_{-2}}{\overset{k_2}{\rightleftharpoons}} 2X_{L(D)} \quad (ii)$$

$$X_L + X_D \overset{k_3}{\longrightarrow} P. \quad (iii)$$

In this scheme, the chiral species X in the two enantiomeric forms X_L and X_D is produced from the achiral substrate S and T directly through reaction (i) and autocatalytically through reactions (ii). With a suitable supply of S and T (to maintain their concentrations at a fixed level), and the irreversible removal of X through reaction (iii), the system can be driven far from thermodynamic equilibrium. The variables of equation (1) for this system are: $\lambda \equiv [S][T]$ and $\alpha \equiv ([X_L]-[X_D])/2$ (where [] denote concentrations). When λ exceeds a certain critical value λ_c, the steady-state value of α switches from zero to either $\alpha > 0$ or $\alpha < 0$ (ref. 14). If the small chiral interaction influences reactions (i) and (ii) so that the rate constants of the L and D enantiomers are unequal, $k_{1L} = k_{1D}(1+g)$, $k_{2L} = k_{2D}(1+g)$, then $C = [k_1 + k_2\beta_c]\lambda_c$, where $\beta \equiv ([X_L]+[X_D])/2$, k_1 and k_2 are rate constants when $g = 0$, and the subscript 'c' denotes values at the critical point. Similar, but more involved, expressions may be obtained for A and B (ref.14).

In contrast to the earlier studies[23-25] that mainly examined such systems for $\lambda > \lambda_c$, we study the system as it slowly evolves through the critical point. The macroscopic steady states (when the fluctuations are ignored) and a sample fluctuating trajectory for the time evolution of α of equation (1) are shown in Fig. 1. When λ is well below λ_c, there is only one steady state, $\alpha \approx Cg/B(\lambda - \lambda_c) \ll 1$, which becomes unstable when λ goes beyond λ_c; for $\lambda > \lambda_c$, the system has two new supercritical branches, $\alpha \approx \pm\sqrt{B(\lambda - \lambda_c)/A}$, to which it can evolve. In the absence of g, the two supercritical branches emerge symmetrically (as a parabola) from the point λ_c and, as the system evolves through the critical point, the fluctuations make both states equally probable. When $g \neq 0$ this is no longer true: one branch is favoured. The effect of g is most marked in the vicinity of the critical point where it separates the two stable supercritical branches by a minimum of $S = (3/2)(4/A)^{1/3}(Cg)^{1/3}$; in contrast, well above and below the critical point, the shift is proportional to Cg. As $Cg \ll 1$, the fractional exponent indicates the enhanced sensitivity of the system near the critical point. However, in this region the α fluctuations are also large. Our aim is to obtain the probability of the system evolving to the favoured branch as λ goes through the critical point at a given rate.

An important factor in calculating this probability is $\varepsilon_1 = Q/VN_A$, which is derived assuming the system is homogeneous over the volume V. We estimate V by considering the homogenization that occurs through diffusion and, for prebiotic considerations, large-scale mixing as might occur in an ocean or a lagoon.

The evolution of the system may be considered in four stages (see Fig. 1). In stages I and IV, the system is well below and well above the critical point respectively; during stages II and III, it is in the vicinity of the critical point where the selection of branches takes place. In stages I and II, there is only one stable steady state and hence the system can be homogeneous over limitlessly large volume. In stage II, if $\alpha \ll 1$ and $\lambda \approx \lambda_c$, α begins to grow slowly, essentially at an average rate Cg. In stage III, however, the system has two possible steady states and hence the volume over which it may be assumed homogeneous has an upper limit. In this stage, in a small volume δV, the autocatalysis of the chemistry (reflected in the term $B(\lambda - \lambda_c)$, $\lambda > \lambda_c$) will cause a fluctuation in α to grow; this growth, however, is curtailed by diffusion and large-scale mixing which transport the excess of α out of δV. As the rate of growth of α

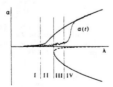

Fig. 1 A sample trajectory of α in equation (1) as λ increases through the critical point (fluctuations exaggerated). Solid line, stable steady states; dashed line, unstable steady state. Stages I-IV are explained in the text.

due to the chemical reactions is proportional to δV, whereas depletion is proportional to the surface area of δV, there is a 'nucleation volume' V_c below which α cannot grow. Within this volume V_c, homogeneity is maintained. Diffusion alone can maintain homogeneity over a length scale $l_D \simeq \sqrt{D/B(\lambda - \lambda_c)}$, D being the diffusion coefficient[26,27]. In our numerical simulation, in stage III, $B(\lambda - \lambda_c) \leq 10^{-11}\,\text{s}^{-1}$, and with $D \sim 10^{-5}\,\text{cm}^2\,\text{s}^{-1}$ we see that $\lambda_D \sim 10^3$ cm. Also, stages II and III are traversed in $\sim 6,000$ yr, during which we may expect homogenization due to large-scale mixing to be on an oceanic scale. We assume this volume V_c to be at least 4×10^9 l (1 km $\times$ 1 km $\times$ 4 m) for our estimate of ε_1. As indicated by the fluctuating trajectory in Fig. 1, by the time stage IV is reached, if the system is already well within the region of attraction of the favoured (upper) branch, it becomes increasingly improbable for a fluctuation in α, occurring at least over a volume V_c, to be large enough to reach the unfavoured (lower) branch.

To obtain the probability for the selection of branches, we study the Fokker–Planck[28,29] equation associated with equation (1), which describes the evolution of the probability density $P(\alpha)$ of α:

$$\frac{\partial P}{\partial t} = -\frac{\partial}{\partial \alpha}\left(-A\alpha^3 + B(\lambda(t) - \lambda_c)\alpha + Cg\right)P(\alpha, t)$$
$$+\left(\frac{\varepsilon}{2}\right)\frac{\partial^2}{\partial \alpha^2}P(\alpha, t) \qquad (2)$$

where $\varepsilon = \varepsilon_1 + (C'\eta)^2$. We suppose, at $t = 0$, λ is well below the critical point and is gradually increasing. Fluctuations in λ have an insignificant role in the process of selection[21], so we ignore them in our theoretical discussion, though not in our numerical simulation of the above model shown in Fig. 2. From equation (2) it follows that, for λ well below λ_c, $P(\alpha, t)$ is essentially a gaussian whose centre is at $\bar\alpha \simeq Cg/B(\lambda_c - \lambda)$, very close to zero (stage I in Fig. 1). As λ increases at a reasonable rate ($\sim 10^{-5}$ M^2 $(10^4\,\text{yr})^{-1}$ for the model) in the vicinity of the critical point, $P(\alpha, t)$ is still a gaussian, but it begins to relax very slowly to the stationary distribution. Thus, in this stage (stages II and III in Fig. 1), $P(\alpha, t)$ is vanishingly small for large α and hence the $-A\alpha^3$ term may be neglected in comparison with the other terms.

For selection, then, we need only consider

$$\frac{\partial P}{\partial t} = -\frac{\partial}{\partial \alpha}\left(B(\lambda(t) - \lambda_c)\alpha + Cg\right)P(\alpha, t) + \left(\frac{\varepsilon}{2}\right)\frac{\partial^2}{\partial \alpha^2}P(\alpha, t) \qquad (3)$$

When $\lambda \simeq \lambda_c$, the term $B(\lambda - \lambda_c)\alpha$, for $\alpha \ll 1$, also becomes small compared with Cg. Here we have a gaussian whose peak is

shifting at a constant rate Cg, but whose width is also increasing because of the 'diffusion term' containing ε. In a time interval T, the drift of the peak will be CgT, while the increase in the width will be $\sqrt{\varepsilon T}$. Thus, for sufficiently large T, even when $Cg \ll \sqrt{\varepsilon}$, CgT can exceed $\sqrt{\varepsilon T}$, which implies that much of $P(\alpha, t)$ will drift into the region $\alpha > 0$ (for $Cg > 0$). This is the heart of the selection mechanism. Now, as λ goes beyond λ_c (stage IV in Fig. 1), the term $B(\lambda - \lambda_c)$ is positive and will cause $P(\alpha, t)$ to spread out rapidly on either side. $P(\alpha, t)$, thus split into two parts, will accumulate around the macroscopic steady states because of the term $-A\alpha^3$; note, however, that by the time this term becomes important, how much of $P(\alpha, t)$ will have evolved into $\alpha > 0$ and $\alpha < 0$ regions will already have been determined. Thus, equation (3) is all we need to calculate the probability of branch selection—an approximation well supported by the numerical solution of the complete equation.

The solution to equation (3), well known[29,30] for constant λ, can be extended to time-dependent λ. With $P(\alpha,0) = \delta(\alpha - \alpha_0)$, the solution $P(\alpha, t|\alpha_0)$ is a gaussian that is drifting and spreading

$$P(\alpha, t|\alpha_0) = \frac{1}{[\pi z(t)]^{1/2}}\exp\left[\frac{(\alpha - \bar\alpha(t))^2}{z(t)}\right] \qquad (4)$$

where

$$\bar\alpha = \alpha_0 \exp(w(t)) + Cg\,\exp w(t)\int_0^t \exp(-w(t'))\,\mathrm{d}t'$$

in which

$$w(t) = \int_0^t B(\lambda(t') - \lambda_c)\,\mathrm{d}t'$$

and

$$z(t) = 2\varepsilon\int_0^t \exp\left(2\int_{t'}^t B(\lambda(t'') - \lambda_c)\,\mathrm{d}t''\right)\mathrm{d}t'$$

For the probability of selection of the $\alpha > 0$ chiral steady state, P_+, we have: $P_+ = \mathrm{Lt}_{t\to\infty}\int_0^\infty P(\alpha, t/\alpha_0)\,\mathrm{d}\alpha$. Because $P(\alpha, t|\alpha_0)$ is a gaussian and $\alpha_0 \approx 0$, we may write

$$P_+(t) = \frac{1}{\sqrt{2\pi}}\int_{-\infty}^N e^{-x^2/2}\,\mathrm{d}x \qquad (5)$$

where

$$N = Cg\left[\exp w(t)\int_0^t \exp(-w(t'))\,\mathrm{d}t'\right]\bigg/\sqrt{z(t)/2} \qquad (6)$$

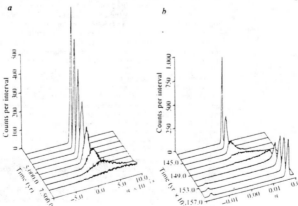

a   b

Counts per interval

Fig. 2 Results of simulation of equation (2) by 5,000 sample trajectories of equation (1), with $\lambda(t) = \lambda_0 + \gamma t + 0.05 \times \lambda_c \times f(t)$, where $\lambda_0 = 0.8 \times 10^{-5}$ M^2, $\gamma = 3.171 \times 10^{-17}$ M^2 s^{-1}, and $f(t)$ is normalized gaussian noise; t goes from 0 to 15,600 yr. Other parameters are as given in the text. Each curve is a histogram of the number of times $\alpha(t)$ fell within an interval $1/400$ of full α scale for the given t. a, $P(\alpha)$ drifting and spreading while λ is near λ_c, which here occurs at $t = 3,000$ yr. b, $P(\alpha)$ spreading and splitting for $\lambda \gg \lambda_c$. Note the change of scale in α by 10^{12} between a and b, because of which the number of trajectories per interval has sharply increased in b.

12

gives the number of standard deviations by which the peak of $P(\alpha, t)$ has drifted from the origin. Now, if we let $\lambda = \lambda_0 + \gamma t$ (λ_0 well below λ_c), for $t \to \infty$, we get

$$N = Cg(\varepsilon/2)^{-1/2}(B\gamma/\pi)^{-1/4} \qquad (7)$$

Thus, equations (7) and (5) give the required probability for the selection of a branch resulting from a chiral interaction.

For the biomolecular context we consider the above model with kinetic constants $k_1 = 5 \times 10^{-5} \, \mathrm{M}^{-1} \mathrm{s}^{-1}$, $k_2 = 2.5 \times 10^{-5} \, \mathrm{M}^{-2} \mathrm{s}^{-1}$, $k_3 = 10^{-3} \, \mathrm{M}^{-1} \mathrm{s}^{-1}$, $k_{-1} = 2.5 \times 10^{-10} \, \mathrm{s}^{-1}$ and $k_{-2} = 1.25 \times 10^{-10} \, \mathrm{M}^{-1} \mathrm{s}^{-1}$. With these values the coefficients of equation (1) are: $A = 1.7 \times 10^{-7} \, \mathrm{M}^{-1} \mathrm{s}^{-1}$, $B = 2.5 \times 10^{-5} \mathrm{M}^{-2} \mathrm{s}^{-1}$, $C = 2.5 \times 10^{-10} \, \mathrm{M} \mathrm{s}^{-1}$ and $\lambda_c = 1.0 \times 10^{-5} \, \mathrm{M}^2$; all concentrations are $\sim 10^{-3}$ M or less during the process of selection and $[\mathrm{X}] \sim 10^{-2}$ M well above λ_c. If we suppose $[\mathrm{S}]$ and $[\mathrm{T}]$ are increasing slowly so that $\lambda \equiv [\mathrm{S}][\mathrm{T}]$ increases from $0.5 \lambda_c$ to $1.5 \lambda_c$ in 10,000 yr, then $\gamma = 3.2 \times 10^{-17} \, \mathrm{M}^2 \mathrm{s}^{-1}$. For $\varepsilon_1 = Q/VN_A$, $Q = (k_1\lambda_c)/2$ for the model [14], and we take $V = 4 \times 10^9$ l as explained above, so that $\varepsilon_1 \simeq 10^{-43} \, \mathrm{M}^2 \mathrm{s}^{-1}$. Using the results of Mason and Tranter for WNC [15], we take $g \simeq 10^{-17}$, which makes $Cg = 2.5 \times 10^{-27}$. For the magnitude of the random chiral influences, we take $C'\eta = 3 \times 10^{-22}$, an estimate explained below. Using these values in equation (7) we get $N \simeq 2.0$, which implies $P_+ \simeq 0.98$, a 98% chance that the enantiomer favoured by WNC will emerge dominant even though the r.m.s. values of the random chiral influences are five orders of magnitude larger. Such sensitivity cannot be realized if the system does not evolve through the critical point.

To check the validity of our approximation of using equation (3) instead of equation (2), we numerically modelled the stochastic equation (1), with $\lambda = \lambda_0 + \gamma t + 0.05 \lambda_c f(t)$, $f(t)$ being a gaussian noise. The results are shown in Fig. 2. For easier graphical visualization, we set $\varepsilon = 3.9 \times 10^{-43}$. Then, according to equation (5), $N = \sqrt{2}$, implying $P_+ = 0.921$ selectivity. Figure 2 is the result of averaging 5,000 trajectories and gives a $P_+ = 0.916$ in good agreement with the analytical result. If λ evolves extremely slowly, the approximation of using equation (3) is not valid. In the limit of infinitely slow λ_D, the results in refs 14, 16 can be used; for the intermediate region no analytical results exist.

For the estimate of the r.m.s. value of the random environmental factors, we take circularly polarized light in the ultraviolet range to be typical. (Hardly any reliable data exist for other effects.) Chirally selective photochemical effects of circularly polarized light are well known [31-33]. In our model, let us assume that chirally selective degradation of X (back reaction of reaction (i)), at a rate $k'[\mathrm{X}]$, occurs with a mean life of about 3,000 yr, that is $k' \simeq 10^{-11} \, \mathrm{s}^{-1}$ for the average solar intensity. (Note, this is also the racemization rate.) For 100% circularly polarized light, the rate of decay of one of the enantiomers is faster by a factor $s \simeq 10^{-3}$ or less for most molecules [32,33], although there are exceptions [34]. Only a fraction $q \simeq 10^{-3}$ of the solar intensity at dawn and dusk is found to be circularly polarized in the infrared frequencies and is at least an order of magnitude smaller for the ultraviolet [35]. The sense of polarization depends on the direction and on the average it is zero. Considering the reduction of q for daylight intensities and attenuating factors in a large body of water, we may take $q_{rms} \simeq 10^{-6}$ with a correlation time $\tau \simeq 10^3$ s. On the evolutionary timescale of $10^{10}-10^{12}$ s considered here, this may be considered white noise of strength $\sqrt{2\tau} \, q_{rms}$ (equivalent to diffusion constant of $(1/2\tau)$ steps per unit time, each of length $2\tau q_{rms}$). Thus, the rate of asymmetric synthesis is $k'[\mathrm{X}_{L(D)}]qs$; when q is a fluctuating quantity it may be represented by a gaussian white-noise term $C'\eta f_2(t)$, with $C'\eta = k'[([\mathrm{X}_L] + [\mathrm{X}_D])/2]s\sqrt{2\tau} \, q_{rms}$. By analogy with Cg, we may define $C' \simeq k'([\mathrm{X}_L] + [\mathrm{X}_D])/2$ and $\eta = s\sqrt{2\tau} \, q_{rms}$. With the above numerical values and $[\mathrm{X}] \simeq 10^{-3}$ M, $C'\eta \simeq 3 \times 10^{-22}$, the value used in our numerical estimates.

We thank Professors I. Prigogine, J. Whitesell, C. Van den Broeck and M. Malek-Mansour for stimulating and useful discussions. D.K.K. thanks the US Department of Energy, Basic Energy Sciences (grant DE-AS05-81ER10947) and G.W.N. thanks the Welch Foundation and the International Paper Company for financial support.

Received 5 November 1984; accepted 23 January 1985.

1. Mason, S. F. Int. Rev. phys. Chem. 3, 217-241 (1983); Nature 311, 19-23 (1984).
2. Elias, W. E. J. chem. Educ. 49, 448-454 (1972).
3. Miller, S. W. & Orgel, L. E. The Origin of Life on Earth (Prentice-Hall, New Jersey, 1974).
4. Fox, S. W. & Dose, K. Molecular Evolution and the Origin of Life (Dekker, New York, 1977).
5. Frank, F. C. Biochim. biophys. Acta 11, 459-463 (1953).
6. Seelig, F. F. J. theor. Biol. 31, 335-361 (1971).
7. Decker, P. J. molec. Evol. 4, 49-65 (1974).
8. Zel'dovich, Ya. B., Saakyan, D. B. & Sobel'man, I. I. Pis'ma Zh. eksp. teor. Fiz. 25, 106-109 (1977) (J. exp. theor. phys. Lett. 25, 94-98].
9. Hegstrom, R. A., Rein, D. W. & Sandars, P. G. H. J. chem. Phys. 73, 2329-41 (1980).
10. Thiemann, W. (ed.) Origins Life 11, 1-194 (1981).
11. Keszthelyi, L. Origins Life 14, 375-382 (1984).
12. Mead, C. A. & Moscowitz, A. J. Am. chem. Soc. 102, 7301-7302 (1980).
13. Peres, A. J. Am. chem. Soc. 102, 7389-7390 (1980).
14. Kondepudi, D. K. & Nelson, G. W. Physica 125A, 465-496 (1984).
15. Mason, S. F. & Tranter, G. E. JCS chem. Commun., 117-119 (1983); Molec. Phys. 53, 1091-1111 (1984).
16. Kondepudi, D. K. & Nelson, G. W. Phys. Rev. Lett. 50, 1023-1026 (1983).
17. Kondepudi, D. K. & Prigogine, I. Physica 107A, 1 (1981).
18. Sattinger, D. H. Lecture Notes in Mathematics No. 762 (Springer, Berlin, 1979).
19. Mangel, M. J. chem. Phys. 69, 3697-3708 (1978).
20. Keizer, J. J. chem. Phys. 69, 2609-2620 (1978).
21. Kondepudi, D. K. in Fluctuations and Sensitivity in Nonequilibrium Systems (eds Horsthemke, W. & Kondepudi, D. K.) 204-213 (Springer, Berlin 1984).
22. Kondepudi, D. K. & Nelson, G. W. Phys. Lett. 106A, 203-206 (1984).
23. Mangel, M. Phys. Rev. A24, 3226-3238 (1981).
24. Morozov, L. L., Kuz'min, V. V. & Gol'danskii, V. I. Origins of Life 13, 69-101 (1983).
25. Morozov, L. L., Kuz'min, V. V. & Gol'danskii, V. I. Pis'ma Zh. exsp. teor. Fiz. 39, 344-345 (1984); JETP Lett. 39, 414-416 (1984).
26. Nicolis, G. & Prigogine, I. Self Organization in Nonequilibirum Systems (Wiley, New York, 1977).
27. Nicolis, G. & Malek-Mansour, M. Phys. Rev. A29, 2845-2853 (1984).
28. Nitzan, A., Ortoleva, P., Deutch, J. & Ross, J. J. chem. Phys. 61, 1056-1074 (1974).
29. Van Kampen, N. G. Stochastic Processes in Physics and Chemistry (North-Holland, Amsterdam, 1981).
30. Uhlenbeck, G. E. & Ornstein, L. S. Phys. Rev. 36, 824-841 (1930).
31. Kuhn, W. & Braun, F. Naturwissenschaften 17, 227-228 (1929).
32. Kagan, H. B. et al Tetrahedron Lett. 27, 2479-2482 (1971).
33. Kagan, H. B., Balavoine, G. & Moradpour, A. J. molec. Evol. 4, 41-48 (1974).
34. Flores, J. J., Bonner, W. A. & Massey, G. A. J. Am. chem. Soc. 99, 3622-3624 (1977).
35. Angel, J. R. P., Illing, R. & Martin, P. G. Nature 238, 389-390 (1972).

II. HOMOCHIRALITY
AND LIFE

The Quest for Chirality

William A. Bonner

Department of Chemistry
Stanford University, Stanford, California 94305

Abstract. The indispensable role played by homochirality and chiral homogeneity in the self-replication of crucial biomolecules is stressed, with the conclusion that life could neither exist nor originate without these chiral molecular attributes. Hypotheses historically proposed for the origin of chiral molecules on Earth are reviewed, including biogenic theories as well as abiotic theories embracing both indeterminate and determinate mechanisms. Indeterminate mechanisms, including autocatalytic symmetry breaking, asymmetric adsorption on quartz and clay minerals, and asymmetric syntheses in chiral crystals, are discussed and evaluated in the context of the prebiotic environment. Abiotic determinate mechanisms based on electric, magnetic and gravitational fields, on circularly polarized light (CPL), and on parity violation effects are summarized, with the emphasis that only CPL has proved practicable experimentally, but that it would be implausible on the primitive Earth. Mechanisms for the amplification of small, indigenous enantiomeric excesses are discussed, with one involving the partial polymerization of amino acids and the partial hydrolysis of polypeptides suggested as potentially viable prebiotically. Aspects of the turbulent, chirality-destructive primeval environment are described, with the conclusion that all of the above mechanisms for the *terrestrial* prebiotic origin of chirality would be non-viable, and that an alternative extraterrestrial source for the accumulation of chiral molecules on primitive Earth must have been operative. A scenario for this is outlined, in which we postulate that asymmetric photolysis of the organic mantles on interstellar grains in molecular clouds by circularly polarized ultraviolet synchrotron radiation from the neutron star remnants of supernovae produces chiral molecules in the grain mantles. These grains with their chiral mantles are ultimately accumulated by Earth as the planet sweeps through the molecular clouds, or by impact delivery to Earth during the prebiotic era after incorporation of the grains into comets or asteroids.

THE ESSENTIAL FUNCTION OF HOMOCHIRALITY

At the outset we must emphasize the crucial dependence of life on molecular chirality, a relationship increasingly appreciated since the time of Pasteur, around the middle of the last century. Today it is common knowledge that the biopolymers supporting life are characterized both by unique homochirality (either all D- or all L-monomer units in the biopolymers) as well as by absolute enantiomeric homogeneity (exclusively D- or L-monomers in each polymer). Thus one finds only L-amino acid monomers forming proteins and only D-sugars in DNA and RNA polymers, with no traces of the "unnatural" enantiomeric monomers in either type of polymer chain.

Why are homochirality and enantiomeric purity of such importance in the molecular structures of biopolymers? Two principal characteristics of life - the synthesis of specific proteins facilitated by DNA and RNA and the replication of DNAs to transcribe genetic information - require a precise, intimate fitting together (complementarity) of the helical polymer strands involved. In 1974 Miller and Orgel (1) first suggested that, although the replication of nucleic acids would be equally possible with either pure D- or pure L-ribotides, it would be impossible with *mixtures* of D- and L-ribotides, and in 1988 Goldanskii and Kuzmin (2) showed with molecular models that the presence of "unnatural" L-sugars in double-stranded ribotide helices disrupted H-bonding between their bases in such ways as to preclude the possibility of complementarity. This rationalized the 1984 observations of Joyce and coworkers (3), who showed that in template - directed nucleotide oligomerizations the assemblage of a complementary strand on a chirally pure template was strongly inhibited if the monomers being incorporated were not themselves chirally pure.We have recently reviewed (4) a number of other studies documenting the importance of chiral purity in facilitating oligomerizations leading to polypeptides and other polymers, as well as in providing stability toward hydrolysis for polypeptides once formed.Today there is increasing acceptance of the 1957 assertion of Terent'ev and Klabunovskii (5) that "life cannot and never could exist without molecular dissymmetry" and of the more recent views of Avetisov *et al.*(6), Goldanskii (7), Goldanskii and Kuzmin (2) (8), and Keszthelyi (9) that homochirality and enantiomeric purity are essential for the self - replication of biopolymers and hence, by implication, for the origin of life. In other words: No Chirality - No Life.

We can now appreciate the fundamental dilemma. Since laboratory syntheses of chiral molecules results in racemic D,L- or (±)-mixtures, how could molecules of a singular chirality arise on Earth, and how did they attain the homochirality and enantiomeric homogeneity necessary for self-replication and life? These questions - *The Quest for Chirality* - are fundamental to the question of the origin of life, and have thus intrigued scientists for well over a century. Unfortunately, they still remain for the most part unanswered. In the sections below, using D- or L- to designate enantiomers of amino acids or sugars and (+)- or (–)- for enantiomers of other types of chiral molecules, we shall summarize briefly the more important experimental and theoretical attempts which have been made to address these questions, and then try to evaluate the suggested answers in the realistic context of the prebiotic environment. This compilation supplements our previous more extensive reviews of these topics (4) (10) (11) (12) (13).

TERRESTRIAL ORIGINS OF CHIRALITY

Biogenic Hypotheses

Biogenic hypotheses for the origin of chiral molecules on Earth presuppose that life was generated in a racemic primordial environment, and that homochirality developed only later to allow for greater biochemical efficiency. Over sixty years ago Mills (14) set the stage for such hypotheses with his peremptory proposal that "the optical activity (*i.e.chirality*) of living matter is an inevitable consequence of its property of growth." Since that time a number of analogous *ad hoc* speculations have been advanced (see Bonner (4) (10) (11) (12)), and as late as 1987 Bada and Miller (15) have maintained that the origin of chiral molecules on Earth "must have occurred at the time of the origin of life or shortly thereafter." We should stress, however, that there is not a shred of experimental evidence supporting such contentions, and that the now - known requirements of homochirality and enantiomeric purity for the self-replication of biopolymers clearly render such biogenic scenarios untenable.

The earlier Panspermia hypothesis, proposed by Arrhenius almost a century ago (see 16), postulates that life was introduced onto Earth by extraterrestrial micro-organisms which, driven through space from other solar systems by "solar winds", ultimately infected our planet with life. While it has collected its fair share of both adherents and detractors in the intervening years (Bonner (4) (12), Marcus and Olsen (17)), the Panspermia hypothesis must from our point of view be regarded as an inherently sterile one, since it merely relegates the origins of chirality and of life to some more remote time and more distant galactic locale.

Finally, we can only logically agree, it would seem, with the recent conclusions of Goldanskii and Kuzmin (2) that a "biogenic scenario of the origin of chiral purity of the biosphere could not, even in principle, be realized in the course of evolution, since without the chiral purity of the medium, the apparatus of self - replication, which is the process of self - reproduction of any organisms, cannot appear: life cannot appear in a racemic medium." Thus some abiogenic processes, whether on Earth or elsewhere, must have provided the causal mechanism(s) leading first to terrestrial chirality and then to the homochirality and enantiomeric homogeneity of terrestrial biopolymers which permitted life to emerge.

Abiogenic Indeterminate Mechanisms

Abiogenic mechanisms which have been proposed historically for the terrestrial generation of chiral molecules fall into two classes, *indeterminate* and *determinate*. In surveying the theoretical and experimental aspects of such mechanisms below, we shall try to evaluate the potential viability of each mechanism within the constraints imposed by the external physical environment. In addition, we should emphasize a further obvious point, namely, that *the mere experimental demonstration that a particular mechanism is effective in generating chiral molecules under contrived and controlled laboratory conditions does not automatically endow that mechanism with either relevance or validity in the prebiotic environment.*

Abiogenic indeterminate mechanisms presuppose that the symmetry - breaking occurring at the molecular level is a totally random process. Thus, like the flip of a coin, there is an equal probability that such mechanisms will produce an excess of either a (+)-enantiomer or a (−)-enantiomer. The indeterminate, chance mechanisms which have been most extensively studied theoretically and/or experimentally are the following.

Spontaneous Symmetry Breaking - Autocatalysis

Over 40 years ago Frank (18) proposed a mathematical scheme for random symmetry breaking in a reacting racemic system where one enantiomeric product (*e.g.* (+)) was a catalyst for further production of itself and an anticatalyst (inhibitor) for the production of the enantiomeric (−)-product. He showed kinetically that such an autocatalytic system was unstable, such that any random fluctuation in the 50:50 population of the (±)-reactant (*e.g.* favoring the (+)-reactant) would result in the

19

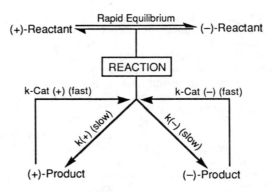

FIGURE 1. Stereospecific Autocatalysis

final dominance of the (+)-product from the favored reactant and the elimination of the enantiomeric (–)-product . Analogous autocatalytic schemes were independently proposed subsequently by Calvin (19) in 1969 and by Seelig (20) in 1971, and these in turn were followed by a plethora of more elaborate mathematical models (see Bonner (11) (12) (21). In general these have involved the kinetic behavior of "far-from-equilibrium" racemic systems which progress to some critical "bifurcation point" where symmetry breaking commences, followed by an inexorable progression of the system into a singular state of enantiomeric homogeneity for the randomly favored product. The only such autocatalytic symmetry breaking mechanism which has been realized experimentally, however, has been Calvin's (19) qualitative scheme for "Stereospecific Autocatalysis," which is illustrated in Figure 1.

Spontaneous Symmetry Breaking on Crystallization. In the recrystallization of racemic *conglomerates*, racemic substances composed of equal numbers of separate crystals of each enantiomer, it sometimes happens that one of the enantiomers may crystallize preferentially. Such spontaneous resolution during crystallization, known since Pasteur's original resolution of sodium ammonium tartrate in 1848, has even to the present era (22) been frequently proposed as a plausible mechanism for the origin of chirality (12). The resolution yield during such crystallizations may be enhanced beyond the obvious 50% maximum, moreover, if the original dissolved enantiomers are able to equilibrate rapidly during the slow crystallization of one of them. The ensuing "2nd order asymmetric transformation" (23) then results in a "total spontaneous resolution" (24), in which the entire racemate may, in principle, crystallize as a single enantiomer. This process of *spontaneous resolution under racemizing conditions* (SRURC) provides the only experimental validation to date of Calvin's stereospecific autocatalysis mechanism, which is shown in the context of SRURC in Figure 2. We see that the reaction involved now becomes the physical process of crystallization, and that the feedback catalyst consists of seed nuclei of whichever enantiomer randomly crystallizes at the bifurcation point.

While several other closely related examples of spontaneous resolution during crystallization have been reviewed recently (12) (21), only three unambiguous realizations of the above SRURC ("enantioselective autocatalysis" (21)) have been recorded during the past half-century. The first was that of Havinga (25) in 1941, who

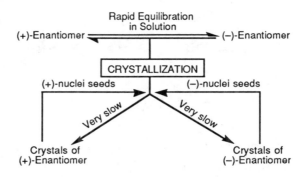

FIGURE 2. Spontaneous Resolution under Racemizing Conditions

found that the crystallization of the quaternary ammonium salt methylethylallyl-anilinium iodide (**I**) from its supersaturated solution in chloroform yielded optically active crystals, sometimes dextrorotatory and sometimes levorotaory, while the mother liquors remained optically inactive. Racemization of the chiral N-atom of **I** occurs through the reversible dissociation of **I** into N-methyl-N-ethylaniline and allyl iodide, as illustrated in Equation (1). Havinga pointed out that his "spontaneous asymmetric synthesis" might provide an appropriate mechanism for the "formation of the first optically active substance", and Calvin (19) cited Havinga's observations as the first example of his "stereospecific autocatalysis."

The second straightforward example of SRURC is provided by the total spontaneous resolution of 1,1'-binaphthyl (**II**), first reported by Pincock and Wilson (26) in 1971. Because of steric hindrance to free rotation about its 1,1'-pivot bond, **II** can exist as two stable enantiomers at room temperature, and optically active samples of **II**, sometimes dextro- and sometimes levorotatory, may results when **II** is allowed to crystallize from the molten state. As illustrated in Equation (2), the racemization of optically active **II**, which permits its SRURC, occurs when it is melted, and the energy barrier to restricted rotation is overcome by the input of thermal energy. Pincock and Wilson (26) emphasized in their original paper that the SRURC of **II** represented an example of Calvin's mechanism for the generation of optical activity.

The third pertinent example of SRURC, now involving enantiomers with only a single asymmetric carbon atom, was described by Okada and coworkers (27) in 1983. They discovered that the bromofluoro-1,4-benzodiazepinooxazole derivative **III** was capable of total spontaneous resolution on recrystallization from methanol, again affording either dextro- and levorotatory crystals which, in turn, underwent rapid racemization when redissolved in methanol. Attributing their observations as due to a 2nd-order asymmetric transformation, Okada *et al.* (28) (29) (30) later studied kinetic and other aspects of this system, and suggested that the rapid equilibration between the enantiomers of **III** occurred *via* the non-chiral benzodiazepinium ion **IV**, as illustrated in Equation (3).

Most of the above examples of SRURC, as well as the somewhat related instances cited elsewhere (12) (21), have been put forward by their authors as models for the prebiotic origin of chirality. Unfortunately, such proposals would appear both unwarranted and unrealistic for several reasons. First, the esoteric molecules under-

$$(1)$$

$$(+)\text{-}\mathbf{I} \qquad\qquad C_6H_5 \qquad\qquad (-)\text{-}\mathbf{I}$$

$$(2)$$

$$(+)\text{-}\mathbf{II} \qquad\qquad (-)\text{-}\mathbf{II}$$

$$(3)$$

$$(+)\text{-}\mathbf{III} \qquad\qquad \mathbf{IV} \qquad\qquad (-)\text{-}\mathbf{III}$$

going SRURC in these studies have generally had no obvious relevance to contemporary biochemical systems, and secondly the physical conditions required for each SRURC were scarcely ones to be anticipated on the primeval Earth. More fundamentally, any solid, crystalline enantiomers produced by SRURC in the aqueous prebiotic environment would inevitably racemize on redissolving in the primitive oceans before undergoing the next stage of their chemical evolution.

Asymmetric Adsorption on Quartz

Quartz crystals found in nature are generally chiral, having either a right (*d-*) or left (*l-*) morphological handedness and a (+) or (−) optical rotation. In 1935 Tsuchida and coworkers (31) in Osaka reported the partial resolution of several racemic cobalt complex salts after passing their solutions through a chromatographic column consisting of *d*-quartz. In 1938 Karagounis and Coumoulos (32), after making similar observations, suggested that such asymmetric adsorption might have occasioned the origin of chiral molecules in nature by "many successive adsorptions and elutions" of racemates on the surfaces of optically active minerals, a view later championed by Bernal (33). During the next thirty years numerous similar claims for the

22

efficacy of quartz as an asymmetric adsorbent were proffered (see Bonner (10)). However, in 1968 Amariglio and coworkers (34) in France were unable to duplicate a number of the earlier observations, and concluded that these positive findings were based on experimental artifacts and on errors inherent in the trivial optical rotations upon which they were based.

In the early 1970s we undertook to reinvestigate these conflicting contentions, using amino acids as "prebiotically realistic" substrates and employing analytical techniques not dependent on the optical rotation criteria criticized by Amariglio *et al.* (34). To evaluate asymmetric adsorption, we used two new experimental proced- ures, namely, radioactivity assays and analytical gas chromatography, to measure the enantiomeric excesses (e.e.s) existing in solutions of initially racemic amino acid derivatives after their exposure to *d*- or *l*-quartz. Provided that scrupulously anhyd- rous solvents were employed, such experiments showed that D-alanine·HCl was preferentially adsorbed by *d*-quartz and L-alanine·HCl by *l*-quartz to the extents of up to 20% (35) (36), and later investigations confirmed these results using other amino acid derivatives (37) (12).

While the above studies indicated that asymmetric adsorption of amino acids on quartz was indeed a verifiable and reproducible phenomenon under proper experi- mental conditions, whatever apparent validity might reside in the earlier (10) or more recent (38) allegations that quartz, either as an asymmetric adsorbent or catalyst, was implicated in the origin of chiral molecules on Earth is nullified by two important considerations. First, the distribution of *d*- and *l*-quartz over the surface of Earth is quite random (39), such that any chirality generated would not be unique, but would be equally distributed between (+)- and (−)-enantiomers. More importantly, the above experiments demonstrating asymmetric adsorption by quartz required care- fully contrived and rigorously maintained anhydrous conditions, with mere traces of moisture nullifying the effect. Since the prebiotic environment was an aqueous one, it is accordingly clear that any direct or indirect mechanism for chirality generation based on asymmetric adsorption by quartz must be summarily dismissed.

Asymmetric Processes on Clay Minerals

In 1970 Degens *et al.* (40) reported that the common clay mineral kaolinite cata- lyzed the asymmetric polymerization of aspartic acid, causing the L-enantiomer to polymerize eight times faster than the D-enantiomer, and a year later Jackson (41) repeated these claims and recounted further that kaolinite also adsorbed phenyl- alanine asymmetrically, with the L-enantiomer more strongly adsorbed at pH 5.8 and the D-enantiomer at pH 2.0. These allegations seemed remarkable to us since clay minerals have no chirality associated with their crystal morphology (42) (43), and we accordingly attempted to verify the observations. Using a variety of alter- native analytical procedures, we found no evidence whatsoever either for the asym- metric adsorption of phenylalanine (44), or for the asymmetric polymerization of aspartic acid (45) by kaolinite, conclusions which were confirmed by others (46) employing yet additional analytical techniques.

Bondy and Harrington (47) later studied the relative adsorptions of ^{3}H-labeled "natural" L-amino acids and D-glucose *versus* their labeled "unnatural" enantiomers by the clay mineral bentonite, assaying the exposed clays for bound radioactivity. Their findings that the "natural" enantiomers were bound 6.5 to 11.3 times more effectively than the "unnatural" ones led to immediate controversy, due to the lack of chirality of the clays. Youatt and Brown (48) ultimately found that the claims were based on an artifact, the adsorption of radioactive decomposition products of the substrates. Using gas chromatography, Friebele and coworkers (49) also found no evidence for stereoselective adsorption of one enantiomer of several racemic amino acids from solutions at different pHs by sodium montmorillonite clay.Thus to date there is no theoretical or experimental evidence of any sort supporting the existence of stereoselective interactions between clays and enantiomeric substrates, despite more recent assumptions to the contrary (see Bonner (12)).

Solid State Lattice-controlled Syntheses in Chiral Crystals

Certain non-chiral organic compounds may, like quartz, crystallize randomly as morphologically chiral crystals. If such a non-chiral substance is able to react in the solid state with itself or with another non-chiral substance intimately incorporated into its chiral crystal lattice, optically active products may result. In solid state asymmetric syntheses of this sort, the chirality of the crystal lattice determines the chirality of the product synthesized within the lattice, with either dextro- or levorotatory products being formed depending on the random chirality of the lattice. An early example of this process is shown in Equation (4), which illustrates the bromination of 4,4'-dimethylchalcone (**V**) to produce one or the other enantiomer of the optically active dibromo product **VI** (50).

Non-chiral Molecule in Chiral Crystal   Chiral Molecule

V   **V I**

A number of analogous solid state lattice-controlled asymmetric syntheses have been described more recently (see (4) (11) (12)), the principles behind them have been extensively reviewed (51) (52) (53), and several of the investigators involved have championed these and several closely related systems (54) as mechanistic models for the origin of chiral molecules in nature. While such stereoselective processes have unquestionably led to the facile synthesis of chiral products, their prebiotic potential, however, would seem rather dubious on several counts: 1) The molecules involved in such syntheses all possess exotic, skillfully selected structures having no obvious relationship to biomolecules, 2) the reaction conditions required prove as a rule to be carefully contrived, very specific and usually critical, and 3) it is hard to see in general how solid state reactions could have had any truly fundamental prebiotic significance on the aqueous primitive Earth.

Abiogenic Determinate Mechanisms

Abiogenic determinate mechanisms postulate that some non-random intrinsic internal or environmentally external physical agent or force, itself chiral by nature, may interact with racemic or prochiral organic substrates in a stereoselective fashion to produce chiral products whose chirality is uniquely predetermined by the chirality of the internl or external agent. Such mechanisms may be classified as either "regional and/or temporal processes," or "universal processes based on parity violation," and we examine them briefly below within the framework of these categories.

Regional and/or Temporal Processes

These processes involve external chiral influences whose inherent chirality may change in direction or magnitude in different topographic or geographical settings on Earth's surface, or in different epochs in Earth's history. The two main processes in this category are the following.

Electric, Magnetic and Gravitational Field Effects. In the middle of the last century Pasteur first experimented with the possibility that magnetic and gravitational fields might be capable of inducing the formation of optically active substances. His experiments were uniformly unsuccessful, however, and Curie later pointed out that such fields were in fact not themselves chiral ones. Nevertheless, between 1939 and the mid-1980s several investigators reported the successful formation of optically active products by conducting syntheses in the presence of electric, magnetic, gravitational, or centrifugal fields, either alone or in combination (see (4) (11) (12)). Such positive claims, based uniformly on trivial optical rotations in the low millidegree range, have sparked a vigorous and sometimes acrimonious dialog as to their experimental and theoretical validity, and to date none of the positive claims has been substantiated independently. A more recent 1986 report of Takehashi and coworkers (55) has alleged that the electrolytic reduction of 2-keto acids (R-CO-COOH) to 2-hydroxy acids (R-CH(OH)-COOH) at a mercury cathode perpendicular to a magnetic field of 0.168 T yielded chiral products in optical yields of up to 25%. We have reinvestigated (56) this remarkable claim and found that the product from such a reduction was totally racemic, even when produced in the vastly more powerful field of 7.03 T. In 1994 Zadel and coworkers (57) made the unbelievable claim that reactions of aldehydes and ketones with Grignard reagents or with lithium aluminum hydride in a magnetic field yielded alcohol products of 100% optical purity. After several unsuccessful attempts (58) (59) to duplicate the results, the claim was subsequently exposed to be fraudulent (60). Thus, while other asymmetric processes based on magnetic fields have been suggested but not yet demonstrated experimentally (12), and while theoretical views are still being advanced as to the possible efficacy of magnetic fields (61), no experimentally substantiated asymmetric reactions have been observed to date under the external influence of either magnetic or other fields of the sorts considered above.

Circularly Polarized Light. Circularly polarized light (CPL) may be regarded as an electromagnetic wave whose electric vector spirals clockwise (to the right, RCPL) or counterclockwise (to the left, LCPL) along its axis of propagation. Being a physical agents thus characterized by "true chirality" (62), RCPL and LCPL are capable of stereoselective photochemical interactions with chiral or prochiral molecules, and over 100 years ago LeBel (63) and van't Hoff (64) proposed that CPL might thus be responsible for the origin of chiral molecules in nature. After earlier investigators had failed, Kuhn and Braun (65) in 1929 conducted the first successful stereoselective reaction implemented by CPL, namely, the asymmetric photolysis of racemic ethyl 2-bromopropionate $((\pm)\text{-}CH_3CH(Br)COOCH_2CH_3)$, and numerous other successful experiments have been reported since that time (see (10) (11) (12)).

CPL-mediated reactions are photochemical, and depend on the *circular dichroism*, of the reactant, that is, on the difference in its molar absorption coefficients $(\Delta\varepsilon)$ for RCPL (ε_R) and LCPL (ε_L) (66). Since the rate of a photochemical reaction depends upon the amount of light absorbed by the reactant, circular dichroism, where the two absorption coefficients are unequal, thus leads to different reaction rates for enantiomeric or prochiral reactants with RCPL *vs.* LCPL, inducing a reaction with a positive asymmetric bias for R- and an equal negative bias for LCPL. Experimentally, CPL has proved capable of producing chiral molecules by any of the following three well-understood mechanisms.

In *Asymmetric Photoequilibrations* the equilibration of two photo-interconvertible enantiomers results in the preponderance of one of them at equilibrium. The process may be illustrated by the photoequilibration of racemic chromium complexes studied by Stevenson and coworkers (67) around 1970, who found that the optical rotations observed while illuminating the reactants with RCPL increased at the outset and then leveled off to a constant value when equilibrium was attained. Equal and opposite results were observed with LCPL. Such equilibrations, however, are both limited in their scope and constrained by the anisotropy factor, $g = \Delta\varepsilon/\varepsilon$, to the trivial optical yields of < 1% or so, and thus would appear to be of little prebiotic relevance.

In *Photochemical Asymmetric Syntheses* a chiral product is synthesized from a prochiral reactant under the influence of R- or LCPL. Such syntheses have been studied mechanistically by Kagan and coworkers in France (68) (69) and by Calvin and coworkers at the University of California, Berkeley (70) in the early 1970s, but they have so far been limited to ring closure reactions similar to that illustrated in Equation (5) for the photochemical asymmetric synthesis of hexahelicene (**VIII**) from the non-chiral 1,2-diarylethylene precursor **VII** with R- and LCPL. The enantiomerism in **VIII** results from its right- or left-handed "screw" structures, and we see that (+)- and (−)-**VIII** are produced with equal and opposite optical rotations. Other examples of such syntheses have been enumerated recently (12). Realistically, however, the esoteric reactants and products, which bear no relationship to current biomolecules, the restrictive reaction conditions, and the g-factor limitations of optical yields again to < 1%, make any actual prebiotic importance of such photochemical asymmetric syntheses quite implausible.

$$[\alpha]^{23}_{436}$$
$$\text{RCPL: } -30.0°$$
$$\text{LCPL: } +30.5°$$

(5)

VII   VIII

In *Asymmetric Photolysis*, because of circular dichroism, CPL induces the preferential photolysis of one enantiomer over the other in a racemic mixture. Thus, if we interrupt the photolysis before completion, we find an excess of the more photostable enantiomer in the undecomposed residue. Such reactions, first demonstrated in 1929 by Kuhn and Braun (65), and subsequently validated by numerous others (see (11) (12)) are not limited by the anisotropy factor g to the negligable optical yields of < 1%, but may in principle provide e.e.s approaching 100% if the substrate g values are high enough and if the photolysis is extensive enough (71). To illustrate, Kagan and coworkers (71) observed an optical purity of 20% in the residue from the 99% photolysis of racemic camphor with ultraviolet light. We (72) later examined the partial photolysis of D,L-leucine using R- and LCPL of 212 nm wavelength. We found that the unphotolyzed residue after 59% photolysis with RCPL had a 2% L > D excess, while that after 75% photolysis with LCPL had a 2.5% D > L excess, again demonstrating equal and opposite stereoselective effects.

Asymmetric photolysis could be potentially important for the origin of chiral molecules on the primitive Earth, and Kagan and coworkers have reemphasized this possibility after their classical experiments (71) in the 1970s. This follows from the facts that asymmetric photolysis is not limited by the anisotropy factor g to the trivial optical yields characteristic of other CPL processes, but can, as shown in the above experiments, provide substantial e.e.s. Nor is it confined to the exotic or biologically irrelevant substrates used in other CPL studies, but may occur with any racemic organic compound having adsorption bands in the visible and/or ultraviolet regions of the spectrum.

What about the availability of CPL at the surface of the prebiotic Earth to engender the production of chiral molecules by any of the above CPL processes? A number of CPL sources and mechanisms have been proposed in recent years (see (11) (12)), but they all suffer from one or more of the same disadvantages: 1) they are extremely weak in intensity, 2) they produce only trivial excesses of RCPL over LCPL (or *vice versa*), and 3) they are subject to reversals over periods of time and/or with different geographical locations or geological features on Earth's surface. Thus even the efficient and general process of asymmetric photolysis would probably prove ineffective in creating chiral molecules in the prebiotic *terrestrial* environment. We shall consider extraterrestrial CPL below.

Universal Processes Based on Parity Violation

In contrast to the above regional and/or temporal processes for the genesis of terrestrial chirality, which may reverse themselves at different times or places during Earth's history, another class of "universal" processes based on parity violation has persisted on Earth (and elsewhere in the galaxy) since the universe came into being. In simplest terms, the principle of parity maintains that natural laws are invariant under spatial reflection, that is, any natural process can also occur as seen reflected in a mirror. Determinate mechanisms based on parity violation stem from the 1956 prediction of Lee and Yang (73) that the parity principle might be violated in "weak interactions" such as those involved in the β-decay of certain radioactive isotopes. This conjecture was verified a year later by Wu and coworkers (74), who found that the electrons emitted during the β-decay of ^{60}Co nuclei were longitudinally polarized with a "left-handed" bias, that is with their spins predominately antiparallel to their direction of propagation. Lee and Yang's Nobel prize-winning prediction soon prompted numerous investigations into the possible implication of parity violation in the origin of biomolecular chirality, since contemporary biomolecules themselves have the characteristic of a parity violation, that is, molecules with mirror image

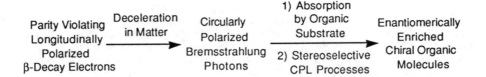

FIGURE 3. The Vester - Ulbricht Hypothesis

molecular structures are absent in our biosphere.

The Vester-Ulbricht (VU) Hypothesis. The first attempt to link parity violation at the nuclear level with that at the biomolecular level was the hypothesis of Vester and Ulbricht in the late 1950s. It was well known that electrons impinging upon matter undergo deceleration, losing some of their energy and thereby emitting a spectrally continuous array of photons called Bremsstrahlung, and it had been recently shown (75) that Bremsstrahlung photons produced during deceleration of longitudinally polarized β-decay electrons were also circularly polarized. Knowing the demonstrated ability of circularly polarized light to facilitate stereoselective photochemical reactions, Ulbricht (76) suggested in 1959 that the circularly polarized Bremsstrahlung photons resulting from the deceleration of β-decay electrons might induce asymmetric syntheses or degradations in organic substrates by CPL-processes similar to those described above. Thus, as summarized in Figure 3, parity violation at the nuclear level might ultimately engender chirality at the molecular level. To test this hypothesis Vester and coworkers (77) (78) undertook careful polarimetric examinations of the products from a number of synthetic and degradative organic reactions conducted in the presence of a variety of β-emitting nuclides (^{32}P, ^{90}Sr, ^{152}Eu, ^{108}Ag) at various radioactivity levels (50 – 2500 mCi), exposure times and temperatures. The products all proved to be optically inactive, however, leading the authors to conclude that more powerful radiation sources and longer exposure times would be necessary to observe stereoselective effects.

The first positive observation regarding the VU hypothesis was that of Garay (79) in Hungary in 1968. He exposed alkaline solutions of D- and L-tyrosine separately to the radiation from 0.36 mCi of dissolved ^{90}SrCl$_2$, and then monitored the 242 nm absorption band characteristic of the substrate at increasing time intervals. After 18 months the intensity of this band for the D-tyrosine solution was diminished to a considerable greater extent than for the L-tyrosine solution, while parallel experiments using non-radioactive ^{88}SrCl$_2$ as a control led to no differences in the absorption spectra of the two enantiomers after the 18-month period. Garay interpreted the apparent stereoselective decomposition of the D-tyrosine as due to an oxidative degradation biased by the ^{90}Sr β-particles or their Bremsstrahlung. While his observations have never been confirmed independently, they encouraged a number of subsequent investigations into the possible efficacy of the VU mechanism, as outlined below

Since we had certain reservations regarding Garay's conclusions and experimental procedures, we undertook in the early 1970s an extensive series of investigations into the VU mechanism. In order to circumvent difficulties inherent in the polarimetric

TABLE 1. Irradiation of Amino Acids with ^{90}Sr-^{90}Y β-Ray Bremsstrahlung

D,L-Amino Acid	Enantiomeric Composition						% Radiolysis
	Irradiated Sample			Control Sample			
	%D	%L	(±)	%D	%L	(±)	
Leucine	50.37	49.63	0.34	50.28	49.72	0.22	47.2
Norleucine	50.14	49.86	0.31	50.23	49.77	0.24	33.6
Norvaline	49.88	50.12	0.15	50.12	49.88	0.19	27.4

or spectral methods of analyses previously employed, we developed and utilized quantitative gas chromatographic (GC) analytical techniques for the determination of both enantiomeric compositions (80) (81) as well as the extents of decomposition (82) (83) characteristic of the products resulting from our stereoselective radiolysis experiments. GC had a unique advantage over polarimetric or spectral analytical methods, in that it could look exclusively at the e.e.s of the residual reactants in the crude residues, regardless of the presence of radiolytic or other contaminants.

Our first experiments (84), intended to supplement those of Garay (79), involved the effects of ^{90}Sr - ^{90}Y Bremsstrahlung on amino acids. A number of racemic and optically active amino acid samples were placed in sealed containers and lowered in-to a 61.7 KCi ^{90}Sr - ^{90}Y β-ray Bremsstrahlung source at Oak Ridge National Laboratory.Samples were retrieved at increasing time intervals and analyzed by GC for their enantiomeric compositions and extents of degradation. The analyses of several of the final D,L-amino acid samples, recovered (85) after 10.9 years following a total radiation dose of *ca.* 2.5×10^9 rads, are shown in Table 1, where we find no evidence whatsoever for stereoselective radiolysis. Comparing the irradiated samples with the non-irradiated control samples, we see that their enantiomeric compositions are identical within experimental error, despite gross radiolyses as high as 47%.

Our next experiments testing the VU hypothesis involved the β-emitting nuclide ^{14}C, an isotope which would have been available at low levels on the primitive Earth

TABLE 2. Self-β-Radiolysis of ^{14}C-Labeled Amino Acids

D,L-Amino Acid	Age Years[a]	Radioactivity mCi/mole	Total Dose Rads x10^{-7}	EnantiomericComposition			% Radiolysis
				%D	%L	(±)	
Alanine	16.9	285	5.05	50.06	49.94	0.85	26.5
Valine	25.8	316	6.51	50.19	49.81	0.20	30.0
Norvaline	24.9	574	11.41	49.94	50.06	0.18	17.4
Leucine	24.0	446	7.63	50.15	49.85	0.22	67.8
Norleucine	24.9	551	9.78	50.10	49.90	0.17	24.1
Aspartic acid	24.1	319	5.40	50.23	49.77	1.02	~50

[a] Between date of preparation and date of analysis.

(86). Here, again using GC to estimate e.e.s and percent radiolysis, we examined (87) a number of ^{14}C-labeled D,L-amino acids which had been synthesized for other reasons at the Lawrence Berkeley Laboratory some 17 to 25 years earlier, and had been undergoing self-β-radiolysis in the solid state during these time intervals. The results of our analyses are shown in Table 2, where we see that all of the samples are totally racemic within experimental error, results agreeing with earlier conclusions of Calvin and coworkers (88). Thus, using ^{14}C-labeled amino acids ,we again find no evidence at all for stereoselective radiolysis, even with gross degradations up to 68%.

Our final experiments involving the VU hypothesis as regards β-ray Bremsstrahlung involved [32]P. These were undertaken to verify a 1976 report by Darge and coworkers (89) claiming, on the basis of the minute optical rotation of 0.7 ± 0.4 millidegree, that the residue remaining after a 12-week exposure of D,L-tryptophan to the β-rays and Bremsstrahlung from 5 mCi of codissolved aqueous ^{32}P-phosphate underwent 33% radiolysis with a remarkable 19% enrichment of the D-enantiomer in the residue. We carefully repeated (90) Darge's experiment, using GC instead of polarimetric and spectral criteria for enantiomeric enrichment and percent radiolysis. As seen in the first line of Table 3, however, we found no evidence at all for the stereoselective β-radiolysis of D,L-tryptophan. In order to confirm these results we subsequently extended (91) such ^{32}P-radiolyses to D,L-leucine, using slightly different experimental conditions. Again, as evident in Table 3, no asymmetric radiolysis was noted.

One might wonder why there has been this consistent failure of the VU mechanism and why stereoselective effects have not been observed with the Bremsstrahlung from β-decay electrons. We have suggested (85) one possible explanation, which is shown in Figure 4. Bremsstrahlung photons embrace a continuous energy spectrum ranging from zero up to the maximum energy of the β-decay electron, with the major portion of the Bremsstrahlung intensity located in the low energy region of the spectrum (92). In contrast, the circular polarization of these Bremsstrahlung photons ranges from zero at the low energy end to almost 100% at the high energy end of the spectrum (93) (94) (95). Thus the low energy majority of the Bremsstrahlung photons, capable of interacting photochemically with organic matter, are of insufficient circular polarization, while the highly polarized high energy photons are of insufficient intensity or unsuitable wavelength to allow for the occurrence of stereoselective interactions.

TABLE 3. Radiolysis of Tryptophan and Leucine with [32]P β-Radiation

D,L-Amino Acid	[32]P mCi	Exposure days	Temp. °C	Enantiomeric Composition		% Radiolysis
				%D	%L	
Tryptophan	5	85	25	50.0 [a]	50.0 [a]	43.5 [a]
Leucine	5	90	-196	49.9 [a]	50.1 [a]	20.6 [a]
Leucine	12.5	124	-196	50.1 [b]	49.9 [b]	25.6 [b]

[a] Average of duplicate experiments, [b] Average of 4 experiments

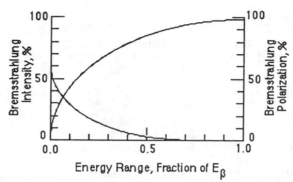

FIGURE 4. β-Decay Bremsstrahlung, Intensity and Circular Polarization

Direct Effects of Chiral Particles. Supplementing the above experiments involving the Bremsstrahlung from β-emitting nuclides, a number of studies have also been undertaken in search for *direct* stereoselective interactions between racemic substrates and a variety of artificially produced spin-polarized subatomic particles, including electrons, protons, positrons and muons (see (11) (12)). Our own initial experiments (96) (97), which investigated the effects of parallel- and antiparallel-spin polarized 120 keV electrons from a linear accelerator on D,L-leucine, seemed at first to indicate the occurrence of a stereoselective effect. However, Hodge and coworkers (98) using an an alternative polarized electron source, were later unable to duplicate our findings, and we concluded that our apparently positive results were due either to artifacts or to an unlucky run of small statistics.

Our next experiments (99) investigated the interaction of parallel- and antiparallel-spin-polarized protons of 0-10 MeV energies on racemic leucine. Irradiations were conducted on samples in tandem, such that the proton beam could pass completely through the first leucine sample and, after attenuation, be finally stopped by the second sample. We see in a number of experiments summarized in Table 4, however, that no stereoselective decomposition at all was evident, whether the proton beam passed completely through the sample or was absorbed by it. The enantiomeric composition of the undecomposed portion of each sample, determined by GC, was still racemic within experimental error.

Similarly, numerous experiments by a number of other investigators have attempted to find stereoselective interactions between enantiomeric substrates and both spin-polarized positrons and muons. Such investigations have often produced controversial claims (see (11) (12)), but have to date uniformly failed to discover any reproducible effects beyond experimental error.

Radioracemization. After examining the 17-25 year old samples of ^{14}C- labeled D,L-amino acids shown in Table 2, we also conducted GC enantiomer analyses on several similar samples which had been resolved and were optically pure at the time of their preparation. Such samples now proved to be partially racemic, suggesting that racemization had accompanied their partial β-radiolysis. This led us to investigate a new phenomenon which we called *radioracemization*, the racemization of an optically active substrate by ionizing radiation without accompanying radiolysis. Since the phenomenon might have implications as regards the VU mechanism, we

TABLE 4. Irradiation of D,L-Leucine with Longitudinally Polarized Protons

Experiment	Spin	Beam	Enantiomeric Composition			% Radiolysis
			%D	%L	(±)	
1	AP[a]	PT[c]	49.99	50.01	-	7
		A[d]	50.12	49.88	-	
2	P[b]	PT	50.01	49.99	0.13	25
		A	50.14	49.86	0.04	
3	AP	PT	50.12	49.88	-	20
		A	50.11	49.89	-	
4	P	PT	50.20	49.80	0.14	20
		A	50.20	49.80	0.11	
5	AP	PT	49.96	50.04	0.15	47
		A	49.98	50.02	0.11	
6	P	PT	50.12	49.88	0.06	50
		A	50.23	49.77	0.19	

[a] Antiparallel, [b] Parallel, [c] Passed through 1st Sample, [d] Absorbed in 2nd Sample

TABLE 5. Radiolysis and Racemization of Amino Acids on γ- Irradiation

Amino Acid	Radiation Dose Rads x 10^{-8}	Enantiomeric Composition			% Racemization	% Radiolysis
		%D	%L	(±)		
L-Alanine	8.1	1.9	98.1	0.1	3.8	38.6
D-2-Amino- butyric acid	8.1	99.2	0.8	0.1	1.6	55.8
L-Norvaline	8.1	1.6	98.4	0.1	3.2	66.1
L-Norleucine	8.1	1.3	98.7	0.1	2.6	63.1
D-Leucine	8.1	97.2	2.8	0.2	5.6	67.9
L-Leucine	8.1	2.5	97.5	0.1	5.0	68.0
D-Leucine	10.2	96.2	3.8	0.6	7.6	96.1
L-Leucine	10.2	6.8	93.2	0.2	13.6	93.2

undertook its systematic examination (100) (101), irradiating several optically pure amino acids with radiation from a 3000 Ci ^{60}Co γ-ray source, then analyzing the irradiated samples for their degradation and enantiomeric composition by GC. Some data pertaining to the radioracemization of amino acids in the solid state are shown in Table 5, where we see that significant racemization of the undecomposed substrate accompanied the radiolysis of each sample.These studies were then extended to the radioracemization of the sodium salts and hydrochloride salts of these amino acids in aqueous solution (100) (101). It was found that the sodium salts were considerably more susceptible to radioracemization, while hydrochloride salts were immune, and rationalizations of these striking differences were offered. We found further (102) (103) that even the non-protein amino acid isovaline, which cannot be racemized under ordinary conditions in aqueous solution, was as susceptible to radioracemization in the solid state as were the protein amino acids, and the geochemical, cosmochemical and paleontological implications of these observations were subsequently elaborated (104) (105) (106). Later studies (107) (108), with the intention of being more realistic prebiotically, investigated the radioracemization of amino acids in the presence of silica and clay minerals. Such mineral surfaces proved generally to render the amino acids even more susceptible to radioracemization.

Early in our above studies we emphasized (100) (101) that radioracemization could only have a deleterious effect on the efficiency of the VU mechanism for generating chiral molecules. Depending upon the relative rate for the formation of an excess of one enantiomer by stereoselective radiolysis of a racemate, compared to the rate of radioracemization of that enantiomer, the latter effect could in principle negate the former, such that no net enantiomeric excess would be generated before gross radiolysis was complete. The implications of radioracemization as regards the use of diagenetic racemization rates of ancient amino acid samples as criteria for geochronological of geothermal calculations have also been stressed (100) (101), with the caveat that radioracemization could make the conclusions reached by such calculations open to question.

Parity Violating Energy Differences. In the above sections we have considered experiments, related to the VU mechanism, which have investigated the possibility of stereoselective interactions between chiral elementary particles resulting from parity violation and enantiomeric molecules. As we have seen, such experiments have failed unambiguously to detect any such interactions. On a more fundamental level, however, parity violation is also believed, by the influence of "parity violating energy differences" (PVEDs), to govern the intrinsic chemical and physical properties of enantiomers themselves. Put in simplest terms, to conserve parity one member of a pair of enantiomers would have to be composed of antimatter. Since this is not the case, *i.e.* both enantiomers consist of matter, there is an intrinsic parity violation in a pair of enantiomers. In 1966 Yamagata (109) first predicted that there should be a small energy difference between enantiomers due to this parity violation, and that this PVED should give rise to minute differences in the physical and chemical properties of two enantiomers. Moreover, he argued, the small reaction rate differences due to PVEDs might be amplified by an "accumulation principle" to allow production of homochiral polymers, thus providing an explanation for the configurational one-handedness of our biosphere. Elaborating on these ideas in the early 1980s, Mason and Tranter (110) (111) (112) and Tranter (113) (114) made *ab initio* calculations purporting to show that the equilibrium populations of "natural" enantiomers (*i.e.* L-amino acids, D-sugars, and their polymers) were favored over their "unnatural" mirror image counterparts to the extents of *ca.* 1 part in 10^{17}. Put into perspective as to magnitude, this excess corresponds to one star out of all the stars in a million of our 100-billion-star Milky Way galaxies, or about one second out of all the time since our Solar System was formed. This minuscule excess is then alleged to be capable of being amplified by some efficient but yet undemonstrated mechanism, into a state of total enantiomeric homogeneity. How might this occur?

In our previous discussion of Spontaneous Symmetry Breaking and Autocatalysis we mentioned the theory that "far-from-equlibium" racemic systems may bifurcate at some critical "bifurcation point", resulting in the production of a single enantiomer. In 1981 Nicolis and Prigogine (115) showed that such systems are profoundly sensitive to environmental asymmetries in the neighborhood of the bifurcation point, which can accordingly influence the system to culminate in a specifically preferred chirality for the enantiomer. A year later Wei-Min (116) suggested that natural asymmetric agents such as CPL or parity violation between enantiomers might provide the requisite environmental asymmetry to induce non-random symmetry breaking, and numerous analogous schemes were soon proposed (see (12)). Amplifying Frank's (18) autocatalytic mechanism for spontaneous symmetry breaking, Kondepudi and Nelson (117) in 1983 suggested an autocatalytic scheme by which the 10^{-17} equilibrium excess of one enantiomer in a racemic mixture due to the PVEDs might be enhanced to a state of enantiomeric purity, and later calculated (118) that such chiral domination could be established within the short span of only 15000 years. Such calculations and conclusions regarding the role of appropriately amplified PVED effects in establishing a unique homochirality on Earth have proved controversial and are by no means universally accepted, and from the early 1980s numerous investigators have argued on theoretical grounds the ineffectuality of PVEDs in determining chirality (See (119)). Unfortunately, the Kondepudi-Nelson autocatalytic amplification mechanism has never been reduced to practice, and other proposed autocatalytic schemes which have been investigated experimentally have been unsuccessful (119). Thus a crucial experimental evaluation of the possibility of generating homochirality as a result of amplified PVED effects has hitherto been lacking.

As mentioned earlier, the only autocatalytic amplification mechanism which has been realized experimentally is Calvin's (19) scheme for "stereospecific autocatalysis" (Figure 1), and that only in the context of "spontaneous resolution under racemizing conditions" (SRURC) (Figure 2).In principle, the stereospecific autocatalysis mechanism should be capable of amplifying any small indigenous enantiomeric excess in a racemic reactant pool, whether due to statistical fluctuations or to PVEDs, into a final state of enantiomeric homogeneity. It thus occurred to us that SRURC, which has led to spontaneous symmetry breaking in several instances (Equations (1), (2), (3)), might provide an ideal amplification system for investigating the possible efficacy of PVED effects in generating homochirality. With this in mind we have reinvestigated the 1,4-benzodiazepinooxazole system (Equation 3) of Okada *et al.*(27) (28) (29). They had previously reported that crystallization of the racemic substrate (±)-**III** under various conditions could lead to three crystalline modifications: optically active α-form crystals, which could also occur as an optically inactive conglomerate, and optically inactive β- or γ-form crystals which were polymorphic modifications of a racemate. Extending their studies, we found at the outset (21) that crystallization of racemic (±)-**III** from methanol solution could afford optically active samples of either (+)-**III** or (−)-**III** of very high optical purity in *quantitative* yield, and that small intentionally contrived enantiomeric excesses of (+)-**III** or (−)-**III** in the racemic substrate could be amplified by digestion in methanol into a state of essentially complete enantiomeric purity in the final crystalline product. This very efficient SRURC involving the total spontaneous resolution of **III** thus appeared to be an excellent system for investigating the possible amplification of PVED effects.

Calvin's scheme, as it applies to the details of the spontaneous resolution of **III**, is shown in Figure 5. As the hot solution of equilibrating (+)- and (−)-**III** is slowly cooled it reaches a "far-from-equilibrium" supersaturated critical point for symmetry

34

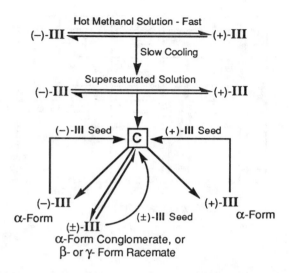

FIGURE 5. Autocatalytic Symmetry Breaking of **III**

breaking, **C.** Here random nuclei of the α-, β-, or γ- crystalline forms of **III** appear, crystallization commences, the far-from-equilibrium condition of super-saturation is relieved, and autocatalysis amplifies whichever crystal form was randomly selected. We may then obtain either the optically inactive α-form con-glomerate crystals, inactive β- or γ-form racemate crystals or optically active (+) or (–) α-form crystals of spontaneously resolved **III**, or a mixture of the various crys-tal forms. The optically inactive crystal forms of **III** prove to be less soluble than the (+)- or (–)-**III** crystals, and spontaneously revert to one of these active forms on stirring in methanol suspension. Our objective was to investigate whether, either in the initial equilibrium or at the critical point **C,** the *ca.* 1 part in 10^{17} PVED excess of the more stable enantiomer of **III** could overcome statistical fluctuations in nucleation and be amplified by this highly efficient autocatalytic mechanism into a singular state of enantiomeric purity. While Okada's previous isolation of either (+)- or (–)-**III** during its spontaneous resolution by crystallization already suggested a negative answer, we felt a more extensive statistical study would be desirable, and accordingly undertook (119) to examine the outcome of repetitive crystallizations of (±)- **III** under racemizing conditions.

In Figure 6 we see graphically the optical rotations of the crude crystalline pro-ducts resulting from 48 sequential slow crystallizations of (±)-**III** from methanol solution. Optically active dextro- or levorotatory crystals were obtained in 52% of the products (25 instances) and optically inactive crystals 48% of the time (23 ins-tances). Of the 25 optically active crystals 52% (13 samples) were dextrorotatory, with specific rotations varying from +3.7° to +262°, while 48% (12 samples) were levorotatory, with rotations varying from –7.7° to –294°. The mean specific rotation for all 48 samples was –1.2°± 115°, and the mean enantiomeric excess (%(+) – %(–) crystals) was –0.4 ± 35%. The overall symmetry of the crystallization results

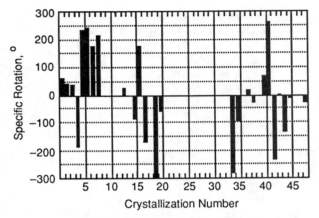

FIGURE 6. Rotations of Crystalline Products

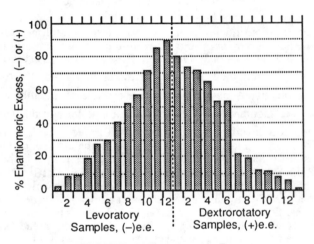

FIGURE 7. Enantiomeric Excesses

is suggested in Figure 7, where we have disregarded the 0.0% e.e.s of the 12 optically inactive samples and arranged the e.e.s of the 12 levorotatory samples in increasing magnitude from left to right and the e.e.s of the 13 dextrorotatory samples from right to left. The data illustrated in Tables 6 and 7 clearly indicate that the spontanous resolution of racemic **III** by the stereoselective autocatalytic amplification mechanism shown in Figure 5 is a random process, and offer no evidence whatsoever for the efficacy or intervention of PVEDs. Had PVEDs been able to supersede random fluctuantioons, we should have consistently obtained either (+)-**III** or (–)-**III**.

The only other statistical study of the distribution of enantiomers resulting from SRURC is that of Pincock et al.(26). They studied the distribution of optical rotations of samples obtained during 200 individual crystallizations of 1,1'-binaphthyl (**II**, Equation (2)) from its melt. They observed rotations ranging from $-218°$ to $+206°$, with a Gaussian distribution of values centered about a mean of $+0.14° \pm 86.4°$, which led them to conclude that this example of Calvin's stereospecific autocatalysis involved a totally random process. Thus here we see also that PVED effects are inconsequential compared to random statistical fluctuations in determining the selection of a specific enantiomer in a system capable of total spontaneous resolution by an efficient autocatalytic amplification mechanism.

Mason (120) and Tranter (121) have further maintained, again on the basis of *ab initio* calculations, that an alleged 1.4% excess of *l*-quartz over *d*-quartz on the surface of Earth is also the inevitable consequence of PVED effects. This frequently repeated (122) allegation is based on a 1962 study (123) reporting 50.7% *l*-quartz and 49.3% *d*-quartz in 16,807 samples collected worldwide. It consistently ignores a more recent and extensive survey (39) which found 50.17% *d*-quartz and 49.83% *l*-quartz in 27,053 samples examined. We therefore conclude that the terrestrial distribution of quartz is quite random, and offers no confirmation whatsoever for the alleged PVED-engendered excess of *l*-quartz.

The most recent suggestion for symmetry breaking based on PVED effects is that of Salam (124), who postulated a Bose condensation phase change which reversed the chirality of "unnatural" enantiomers into their more stable "natural" configurations below some critical low temperature. However, recent careful experiments by Figureau et al.(125), in which racemic cystine was subjected to 0.6 - 77 K temperatures for various time periods, have so far failed to detect the predicted configurational phase change.

Evaluating all of the above mechanisms for chirality generation based on parity violation, it is immediately apparent that there is to date neither uncontested theoretical documentation, observational verification, nor reproducible experimental evidence to support the contention that causal relationships of any kind exist between the homochirality of terrestrial biomolecules and parity violation at the elementary particle level, whether through the auspices of β-decay or PVEDs. We must, I think, agree for now with the earlier conclusions of Goldanskii and Kuzmin (2) that all mechanisms based on parity violation are incapable of leading to symmetry breaking or the genesis of chiral molecules, and therefore would have been totally ineffectual on the prebiotic Earth. It would thus appear that of all the Abiogenic Determinate Mechanisms considered above, only the efficient and well-established process of Asymmetric Photolysis might be viable, but for reasons already indicated even this mechanisms would appear implausible on the primitive Earth.

The Amplification of Enantiomeric Excesses

While the above mechanisms for abiotic chirality generation would, as we have seen, appear implausible on the prebiotic Earth, nevertheless several of them *have* been shown experimentally to be capable of generating small enantiomeric excesses. If any of these by chance *were* operative, it is clear that additional abiotic mechanisms for the subsequent amplification of such small e.e.s must have been available to culminate in the high degree of homochirality and chiral purity necessary for the emergence of self-replicating biopolymers. Physical processes capable of enhancing enantiomeric purity have been outlined by Wagener (126) and DeMin *et al.* (127) (See also Bonner (11) (12)), and here we shall summarize briefly those mechanisms which have been proposed experimentally or theoretically for the amplification of e.e.s in the prebiotic era, and comment on their potential prebiotic viability.

Kinetic Resolutions

When an enantiomerically impure substance reacts incompletely with itself or with a second chiral compound, either the product or the unreacted starting material will have an e.e. greater than it had before the reaction occurred. Enantiomeric enhancement by this process of "kinetic resolution" has recently been reviewed by Kagan and Fiaud (128), who emphasized its relevance to the problem of chirality amplification. For its success, however, the process has been demonstrated only under restricted and contrived experimental conditions, which make its potential viability in the aqueous prebiotic environment rather dubious. The same restrictions apply to the novel and efficient enantiomeric amplification scheme of Lahav *et al.* (129) involving topochemical solid-state photodimerization reactions.

Partial Evaporation or Precipitation

If there exists a difference in solubility between a racemate and its individual enantiomeric constituents, enantiomeric enrichment may occur during the partial evaporation of a solution or the partial precipitation of a solid from a solution where the initial concentrations of enantiomers are unequal. Such phenomena have been studied experimentally and theoretically in only a few instances (130) (131), which is rather unfortunate, since these processes might be intuitively construed as potentially viable on the prebiotic Earth. Addadi *et al.* (132) have recently proposed a scheme for the generation and amplification of optical activity involving the enantioselective occlusion of enantiomers on the opposite faces of growing centrosymmetric crystals such as glycine, and Thiemann and Teutsch (133) have invoked lyotropic liquid crystals as a potential means of generating and amplifying chirality. While interesting in themselves, these mechanisms must unfortunately be discounted as to prebiotic relevance due to their severe restrictions in scope and the specialized experimental conditions required for their implementation.

Autocatalysis

The models and schemes discussed above for autocatalytic spontaneous symmetry breaking would, of course, equally well provide mechanisms for the amplification of small e.e.s, and indeed such schemes have been relied upon theoretically to amplify the alleged effects of PVEDs (117). As indicated earlier, however, the only model which has been reduced to practice experimentally has been the highly efficient amplification occurring during crystallizations subject to "2nd order asymmetric transformations," *i.e.* spontaneous resolutions under racemizing conditions (Figure 2), but the implausibility of such an amplification mechanism in the aqueous prebiotic environment has also been stressed above. Attempts to develop solid-state photodimerization and photopolymerization reactions of crystalline divinylbenzene monomers into autocatalytic schemes for generating and amplifying chirality (134) (135) have not only involved specific reactions and experimental conditions which were

38

quite unrealistic prebiotically, but have so far led only to systems which showed "negative feedback" steps which precluded amplification. Other unsuccessful experimental attempts to develop autocatalytic amplification schemes have been reviewed recently (12) (21).

Partial Polymerization - Partial Hydrolysis

In 1957 Wald (136) suggested that the chiral α-helix secondary structure of a growing polypeptide chain might dictate the subsequent selection of chirally similar amino acid monomer units being incorporated into the polymer chain as it grows. This hypothesis quickly received experimental verification in the 1960s in numerous model experiments in which amino acid N-carboxy anhydrides (NCAs) (**IX**) underwent base-catalyzed polymerization into polypeptides (**X**) (Equation (6)) (See (11) (12). In the mid-1970s we undertook a series of studies specifically designed to investigate the possibility of chirality amplification of amino acids by such a mechanism.

$$\underset{\textbf{IX}}{\overset{\displaystyle R\text{-CH-CO}}{\underset{\displaystyle NH\text{-CO}}{\big|\hspace{-0.3em}\diagdown\hspace{-0.3em}O}}} \quad \xrightarrow[\text{2. } H_2O]{\text{1. Base cat.}} \quad \underset{\textbf{X}}{H_2N\text{-}\overset{R}{\underset{|}{C}}H\text{-CO-(NH-}\overset{R}{\underset{|}{C}}H\text{-CO)}_n\text{-NH-}\overset{R}{\underset{|}{C}}H\text{-COOH}} + CO_2 \quad (6)$$

Our experimental approach involved the partial polymerization of enantiomerically unequal mixtures of an amino acid NCA (**IX**), followed by G. C. determination of the e.e.s in the resulting polymer (**X**) and in the unreacted amino acid NCA monomer. As seen in Table 6, we found (137) that polymerizations to the extents of *ca.* 50% of D ≠ L mixtures of leucine NCAs generally led to an e.e. increase in the polyleucine polymers and a comparable decrease in the e.e.s of the unreacted leucine NCA. Thus e.e. enrichments between 3.6 and 13.9% of the enantiomer initially in excess occurred during the partial polymerization of enantiomerically unequal mixtures of leucine NCAs, which form a polymer having a helical secondary structure.

TABLE 6. Partial Polymerization of D ≠ L Mixtures of Leucine NCA

| Initial Leucine NCA | | Polymerization | Polymer | | Unreacted NCA | |
Excess of	% E.e.[a]	% Completion	%E.e.[a]	%Δ[b]	% E.e.[a]	%Δ[b]
D	8.7	53	12.3	3.6	4.5	−4.2
L	9.0	54	12.4	3.4	4.9	−4.1
L	27.0	56	39.5	12.5	18.3	−8.7
D	31.2	52	42.1	10.9	21.8	−9.4
L	50.5	56	62.9	12.4	42.4	−8.1
L	69.6	54	83.5	13.9	58.7	−10.9

a. Absolute value of %D − %L.   b. E.e.of product − E.e.of Initial NCA .

TABLE 7. Partial Hydrolysis of D ≠ L Leucine Polymers

Initial Leucine Polymer Excess of	% E.e.[a]	Hydrolysis % Completion	Unhydrolyzed Residual Polymer % E.e.[a]	%Δ[b]	Recovered Leucine % E.e.[a]
L	45.4	10.4	49.5	4.1	31.2
L	45.4	16.9	50.1	4.7	30.5
L	45.4	27.0	54.9	9.5	39.2
D	41.4	57.0	51.5	10.1	39.5

a. Absolute value of %D – %L.   b. Unhydrolyzed polymer e.e.– Initial polymer e.e.

We also found (138) that the partial hydrolysis of peptide polymers such as those in Table 6 could lead to still greater e.e. enhancement for the amino acids in the residual unhydrolyzed polymer. Note in the second column of Table 7 the e.e. of the D ≠ L leucine in the original polymer, and in the fourth column the e.e. increase for the leucine in the unhydrolyzed portion of the polymer - some 4 to 10%, depending upon the extent of hydrolysis. Note also in the last column the anticipated decrease in the e.e. of the leucine from the hydrolyzed portion of the polymer.

The data in Tables 6 and 7 suggested that the two processes, partial polymerization and partial hydrolysis, be combined sequentially to asses the overall possibility for e.e. enhancement. In this experiment (139) a D ≠ L mixture of leucine NCAs consisting of 65.6% D- and 34.4% L- isomer (e.e.= 31.2%) was polymerized to the extent of 52% to yield a polyleucine having an enantiomeric composition of 72.7% L- and 27.3% D-leucine (e.e.= 45.4%, a 14.2% e.e. increase). This same polymer was then subjected to partial hydrolysis to the extent of 27%, affording a residual unhydrolyzed polymer composed of 77.5% L- and 22.5% D-leucine (e.e.= 55.0%, a 9.6 % e.e. increase). We thus see an impressive 23.8 % overall increase in the e.e. of the original monomer in just two sequential steps of partial polymerization - partial hydrolysis.

The encouraging results in the above model experiments led us to propose (139) the scheme shown in Figure 8 for the abiotic amplification of low e.e.s in original amino acid monomer mixtures into high e.e.s in final polypeptide polymers. The

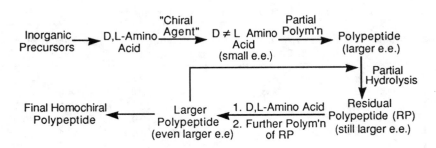

FIGURE 8. Model for the Abiotic Genesis of Homochiral Polypeptides

mechanism involves a repetitive cyclic sequence of partial polymerization - partial hydrolysis steps, which could in principle be driven by environmental dry - wet cycles on the primitive Earth. While the amino acid derivatives (NCAs) used in the above amplification experiments were clearly unrealistic prebiotically, the principles demonstrated in the model system showed nevertheless that such processes might be prebiotically reasonable. In addition, they would have the distinct advantage of leading directly to homochiral polypeptides on the prebiotic Earth.

ENVIRONMENTAL CONSTRAINTS ON PREBIOTIC CHIRALITY

Pace (140) has recently emphasized that realistic considerations of "constraints that might be imposed by the physical setting" have generally been excluded from speculations regarding the origin of life. As implied in our discussions above, the same regrettable assessment might also be applied to all hypotheses and experiments to date which pertain to the origin and amplification of molecular chirality. Let us consider briefly the principal constraints to chirality in the prebiotic environment and the dilemma in which they leave us.

A major hazard to prebiotic chiral molecules, of course, was racemization. In 1979 Keszthelyi *et al.* (141) stressed that prebiotic processes yielding chiral molecules must occur in a much shorter time interval than that of their racemization. Bada and Miller (15) later estimated that the racemization half-lives of amino acids are typically $10^5 - 10^6$ years at ambient temperatures, and it is well known that racemization rates are markedly enhanced in solution by metal ions, alkaline conditions and ionizing radiation (103). Thus chiral amino acids (or any other prebiotic monomers) must have enjoyed only a precarious existence indeed on the aqueous prebiotic Earth. An even greater nemesis to prebiotic chirality was the primeval environment itself. Contrary to affording a congenial locale for the life-nurturing "warm little pond" of Darwin or the gradually evolving, optically inactive (1), dilute "primordial soup" postulated in the Haldane-Oparin scenario (142), primitive Earth was a planet characterized by "rampant volcanism, scorching heat, and a murderous bombardment from comets and asteroids" (143) during the first 700 million or so years after its formation 4.5 Gyr ago. Such catastrophic impacts are now thought to have frustrated the origin of life during this tumultuous period by sterilizing Earth's surface and vaporizing its oceans, such that only after impact frequencies had vastly diminished could life begin to evolve. Over twenty years ago Sagan (144) argued that such conditions could shorten the period available for the origin of life to "only a few hundred million years - perhaps much less," and later authors (145) (146) (147) have concluded that intervals between annihilating impacts became long enough for life to emerge and proliferate only about 3.8 Gyr ago, and that the maximum time needed for the origin of life could be as little as 2.5 to 133 million years (148), depending upon whether $^{12}C/^{13}C$ ratios in sedimentary rocks or the oldest microfossils are used as criteria for the earliest life. Such conclusions clearly jeopardize the comfortable earlier paradigm presupposing the successive accumulation of prebiotic reactants in a "primordial soup," followed by their gradual evolution into self-replicating biomolecules (147). Now, if the above time constraints are valid and if replicating biopolymers are a prerequisite for the emergence of life, how much shorter then must have been the time available for the origin of chirality and its amplification into the homochirality and chiral purity necessary to permit molecular self-replication? And if the numerous mechanisms which we have discussed for the *terrestrial* genesis of chirality are either impotent *per se* or would be non-viable in the hostile and turbulent prebiotic environment, what alternative remains? We are forced to conclude, it would seem, that the chiral molecules available on Earth within the brief time interval available must have been provided *repetitively* from some exogenous, *extraterrestrial* source. Let us now consider how this might have occurred.

THE EXTRATERRESTRIAL ORIGIN OF CHIRALITY

Following the 1961 suggestion of Oró (149) that organic matter might have been transported from outer space to Earth by comets, it is now generally accepted that infalling exogenous organic material, provided by cometary impacts, supplemented terrestrially formed organic molecules in providing the precursors of biomolecules on the prebiotic Earth (150) (151) (152) (153). In 1980 Khasanov and Gladyshev (154) first suggested that *chiral* molecules might also have been formed in outer space by some unspecified "conjoint stereospecific effect of a magnetic and other molecule-orienting field," and three years later several of us (155) proposed a detailed scheme, a *Cosmic Connection*, so to speak, for the extraterrestrial production of chiral molecules in outer space and their subsequent transport to Earth. We shall now examine this hypothesis in greater detail.

The starting point of this scenario is the formation of a neutron star after the occurrence of a supernova. It is well known (156) (157) that when stars of a suitable mass (*ca.* 8 - 10 solar masses) have "used up all of their nuclear fuel," that is, when the nuclear fusions sustaining the star consume rather than release energy, the star undergoes an abrupt gravitational collapse. This precipitous implosion, occurring within a timescale of seconds to hours, produces a shock wave which rebounds from the core, ejecting the outer mantle of the star in a colossal explosion known as a Type II supernova, forming a rapidly expanding nebula of debris, compressing the interstellar medium by releasing prodigious amounts of energy, and finally leaving a neutron star remnant. Supernovae are by no means rare events in our galaxy, appearing once every 50 years or so. This suggests that some 94-million supernovae might have occurred since our Solar System was formed, which itself has been postulated to be the result of a supernova which happened some 4.7 Gyr ago (158).

The neutron star remnant of a supernova is a small object (*ca.*10 km diameter), having an incredible density (*ca.* 10^8 g/cm^3) and an intense magnetic field (*ca.* 10^{13} Gauss, compared to Earth's 0.5 Gauss), which continuously emits and is surrounded by a plasma of electrons and protons. In conserving the angular momentum of the original star, the neutron star rotates extremely rapidly. One of the faster ones known rotates at 1.5 milliseconds per revolution, or 40,000 r.p.m.This rapidly rotating magnetic field around the star creates colossal electric fields, which in turn accelerate the plasma electrons around the core to near relativistic velocities. Now, restricted by the intense magnetic field and thus forced to move in circular orbits around the equatorial plane of the neutron star, these circularly accelerated electrons thereupon emit synchrotron radiation (159), just as they do under analogous laboratory conditions.

Synchrotron radiation is known to be spectrally continuous from X-rays to radio frequencies, and to be plane polarized in the plane of the orbiting electrons, but elliptically and circularly polarized (of opposite handedness) above and below the plane of the orbit (169). This situation around a neutron star would in principle result in vast domains of space "above" and "below" the equator of the neutron star which would be bathed in elliptically and circularly polarized synchrotron radiation of opposite handedness, including *circularly polarized light* of all photochemically pertinent wavelengths.

Let us now shift our attention to the recent suggestions of Greenberg (160) (161) regarding the birth and evolution of interstellar grains having outer mantles containing complex organic molecules. Greenberg proposed that submicroscopic silicate "seedling" particles in dense interstellar molecular clouds at low interstellar temperatures would condense out upon their surfaces "ice" mantles comprised of frozen H_2O, NH_3, CH_4, CO, CO_2, CH_2O and other small molecules present in the molecular clouds. The molecules in these "ices" then become photolyzed by ultraviolet radiation into free radicals, which remain trapped in the ices. On subsequent slight warming the trapped free radicals eventually recombine in a variety of ways to yield larger organic molecules. Repetition and extension of such events lead ultimately to interstellar grains coated with outer mantles containing more complex and even polymeric organic molecules. Greenberg and coworkers (162) have supported this hypothesis by laboratory simulations in which the above interstellar gas molecules were condensed as "ices" on a solid surface, irradiated with ultraviolet light and allowed to warm. The complex mixture of products formed was shown by GC-MS analysis to contain a number of identifiable organic constituents, some of which could exist as optically active enantiomers.

We have combined these concepts (155) (163) (164) (165) into the unified hypothesis for the extraterrestrial production of chiral molecules illustrated in Figure 9. In this scenario the ultraviolet CPL components of the synchrotron radiation from a neutron star photolyze the racemic constituents of the organic mantles on interstellar grains to form chiral components in the mantles by the familiar and effective process of asymmetric photolysis, which Greenberg (166) has recently found to be workable even at interstellar temperatures. Earth could then accrete these grains with their partially chiral mantles as the Solar System passes through the interstellar clouds while

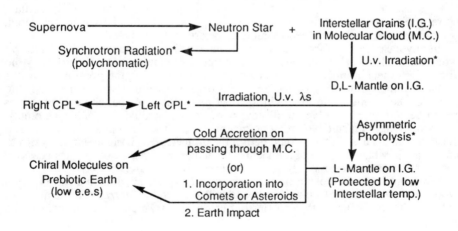

FIGURE 9. The "Cosmic Connection" to Terrestrial Chirality

rotating every 110-million years about the center of our galaxy, thus accumulating vast quantities of exogenous organic matter with certain constituents having a uniform chirality, unracemized since their formation due to the cold interstellar temperatures. Another plausible mode of delivery would be the incorporation of such interstellar grains with their chiral mantles into comets and asteroids, followed by impact delivery to Earth during its turbulent prebiotic era. Greenberg *et al.* (167) have recently concluded that the fluffy morphological ice structure of comets would allow them to impact Earth without the destruction of their organic constituents. Chyba (168) has suggested that Earth acquired its oceans as late-accreting veneers by cometary impact, and others have concluded more recently that comets provided all the molecules now comprising our biosphere (151), as well as amino acids (153) and other complex organic molecules essential for the origin of life (152). It is thus reasonable to speculate that the primitive oceans might have been well stocked with uniformly chiral molecules, produced as described in Figure 9, then delivered to Earth by *repetitive* cometary impacts during its tumultuous prebiotic era.

What objective evidence exists to indicate that the hypothesis in Figure 9 is plausible and potentially realistic? The steps in Figure 9 which have had direct observational and/or experimental verification are indicated by asterisks. Thus light from supernova remnants has been found to include polychromatic synchrotron radiation (159), and synchrotron radiation is known to contain right- and left-CPL components (169). The laboratory simulations of the production of organic mantles on interstellar grains has already been mentioned (162), and the asymmetric photolysis of D,L- substrates with circularly polarized ultraviolet radiation to produce optically active products is a well established phenomenon (71) (72) (164). The only currently undocumented steps in the scenario in Figure 9 are the accretion of interstellar grains with chiral mantles as Earth passes through interstellar molecular clouds, and the incorporation of such grains into comets or asteroids followed by impact delivery to Earth. As regards the latter, the intriguing but unconfirmed report of Engel *et al.* (170) that L-alanine exceeds D-alanine in the Murchison meteorite to the extent of *ca.* 18%, with both enantiomers having a ^{13}C content consistent with an extraterrestrial origin, would if eventually corroborated lend independent support to the extraterrestrial genesis of chiral molecules. Meanwhile, the question whether comets contain chiral molecules among their inventory of organic constituents will be answered, perhaps, only after we are able to sample cometary material directly.

In any case, the uniformly chiral molecules of low e.e., conveyed to Earth by whatever mode of delivery, could then undergo subsequent e.e enhancement by the above (or other) amplification mechanisms, thus providing the possibility for the global symmetry breaking of the terrestrial organic medium which finally set the stage for the origin of life. It would thus seem that the principal chirality problems yet to be addressed involve the discovery of other plausible means for e.e amplification, along with suitable prebiotic mechanisms for molecular sequestration and encapsulation. These in turn would permit the protected e.e. enrichment processes to culminate in whatever state of homochirality and enantiomeric purity was necessary for the rudimentary self-replication leading to the origin of life.

We should note in closing that the supernova hypothesis illustrated in Figure 9 would allow that the molecular chirality associated with *extraterrestrial* life need not necessarily be similar to that on Earth, but might vary randomly throughout the galaxy.

REFERENCES

1. Miller, S. L., and Orgel, L. E., *The Origins of Life on Earth*, Englewood Cliffs, N.J.: Prentice-Hall, Inc., 1974, pp.166–174.
2. Goldanskii, V. I., and Kuzmin, V. V., *Z. Phys. Chem.* **269**, 216–274 (1988).
3. Joyce, G. F., Visser, G. M., van Boeckel, C. A. A., van Bloom, J. H., Orgel, L. E., and van Westrenen, J., *Nature* **310**, 602–604 (1984).
4. Bonner, W. A., *Origins Life Evol. Biosphere* **25**, 175–190 (1995).
5. Terent'ev, A. P., and Klabunovskii, E. I., *The Origin of Life on Earth*, Clark, F., and Synge, R. L. M. (eds.), New York: Pergamon Press, 1957, pp. 95–105.
6. Avetisov, V. A., Goldanskii, V. I., and Kuzmin, V. V., *Physics Today* **44**, 33–41 (1991).
7. Goldanskii, V. I., *Wiss. Fortschr.* **38**, 188–190 (1988).
8. Goldanskii, V. I. and Kuzmin, V. V., *Nature* **352**, 114 (1991).
9. Keszthelyi, L., *BioSystems* **20**, 15–19 (1987).
10. Bonner, W. A., "Origins of Molecular Chirality," in *Exobiology*, Ponnamperuma, C. (ed.), Amsterdam: North-Holland Publ. Co., 1972, pp. 117–181.
11. Bonner, W. A., "Origins of Chiral Homogeneity in Nature," in *Topics in Stereochemistry*. *Vol. 18*, Eliel, E. L. and Wilen, S. H. (eds.), New York: John Wiley & Sons, 1988, pp. 1–96.
12. Bonner, W. A., *Origins Life Evol. Biosphere* **21**, 59–111 (1991).
13. Bonner, W. A., *Origins Life Evol. Biosphere* **21**, 407-420 (1992).
14. Mills, W. H., *Chem. and Ind.* **51**, 750–759 (1932).
15. Bada, J. L., and Miller, S. L., *BioSystems* **20**, 20–26 (1987).
16. Arrhenius, S., in *The Quest for Extraterrestrial Life*, Goldsmith, D. (ed.), California: University Science Books, 1980, pp. 32-33.
17. Marcus, J. N., and Olsen, M. A., "Biological Implications of Organic Compounds in Comets," in *Comets in the Post-Halley Era*, Newburn, R. L., Neugebaer, M., and Rahe, J. (eds.), Dordrecht: Kluwer Acad. Publ., 1991, pp. 439–462.
18. Frank, F. C., *Biochim. Biophys. Acta* **11**, 459–463 (1953)
19. Calvin, M., *Chemical Evolution*, Oxford: Oxford University Press, 1969, pp. 149–152.
20. Seelig, F. F., *J. Theor. Biol.* **31**, 355–361 (1971).
21. Bonner, W. A., *Origins Life Evol. Biosphere* **24**, 63–78 (1994).
22. Bernal, I., *Inorg. Chim. Acta* **96**, 99–110 (1985).
23. Harris, M. M., *Progr. Stereochem.* **2**, 159 (1958).
24. Jacques, J., Collet, A., and Wilen, S. H., *Enantiomers, Racemates and Resolutions*, New York: Wiley, 1981, pp. 369–373, 430–434.
25. Havinga, E., *Chem. Weekblad* **13**, 642 (1941); *Biochim. Biophys. Acta* **13**, 171–174 (1952).
26. Pincock, R. E., and Wilson, K. R., *J. Am. Chem. Soc.* **93**, 1291–1292 (1971); *J. Chem. Educ.* **50**, 455–457 (1973).
27. Okada, Y., Takebayashi, T., Hashimoto, M., Kasuga, S., Sato, S., and Tamura, C., *J. Chem. Soc. Chem. Commun.* 784–785 (1983).
28. Okada, Y., and Takebayashi, T., *Chem. Pharm. Bull.* **36**, 3787–3792 (1988).
29. Okada, Y., Takebayashi, T., and Sato, S., *Chem. Pharm. Bull.* **37**, 5–8 (1989).
30. Okada, Y., Takabayashi, T., Seiichi, A., and Sato, S., *Heterocycles* **31**, 1923-1926 (1990).
31. Tsuchida, R., Kobayashi, M., and Nakamura, A., *J. Chem. Soc. Japan* **56**, 1339–1345 (1935).
32. Karagounis, G., and Coumoulos, G., *Nature* **142**, 162–163 (1938).
33. Bernal, J. D., *The Physical Basis of Life*, London: Routledge and Paul, 1951, pp. 32–39.
34. Amariglio, A., Amariglio, H., and Duval, X., *Helv. Chim. Acta* **51**, 2110–2132 (1968); *Ann. Chim.* **3**, 5–25 (1968).
35. Bonner, W. A., Kavasmaneck, P. R., Martin, F. S., and Flories, J. J., *Science* **186**, 143–144 (1974); *Origins of Life* **6**, 367–376 (1975).
36. Bonner, W. A., and Kavasmaneck, P. R., *J. Org. Chem.* **41**, 2225–2226 (1976).

37. Kavasmaneck, P. R., and Bonner, W. A., *J. Am. Chem. Soc.* **99**, 44–50 (1977).
38. Klabunovskii, E. I., *Origins of Life* **12**, 401–404 (1982).
39. Frondel, C., *Amer. Minerol.* **63**, 17–27 (1978).
40. Degens, E. T., Matheja, J., and Jackson, T., *Nature* **227**, 492–493 (1970).
41. Jackson, T. A., *Experientia* **27**, 242–243 (1971); *Chem. Geol.* **7**, 295–306 (1971).
42. Brindley, G. W., in *The X-Ray Identification and Crystal Structures of Clay Minerals*, Brown, G. (ed.), London: Minerological Society, 1961, p.55.
43. Grim, R. E., *Clay Minerology*, New York: McGraw-Hill, 1968, pp. 57–69.
44. Bonner, W. A., and Flores, J. J., *Curr. Mod. Biol.* **5**, 103–115 (1973).
45. Flores, J. J., and Bonner, W. A., *J. Mol. Evol.* **3**, 49–56 (1974).
46. McCullogh, J. J., and Lemmon, R. M., *J. Mol. Evol.* **3**, 57–61 (1974).
47. Bondy, C. S., and Harrington, M. E., *Science* **203**, 1243–1244 (1979).
48. Youatt, J. B., and Brown, R. D., *Science* **212**, 1145–1146 (1981).
49. Friebele, E., Shimoyama, A., Hare, P. E., and Ponnamperuma, C., *Origins of Life* **11**, 173–184 (1981).
50. Penzien, K., and Schmidt, G. M. J., *Angew. Chem. Int. Ed. Engl.* **8**, 608–609 (1969).
51. Addadi, L., and Lahav, M., "An Absolute Asymmetric Synthesis of Chiral Dimers and Polymers with Quantitative Enantiomeric Yield," in *Origins of Optical Activity in Nature*, Walker, D. C. (ed.), New York: Elsevier, 1979, pp. 179–192.
52. Addadi, L., Cohen, M. D., and Lahav, M., "Syntheses of Chiral Non-racemic Dimers and Polymers via Topochemical Reactions in Chiral Crystals; An Example of 'Absolute' Asymmetric Synthesis," in *Optically Active Polymers*, Selegny, E. (ed.), Dordrecht, Holland: Kluwer Acad. Publ., 1979, pp. 183–197.
53. Green, B. S., Lahav, M., and Rabinovich, D., *Acc. Chem. Res.* **12**, 191–197 (1979).
54. Addadi, L., Berkovitch-Yellin, Z., Weissbuch, I., van Mil, J., Shimon, L. J. W., Lahav, M., and Leiserowitz, L., *Angew. Chem. Int. Ed. Engl.* **24**, 466–485 (1985).
55. Takahashi, F., Tomii, K., and Takahashi, H., *Electrochim. Acta* **31**, 127–130 (1986).
56. Bonner, W. A., *Electrochim. Acta* **35**, 683–684 (1990); *Origins Life Evol. Biosphere* **20**, 1–13 (1990).
57. Zadel, G., Eisenbraun, C., Wolff, G. J., and Breitmaier, E., *Angew. Chem. Int. Ed. Engl.* **33**, 454–456 (1994).
58. Feringa, B. L., Kellog, R. M., Hulst, R., Zondervan, C., and Kruizinga, W. H., *Angew. Chem. Int. Ed. Engl.* **33**, 1458–1459 (1994).
59. Kaupp, G., and Marquardt, T., *Angew. Chem. Int. Ed. Engl.* **33**, 1459–1461 (1994).
60. Breitmaier, E., *Angew. Chem. Int. Ed. Engl.* **33**, 1461 (1994).
61. Barron, L. D., *Science* **266**, 1491–1492 (1994).
62. Barron, L. D., *Chem. Phys. Lett.* **123**, 423–427 (1986).
63. LeBel, J. A., *Bull. Soc. Chim. Fr.* **22**, 337–347 (1874).
64. Van't Hoff, J. H., *The Arrangement of Atoms in Space*, 2nd Ed., Braunschweig, 1894, p. 30.
65. Kuhn, W., and Braun, E., *Naturwissenschaften* **17**, 227–228 (1929).
66. Buchardt, O., *Angew. Chem. Int. Ed. Engl.* **13**, 179–185 (1974).
67. Stevenson, K. L., and Verdieck, J. F., *J. Am. Chem. Soc.* **90**, 2974–2975 (1968); *Mol. Photochem.* **1**, 271–288 (1970).
68. Kagan, H., Moradpour, A., Nicoud, J. F., Balavoine, G., Martin, R. H., and Cosyn, J. P., *Tetrahedron Lett. (No. 27)*, 2479–2482 (1971).
69. Kagan, H. B., Balavoine, G., and Moradpour, A., *J. Mol. Evol.* **4**, 41–48 (1974).
70. Bernstein, W. J., Calvin, M., and Buchardt, O., *J. Am. Cem. Soc.* **94**, 494–498 (1972); *Tetrahedron Lett. (No. 22)*, 2195–2198 (1972).
71. Balavoine, G., Moradpour, A., and Kagan, H. B., *J. Am. Chem. Soc.* **96**, 5152–5158 (1974).
72. Flores, J. J., Bonner, W. A., and Massey, G. A., *J. Am. Chem. Soc.* **99**, 3622–3625 (1977).
73. Lee, T. D., and Yang, C. N., *Phys. Rev.* **104**, 254–257 (1956).

74. Wu, C. S., Ambler, E., Hayward, R. W., Hoppes, D. D., and Hudson, R. P., *Phys. Rev.* **105**, 1413–1415 (1957).
75. Goldhaber, M., Grodzins, L. and Sunyar, A. W., *Phys. Rev.* **106**, 826–828 (1957).
76. Ulbricht, T. L. V., *Quart. Rev.* **13**, 48–60 (1959).
77. Vester, F., Ulbricht, T. L. V., and Krauch, H., *Naturwissenschaften* **46**, 68 (1959).
78. Ulbricht, T. L. V., and Vester, F., *Tetrahedron* **18**, 629–637 (1962).
79. Garay, A., *Nature* **219**, 338–340 (1968).
80. Bonner, W. A., *J. Chromatogr. Sci.* **10**, 159–164 (1972).
81. Bonner, W. A., Van Dort, M. A., and Flores, J. J., *Anal. Chem.* **46**, 2104–2107 (1974).
82. Bonner, W. A., *J. Chromatogr. Sci.* **11**, 101–104 (1973).
83. Bonner, W. A., and Blair, N. E., *J. Chromatogr.* **169**, 153–159 (1979).
84. Bonner, W. A., *J. Mol. Evol.* **4**, 23–39 (1974).
85. Bonner, W. A., and Liang, Y., *J. Mol. Evol.* **21**, 84–89 (1984).
86. Noyes, H. P., Bonner, W. A., and Tomlin, J. A., *Origins of Life* **8**, 21–23 (1977).
87. Bonner, W. A., Lemmon, R. M., and Noyes, H. P., *J. Org. Chem.* **43**, 522–524 (1978).
88. Bernstein, W. J., Lemmon, R. M., and Calvin, M., "An Investigation of the Possible Differential Radiolysis of Amino Acid Optical Isomers by [14]C Betas," in *Molelcular Evolution, Prebiological and Biological*, Rolfing, D. L., and Oparin, A. I. (eds.), New York: Plenum, 1972, pp. 151–155.
89. Darge, W., Laczko, I., and Thiemann, W., *Nature* **261**, 522–524 (1976).
90. W. A. Bonner, and N. E. Blair, *Nature* **281**, 150–151 (1979).
91. N. E. Blair, and W. A. Bonner, *J. Mol. Evol.* **15**, 21–28 (1980).
92. Wyard, S. J., *Nucleonics* **13**, 44–45 (1955).
93. McVoy, K. W., *Phys. Rev.* **106**, 828–829 (1957).
94. Fronsdal, C., and Überall, H., *Phys. Rev.* **111**, 580–586 (1958).
95. Schopper, H., and Galster, S., *Nuclear Phys.* **6**, 125–131 (1958).
96. Bonner, W. A., Van Dort, M. A., and Yearian, M. R., *Nature* **258**, 419-421 (1975).
97. Bonner, W. A., Van Dort, M. A., Yearian, M. R., Zeman, H. D., and Li, G. C., *Israel J. Chem.* **15**, 89-95 (1976/77).
98. Hodge, L. A., Dunning, F. B., Walters, G. K., White, R. H., and Schroepfer, Jr., G. J., *Nature* **280**, 250-252 (1979).
99. Lemmon, R. M., Conzett, H. E., and Bonner, W. A., *Origins of Life* **11**, 337-341 (1981).
100. Bonner, W. A., and Lemmon R. M., *J. Mol. Evol.* **11**, 95-99 (1978).
101. Bonner, W. A., and Lemmon R. M., *Bioorg. Chem.* **7**, 175-187 (1978).
102. Bonner, W. A., Blair, N. E., and Lemmon, R. M., *J. Am. Chem. Soc.* **101**, 1049 (1979).
103. Lemmon, R. M. and Bonner, W. A., "Radioracemization of Amino Acids," in *Origins of Optical Activity in Nature*, Walker, D. C. (ed.), New York: Elsevier, 1979, pp. 47-53.
104. Bonner, W. A., Blair, N. E., and Lemmon, R. M., *Origins of Life* **9**, 279-290 (1979).
105. Bonner, W. A., Blair, N. E., Lemmon, R. M., Flores, J. J., and Pollock, G. E., *Geochim. Cosmochim. Acta* **43**, 1841-1846 (1979).
106. Bonner, W. A., Blair, N. E., and Lemmon, R. M., "The Radioracemization of Amino Acids by Ionizing Radiation: Geochemical and Cosmochemical Implications", in *Biogeochemistry of Amino Acids*, Hare, P. E., Hoering, T. C., and King, Jr., K. (eds.), New York: Wiley, 1980, pp. 357-374.
107. Bonner, W. A., and Lemmon, R. M., *Origins of Life* **11**, 321-330 (1981).
108. Bonner, W. A., Hall, H., Chow, G., Liang, Y., and Lemmon, R. M., *Origins of Life* **15**, 103-114 (1985).
109. Yamagata, Y., *J. Theor. Biol.* **11**, 495-498 (1966).
110. Mason, S. F., and Tranter, G. E., *Chem. Phys. Lett.* **94**, 34-37 (1983).
111. Mason, S. F., and Tranter, G. E., *Mol. Phys.* **53**, 1091-1111 (1984).
112. Mason, S. F., and Tranter, G. E., *Proc. Roy. Soc. London A* **397**, 45-65 (1985).

113. Tranter, G. E., *Chem. Phys. Lett.* **120**, 93-96 (1985).

114. Tranter, G. E., *J. Chem. Soc. Chem. Commun.*, 60-61 (1986).

115. Nicolis, G., and Prigogine, I., *Proc. Natl. Acad. Sci. U. S. A.* **78**, 659-663 (1981).

116. Wei-Min, L., *Origins of Life* **12**, 205-209 (1982).

117. Kondepudi, D. K., and Nelson, G. W., *Phys. Rev. Lett.* **50**, 1023-1026 (1983).

118. Kondepudi, D. K., and Nelson, G. W., *Nature* **314**, 438-441 (1985).

119. W. A. Bonner, *Origins Life Evol. Biosphere*, (In Press).

120. Mason, S. F., *Nature* **311**, 19-23 (1984); *Nuov. J. Chem.* **10**, 739-747 (1986).

121. Tranter, G. E., *Nature* **318**, 172-173 (1985); *BioSystems* **20**, 37-48 (1987).

122. MacDermott, A. J., and Tranter, G. E., *Croat. Chem. Acta* **62**, 165-187 (1989).

123. Palanche, C., Berman, H., and Frondel, C., *Dana's System of Minerology.*, *7th Ed.,Vol. III*, New York: Wiley, 1962, pp. 16-17.

124. Salam, A., *J. Mol. Evol.* **33**, 105-113 (1991).

125. Figureau, A., Duval, E., and Boukenter, A., preprint *LYCEN* 9244, (November,1992).

126. Wagener, K., *J. Mol. Evol.* **4**, 77-84 (1974).

127. DeMin, M. Levy, G., and Micheau, J. C., *J. Chim. Phys.* **85**, 603-619 (1988).

128. Kagan, H. P., and Fiaud, J. C., "Kinetic Resolutions," in *Topics in Stereochemistry, Vol. 18*, Eliel, E. L., and Wilen, S. H. (eds.), New York: John Wiley & Sons, 1988, pp. 249-330.

129. Lahav, M. Laub, F., Gati, E., Leiserowitz, L., and Ludmer, Z., *J. Am. Chem. Soc.* **98**, 1620-1622 (1976).

130. Morowitz, H. J., *J. Theor. Biol.* **25**, 491-494 (1969).

131. Thiemann, W., *J. Mol. Evol.* **4**, 85-97 (1974).

132. Addadi, L. Berkovitch-Yellin, Z., Weissbuch, I., van Mil, J., Shimon, L. J. W., Lahav, M., and Leiserowitz, L., *Angew. Chem. Int. Ed. Engl.* **24**, 466-485 (1985).

133. Thiemann, W., and Teutsch, *Origins Life Evol. Biosphere* **20**, 121-126 (1990).

134. Addadi, L., van Mil, J., Gati, E., and Lahav, M., *Origins of Life* **11**, 107-118 (1981).

135. van Mil, J., Addadi, L., Gati, E., and Lahav, M., *J. Am. Chem. Soc.* **104**, 3429-3434 (1982).

136. Wald, G., *Ann. N. Y. Acad. Sci.* **69**, 353-368 (1957).

137. N. E. Blair, and W. A. Bonner, *Origins of Life* **10**, 255-263 (1980).

138. N. E. Blair, F. M. Dirbas, and W. A. Bonner, *Tetrahedron* **37**, 27-29 (1981).

139. N. E. Blair, and W. A. Bonner, *Origins of Life* **11**, 331-335 (1981).

140. Pace, N. R., *Cell* **65**, 531-533 (1991).

141. Keszthelyi, L., Czege, J., Fajszi, C., Posfai, J., and Goldanskii, V. I., "Racemization and the Origin of Asymmetry of Biomolecules," in *Origins of Optical Activity in Nature*, Walker, D. C. (ed.), New York: Elsevier, 1979, pp. 229-244.

142. Shapiro, R., *Origins - a Skeptics Guide to the Creation of Life on Earth*, New York: Summit Books, 1986, pp. 98-116.

143. Waldrop, M. M., *Science* **250**, 1078-1080 (1990).

144. Sagan, C., *Origins of Life* **5**, 497-505 (1974).

145. Maher, K. A., and Stevenson, D. J., *Nature* **331**, 612-614 (1988).

146. Sleep, N. H., Zahnle, K. J., Kasting, J. F., and Morowitz, H. J., *Nature* **343**, 139-142 (1989).

147. Oberbeck, V. R., and Vogleman, G., *Origins Life Evol. Biosphere* **20**, 181-195 (1990).

148. Oberbeck, V. R., and Vogleman, G., *Origins Life Evol. Biosphere* **19**, 549-560 (1989).

149. Oró, J., *Nature* **190**, 389-390 (1961).

150. Chyba, C., and Sagan, C., *Nature* **355**, 125-132 (1992).

151. Delsemme, A., *Origins Life Evol. Biosphere* **21**, 279-298 (1992).

152. Huebner, W. F., and Boice, D. C., *Origins Life Evol. Biosphere* **21**, 299-315 (1992).

153. Oberbeck, V. R., and Aggarwal, H., *Origins Life Evol. Biosphere* **21**, 317-338 (1992).

154. Khasanov, M. M., and Gladyshev, G. P., *Origins of Life* **10**, 247-254 (1980).

155. Rubenstein, E., Bonner, W. A., Noyes, H. P., and Brown, G. S., *Nature* **306**, 118 (1983).
156. Clark, D. H., and Stephenson, F. R., *The Historical Supernovae*, New York: Pergamon, 1977, pp. 1-113.
157. Bethe, H. A., and Brown, G., *Sci. Am.* **252**, 60-68 (1985).
158. Clayton, D. D., "Supernovae and the Origin of the Solar System," in *Supernovae: A Survey of Current Research*, Rees, M. J., and Stoneham, R. J. (eds.), Boston: Reidel, 1982, pp. 535-564.
159. Oort, J. H., *Sci. Am.* **196**, 53-60 (1957).
160. Greenberg, J. M., "The Largest Molecules in Space: Interstellar Dust," in *Cosmochemistry and the Origin of Life*, Ponnamperuma, C. (ed.), Boston: Reidel, 1983, pp. 71-112.
161. Greenberg, J. M., *Sci. Am.* **250**, 124-135 (1984).
162. Briggs, R., Ertem, G., Ferris, J. P., Greenberg, J. M., McCain P. J., Mendoza-Gomez, C. X., and Schutte, W., *Origins Life Evol. Biosphere* **22**, 287-307 (1992).
163. Bonner, W. A., and Rubenstein, E., *BioSystems* **20**, 99-111 (1987).
164. Bonner, W. A., and Rubenstein, E., "Photochemical Origins of Biomolecular Chirality", in *Prebiological Self Organization of Matter*, Ponnamperuma, C., and Eirich, F. R. (eds.), Hampton, Virginia: A. Deepak Publ., 1990, pp. 35-50.
165. Bonner, W. A., *Chem. Ind. (London) (No. 17)*, 640-644 (1992).
166. Greenberg, J. M., Private communication, (1992).
167. Greenberg, J. M., Zhao, N., and Hage, J., *Ann. Phys. Fr.* **14**, 103-131 (1989).
168. Chyba, C., *Nature* **343**, 129-133 (1990).
169. Winick, H., "Properties of Synchrotron Radiation", in *Synchrotron Radiation Research*, Winick, H., and Doniac, S. (eds.), New York: Plenum, 1980, pp. 11-20.
170. Engel, M. H., Macko, S. A., and Silfer, J. A., *Nature* **348**, 47-49 (1990).

FORMALDEHYDE AS HYPOTHETICAL PRIMER OF BIOHOMOCHIRALITY

Vitalii I. Goldanskii

ABSTRACT

One of the most intriguing and crucial problems of the prebiotic evolution and the origin of life is the explanation of the origin of biohomochirality. A scheme of conversions originated by formaldehyde (FA) as hypothetical primer of biohomochirality is proposed. The merit of FA as executor of this function is based -inter alia - on the distinguished role of FA as one of earliest and simplest molecules in both warm, terrestrial and cold, extraterrestrial scenarios of the origin of life. The confirmation of the role of FA as primer of biohomochirality would support the option of an RNA world as an alternative to the protein world.

The suggested hypothesis puts forward for the first time a concrete sequence of chemical reactions which can lead to biohomochirality. The spontaneous breaking of the mirror symmetry is secured by the application of the well-known Frank scheme (combination of autocatalysis and "annihilation" of L and D enantiomers) to the series of interactions of FA "trimers" (i.e. $C_3H_6O_3$ compounds) of (aaa), (apa) and (app) types, where the monomeric groups (a) mean "achirons" ($a=CH_n$, $n \geq 2$ and $C=M$, $M=C,O$) and (p) mean "prochirons" ($p=HC^*OM$, $M=H, C$)

N.N. Semenov Institute of Chemical Physics of the Russian Academy of Sciences, Kosygin Street 4, Moscow, 117334, Russia

One of most mysterious and significant problems connected to the prebiotic evolution and the origin of life is the problem of the origin of biohomochirality i.e. of the homochirality of the bioorganic world (exclusively left (L) enantiomers of natural aminoacids, exclusively right (D) enantiomers of natural sugars).

I suggest a hypothesis which treats formaldehyde (FA) as the primer of biohomochirality. I fully understand that this hypothesis needs to be experimentally proven, but the outstanding importance of the problem makes it worthwhile to present any considerations on that subject to the judgment of broad scientific circles.

The formaldehyde hypothesis of the origin of biohomochirality is particularly attractive since FA is generally recognized as one of the earliest simple molecules both on the Earth and in space, i.e. in both terrestrial, warm and extraterrestrial, cold scenarios of the origin of life. Polymerization of FA even near the absolute zero (at 4.2 K) was observed experimentally at laboratory conditions (1). This experiment led to the discovery of molecular tunneling and the formulation of the possibility of a "cold prehistory of life"(1,2)

The confirmation of the role of FA as a primer of biohomochirality could bring an additional strong argument in favor of the concept of an "RNA world"(3) which we have already earlier(4) considered as more rational than the traditional concept of a "protein world".

Turning now to FA hypothesis let us use the following designations.

Let us call achirons (a) the monomeric groups with nonasymmetric C atom - CH_n ($n \geq 2$) and $C = M$ ($M = C,O$), and prochirons (p) the monomeric groups with asymmetric C^* atom - HC^*OM ($M = H,C$).

An important intermediate step in the origin of biohomochirality is the formation of "trimers" of FA, (i.e. $C_3H_6O_3$ compounds); achiral (aaa): ketotriose dioxyacetone $HOCH_2 - (C=O) - CH_2OH$ (a) β-oxypropionic acid $HOCH_2 - CH_2 - COOH$ (b) metoxyacetic acid $CH_3O - CH_2 - COOH$ (c), mono-ethyl ester of carbonic acid $C_2H_5O\text{-}COOH$ (d) (it is left aside the heterocyclic trioxane $(CH_2O)_3$); chiral (apa) : aldotriose glyceroaldehyde (GA) $HOCH_2 - C^*H(OH) - COH$ (e), monosemiacetal of

51

oxalic aldehyde $CH_3O - C^*H(OH) - COH$ (f), α-oxypropionic acid $CH_3 - C^*H(OH) COOH$ (g) and heterocyclic 2-oxy-dioxalan $C^*H(OH)-CH_2-O-CH_2-O$ (h).

"Trimers" (app) with two asymmetric C^* atoms, e.g. heterocyclic 2.4-dioxyoxetane $C^*H(OH)-O-C^*H(OH)$ (i) can contain both C^* atoms of the

same (LL, DD) or opposite (LD) chirality. If in the latter case both C^* atoms have identical environment, the corresponding (app) (LD) compound is optically inactive.

It should be noted that among the methods of obtaining the above mentioned FA "trimers" (in particular, the chiral (apa) trimer GA) the condensation of FA is conspicuous.

As a next stage of the formation of biohomochirality the interaction of two "trimers", without or with the formation of hexameric products (i.e. $C_6H_{12}O_6$ compounds) is considered.

The possibility of mutual transformation (aaa) $\leftrightarrow$ (apa) via the exchange of H and OH by two monomeric groups, e.g. the possibility of autocatalytic processes of the type (aaa) + (apa) $\Rightarrow$ (apa) + (apa) is significant, either in the direct interactions of "trimers", or as some heterogeneous process, or with the participation of "hexamers" as peculiar matrices.

The mentioning of this latter variant is stimulated by the brilliant works of von Kiedrowsky(5,6) on the autocatalytic self-replication of hexanucleotides of GCCGGC type with the participation of complementary trinucleotides CGG and CCG.

No complementarity is known for the FA oligomers. However, one should keep in mind that chiral purity of sugar groups is a necessary condition for the complementarity of oligonucleotides(7) and the appearance of such chiral purity had to precede the formation of double RNA helices.

A well-known example of the dimerization of trioses is the condensation of GA in alkaline media with the formation of hexoses. It is assumed that the autocatalytic reaction (aaa) A + (apa) L $\Rightarrow$ (apa) L + (apa) L (same for (apa) D) dominates over the "erroneous" autocatalysis of the type (aaa) A + (apa) L $\Rightarrow$(apa)L + (apa) D (here and below A means an achiral substrate).

However the sole autocatalysis can not secure by itself the origination of biohomochirality. The detailed analysis of this problem (8,9) leads to the conclusion that neither gradual, evolutionary increase of chiral polarization of the system $\eta = \frac{[L]-[D]}{[L]+[D]}$ $(0 < \eta \leq 1)$ can result in the chiral purity ($|\eta| = 1$) - it can be achieved only jumpwise, via a peculiar phase transition, passage through the critical point, i.e. the dynamical equation for $\dot{\eta}$ should be of bifurcational type. The racemic state ($[L] = [D]$, $\eta = 0$) is stable in the subcritical region, while above the critical point it loses stability, and there appear two branches of the solution of dynamic equation- positive and negative η, in the limit of very high overcriticalities, $\eta = +1$ and $\eta = -1$, i.e. chirally pure L and D states.

The simplest and quite general scheme of conversions able to lead to the spontaneous breaking of mirror symmetry was suggested by Frank(10). In the famous Frank scheme, along with the autocatalysis the deracemizing process of "annihilation" of optical antipodes $L + D \Rightarrow A + A$ is considered. Frank took into account also the "self-destruction" of enantiomers: $L + L \Rightarrow A + A$ (same for D) and showed that the possibility of spontaneous breaking of mirror symmetry still remains if the rate constant of "annihilation" exceeds such constants for "self-destruction": $K_{LD} > K_{LL} = K_{DD}$. (We don't touch here the role of the so-called advantage factors (8,9) AF, i.e. of some inequalities in the energies of L and D enantiomers and in their reactions rates and omit also the problem of the choice of certain sign of biohomochirality - whether it is random or predetermined by some AF).

The dynamic equation which corresponds to the Frank scheme is $\dot{\eta} = \alpha (\eta - \eta^3)$, where α depends on the set of characteristics of the reactants and the external conditions.

Thus, in the suggested FA hypothesis one should find a place for the "annihilation" reactions   apa (L) + apa (D) $\Rightarrow$ aaa (A). The role of such reactions can be played e.g. by the exchange of H- and OH-groups between L and D enantiomers of GA: $GA(L) + GA(D) \Rightarrow 2\ HOCH_2 - CH_2 - COOH$ or by the formation of "achiral" hexameric complexes with hydrogen bonds where L and D asymmetric C* atoms have identical environment, e.g.

REFERENCES

1.   V. I. Goldanskii, M.D. Frank-Kamenetskii, I.M. Barkalov, Science, 182, 1344 (1973).
2.   V. I. Goldanskii, Nature, 279, 109 (1979)
3.   G. F. Joyce, Nature, 338, 217 (1989)
4.   V.I. Goldanskii, Europ. Rev. 1, 137 (1993).
5.   G. von Kiedrowski, Angew. Chem. Int. Edn. 25, 932 (1986)
6.   D. Sievers and G. von Kiedrowski, Nature, 369, 221 (1994).
7.   V. I. Goldanskii, V.A. Avetisov, V.V. Kuz'min, FEBS Lett. 207, 181 (1986).
8.   V. I. Goldanskii and V.V. Kuz'min, Sov. Phys. Uspekhi, 32, 1 (1989).
9.   V.A. Avetisov, V.I. Goldanskii, V.V. Kuz'min, Phys. Today, 44 (7), 33 (1991).
10.   F.C. Frank, Biochim. Biophys. Acta, 11, 459 (1953).

Peptide Nucleic Acid (PNA). Implications for the origin of the genetic material and the homochirality of life

Peter E. Nielsen[*]

Center for Biomolecular Recognition, Department of Medical Biochemistry and Genetics, Laboratory B, The Panum Institute, Blegdamsvej 3c, DK-2200 N, Copenhagen,Denmark

Abstract. PNA is a pseudopeptide DNA mimic in which the natural nucleobases have been retained, but the backbone consists of N-(2-aminoethyl)glycine units to which the nucleobases are attached via methylene carbonyl linkers. The finding that PNA forms Watson-Crick-like helices with complementary DNA, RNA or PNA combined with the fact PNA is held together by amide bonds has made PNA of interest as a model for a primordial genetic material. Futhermore, the PNA backbone is achiral, while preferred chirality can be induced in PNA-PNA double helices by attached chiral ligands, thereby providing a new way of "chiral amplification". Finally, it has been demonstrated that PNA-template directed synthesis of RNA and PNA is feasible.

INTRODUCTION

Almost all recent discussion of the origin of life and of the origin of its homochirality has ex- or implicitly assumed the creation of an RNA-world as the crucial step in establishing a genetic material (1-3). Although the RNA world represents a very attractive and plausible scenario, recent results with a pseudopeptide DNA mimic, PNA (Peptide Nucleic Acid), have made it pertinent to consider other chemical backbones than the (deoxy)ribosephosphate backbone of RNA (DNA) as possibilities in a prebiotic genetic material (4-6). In this context peptide-like backbones are especially attractive in view of their chemical stability and the probability of their accumulation in a "primordial soup" environment.

PEPTIDE NUCLEIC ACID (PNA)

PNA is a pseudopeptide DNA mimic in which the natural nucleobases have been retained, but the backbone consists of N-(2-aminoethyl)glycine units to which the nucleobases are attached via methylene carbonyl linkers (Figure 1). PNA has attracted wide attention within medicinal chemistry, diagnostics and molecular biology due to its DNA- and RNA hybridization properties (7-12).

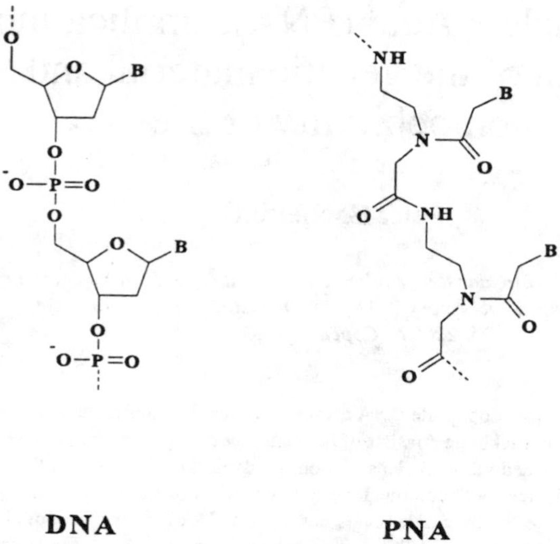

FIGURE 1. Chemical structures of PNA and DNA.

Especially three features also make PNA of interest when discussing the origin of the genetic material and the homochirality of life. First, PNA is a pseudopeptide, *i.e.* a type of compound for which a prebiotic synthesis may be easier to imagine (13) than for ribonucleosides and -tides. Second, complementary PNA oligomers sequence selectively form duplexes according to the Watson-Crick base pairing rules (5,14). Third, the PNA molecule itself and its building blocks are achiral, whereas PNA-PNA helical duplexes are chiral and a preferred helicity can be induced by a chiral ligand such as a chiral amino acid attached to one of the PNA strands (14).

PNA-PNA DUPLEXES

PNA is a polymer containing a linear array of the four nucleobases adenine, cytosine, guanine and thymine and as such formally carrries genetic information. In order for PNA to function as a hypothetical genetic material in a biological sense, the sequence information must be transferable to a new copy of PNA for proliferation as well as to functional molecules that can execute the instructions embedded in the base sequence.

Thus for a replication process it is essential that PNA is able to recognize a complementary PNA strand sequence specifically. The results presented in Figure 2 shows that this is indeed the case. As judged from these thermal stability (Tm) measurements PNA recognizes PNA as well as it recognizes DNA and therefore as well as DNA recognizes DNA. No detailed structural information concerning PNA-PNA duplexes is yet available, but the CD data leave little doubt that these are helical and most probably resemble DNA-helices. Since the PNA molecules are inherently achiral, a racemic mixture of right- and left-handed PNA-PNA helices must exit in solution. However, a single chiral amino acid attached to one end of one of the PNA

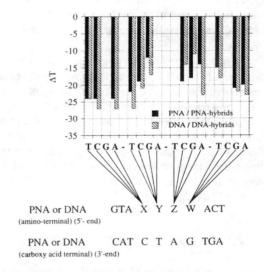

FIGURE 2. Differences in thermal stabilities of PNA-PNA and PNA-DNA duplexes upon introducing single base pair mismathes.

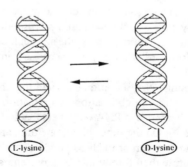

FIGURE 3. Chiral induction in PNA-PNA helices

57

strands is sufficient to significantly shift the equilibrium towards one of the chiral forms (5,4) (Figure 3). This "chiral communication" is critically dependent on the connection between the amino acid and the PNA oligomers as well as on the identity of the first base pair, which in this case should be a G-C base pair (14). These results demonstrate that secondary structure chirality in "macromolecules" is feasible by induction via a chiral ligand that is not an integral part of the macromolecule.

As mentioned above PNA template directed synthesis of another PNA strand is a prerequisite for a hypothetic prebiotic scenario in which PNA constitutes the genetic material to be replicated in the course of proliferation. Recent results using a PNA C_{10} template on which PNA-G_2 units were oligomerized have demonstrated the feasibility of such reactions (6).

Even if one can present arguments for the possibilities of a PNA-prebiotic world, this scenario obviously would require a genetic take-over by the present day genetic material DNA or probably more likely RNA. The feasibility of such a genetic take-over was likewise demonstrated by the chemical synthesis of a G_{10} RNA oligomer. Finally, it was shown that a DNA template (dC_{10}) could direct the synthesis of a PNA G_{10} oligomer.

In order for PNA to function hypothetically as a genetic material in a biological sense, the information must be transferable to a new copy for proliferation as well as to functional molecules that can execute the instructions embedded in the base sequence. oligomer on a PNA C_{10} template.

A PREBIOTIC GENETIC MATERIAL

These results naturally do not show that PNA was the prebiotic genetic material. In fact I do not even wish to propose the *N*-(2-aminoethyl)glycine structure as prebiotic. I think that the most important lesson to be learned from the results presented here is that oligomers/polymers with some of the crucial properties of a biologically relevant genetic material may have a backbone based on (pseudo)peptide chemistry. Further studies, including ones attempting to synthesize such compounds under conditions that immitate the "primordial soup", are necessary in order to evaluate possible candidate structures.

Specifically discussing PNA, it could be argued that the thermal stability of PNA-PNA duplexes is much too high to allow a replication process to take place. Thus backbone structures that result in PNA oligomers that hybridize less efficiently could be favoured (Figure 4). On the other hand, one may argue that life originated near hot springs either in the ocean or in "land poddles" and that a very thermally stable genetic duplex structure therefore would be a prerequisite for the origin of life.

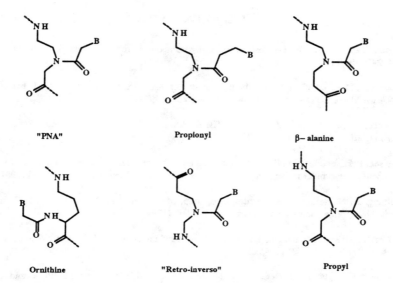

"PNA"   Propionyl   β– alanine

Ornithine   "Retro-inverso"   Propyl

FIGURE 4. Chemical structures of alternative PNA backbones (cf. References 15-17)

IMPLICATIONS FOR THE ORIGIN OF HOMOCHIRALITY

Chiral induction in an inherently achiral genetic material naturally does not explain how life became homochiral. However, it presents a mechanism by which a single event, attachment of a single chiral ligand to a polymer, results in an amplification of the chiral expression. Thus the primordial genetic material and life could have started achiral/racemic and stepwise have drifted towards homochirality as more and more chiral ligands cooperatively stabilize the one chiral form until finally chiral building blocks of the genetic material itself (e.g., in the form of RNA) took over.

Finally, one could imagine that an achiral template like PNA might direct the synthesis of a chiral complement (as e.g. RNA) provided that the PNA during synthesis was attached to a chiral support or if the catalyst was chiral.

ACKNOWLEDGEMENTS

This work was supported by the Danish National Research Foundation.

REFERENCES

1. Ertem, G. and Ferris, J. P., Synthesis of RNA oligomers on heterogeneous templates, *Nature* **379**, 238-240 (1996).

2. Robertson, M. P. and Miller, S.L., Prebiotic synthesis of 5-substituted uracils: a bridge between the RNA world and the DNA-protein world, *Science* **268** (1995).

3. Gesteland, R. F. and Atkins, J. F. (eds), The RNA world, *Cold Spring harbor Lab. Press*, Cold Spring Harbor (1993).

4. Nielsen, P. E., Egholm, M., Berg, R. H. & Buchardt, O., Sequence selective recognition of DNA by strand displacement with a thymine-substituted polyamide, *Science* **254**, 1497-1500 (1991).

5. Wittung, P., Nielsen, P. E., Buchardt, O., Egholm, M. and Nordén, B., DNA-like double helix formed by peptide nucleic acid, *Nature* **368**, 561-563 (1994).

6. Böhler, C., Nielsen, P. E. and Orgel, L. E., template switching between PNA and RNA oligonucleo tides, *Nature* **376**, 578-581 (1995).

7. Buchardt, O., Egholm, M., Berg, R. and Nielsen, P. E., Peptide Nucleic Acids (PNA) and their potential applications in medicine and biotechnology, *Trends Biotechnology* **11**, 384-386 (1993).

8. Nielsen, P. E., Egholm, M., and Buchardt, O., Peptide Nucleic Acids (PNA). A DNA mimic with a peptide backbone, *Bioconjugate Chemistry* **5**, 3-7 (1994).

9. Nielsen, P. E., Egholm, M. and Buchardt, O., sequence specific transcription arrest by PNA bound to the template strand, *Gene* **149**, 139-145 (1994).

10. Hanvey, J. C., Peffer, N. C., Bisi, J. E., Thomson, S. A., Cadilla, R, Josey, J. A., Ricca, D. J., Hassman, C. F., Bonham, M. A., Au, K. G., Carter, S. G., Bruckenstein D. A., Boyd, A. L., Noble S. A. and Babiss, L. E. Antisense and antigene properties of peptide nucleic acids, *Science* **258**, 1481-1485 (1992).

11. Nielsen, P. E. and Ørum, H., "Peptide Nucleic Acid (PNA) as New Biomolecular Tools" in *Molecular Biology: Current Innovations and Future Trends* (H. Griffin, ed.) *Horizon Scientific Press, UK,* p.73-86 (1995).

12. Ørum, H., Jørgensen, M., Koch, T., Nielsen, P. E., Larsson, C. & Stanley, C., Sequence specific purification of nucleic acids by PNA controlled hybrid selection, *Biotechniques* **19**, 472-480 (1995).

13. Nielsen, P. E. Peptide Nucleic Acid (PNA) A model structure for the primordial genetic material. *Origins of Life* **23**, 323-327 (1993).

14. Wittung, P., Lyng, R., Eriksson, M., Nielsen, P. E. and Nordén, B., Induced Chirality in the PNA-PNA Duplex, *J. Amer. Chem. Soc.* **117**, 10167-10173 (1995).

15. Hyrup, B., Egholm, M., Nielsen, P.E., Wittung, P., Nordén, B. and Buchardt, O., Structure-Activity Studies of the Binding of Modified Peptide Nucleic Acids (PNA) to DNA (1994) *J. Amer. Chem. Soc.* **116**, 7964-7970.

16. Krotz, A.H., Buchardt, O. and Nielsen, P.E., Synthesis of "Retro-Inverso" Peptide Nucleic Acids: 2. Oligomerization and stability. (1995) *Tetrahedron Lett*. **36**, 6941-6944.

17. Petersen, K.H., Buchardt, O. and Nielsen, P.E. (submitted).

III. MODELS OF PHYSICAL
CHIRAL SYMMETRY BREAKING

SELECTION OF HANDEDNESS IN PREBIOTIC CHEMICAL PROCESSES

Dilip K. Kondepudi
Wake Forest University, Winston-Salem, NC 27109

ABSTRACT

We see chiral asymmetry in nature at all levels: from elementary particles to living beings. This naturally makes us wonder if these asymmetries are interrelated. Is it possible that the particular asymmetry we see in life's chemistry is a consequence of the chiral asymmetry (parity violation) at the level of electro-weak interactions ? Here we present a theory that relates the strength of a chiral asymmetry and random chiral fluctuations to the probability that molecules with a particular handedness will dominate in a symmetry breaking transition. This theory tells us that, under reasonable prebiotic conditions, the molecular chiral asymmetry could be determined by chiral asymmetries as small as those due to weak neutral currents.

INTRODUCTION

The discovery of weak neutral currents (WNC) gave us a new and unified perspective of chiral asymmetry in nature. Chiral asymmetry is not just a curious aspect of life on this planet, but it is a general property of all matter: elementary particle, atoms, biomolecules and almost all living organisms exhibit chiral asymmetry[1]. The obvious chiral asymmetry in morphological and functional aspects of living beings (such as spiral sea shells being predominantly dextral, about 90% of the people in every continent and culture being right-handed) have been noted many centuries ago. It was in the second half of nineteenth century, however, that Louis Pasteur discovered the chiral asymmetry in the *chemistry* of living organisms. Today we know the universal asymmetry of biomolecules in more precise terms than did Louis Pasteur: as far as we know, proteins are exclusively made of L-amino acids and the sugars in DNA and RNA are all of the D form. The chemistry of life crucially depends on this asymmetry.

In spite of all the advances in biochemistry and in our understanding of the chemistry of life's evolution, processes that led to the chiral asymmetry of proteins and DNA remain elusive. Though there have been many theoretical speculations, due to lack of experimental support we are still without any consensus on this issue[2]. We can not even say with certainty if this asymmetry gradually developed after living cells came into existence or if it arose before, through some prebiotic process. There is no doubt, however, that if chiral asymmetry of the chemical building blocks of life arose under prebiotic conditions, the evolution of life would have been greatly facilitated. Some even consider chiral asymmetry a prerequisite for the evolution of life[3].

In the following sections we will first give a brief description of chemical systems that spontaneously or kinetically break chiral symmetry. Then we will present a theory that describes how systems that spontaneously break chiral symmetry will respond to small chiral asymmetries. Finally, we will discuss how

Reprinted from AIP Conference Proceedings vol. 300, edited by A.K. Mann and D.B. Cline, pp. 491-498.
© 1994 American Institute of Physics.

this theory can be applied to a prebiotic situations and see under what conditions an asymmetry as small as that due to weak neutral currents will be able to determine the particular handedness of the molecules that will dominate.

CHIRAL SYMMETRY BREAKING IN CHEMICAL SYSTEMS: SENSITIVITY TO SMALL CHIRAL ASYMMETRIES

The phenomenon of symmetry breaking is general and well known in physics and chemistry. Equilibrium phase transitions are often associated with spontaneous breaking of symmetry: the final state of the system is less symmetric than the underlying interactions. Thus, in a magnetic transition a direction of magnetization is spontaneously created, though the spin-spin interactions are isotropic. A similar breaking of symmetry also appears in non-equilibrium transitions in chemical systems; the interplay between isotropic diffusion and chemical reactions, for example, can produce states that are anisotropic [4-6]. Theoretical descriptions of these equilibrium and non-equilibrium processes have much in common [5, 7].

From a thermodynamic view point, asymmetric chiral states can only occur in non-equilibrium systems. In some cases, the theory of transitions to chirally asymmetric states bares a close resemblance to the theory of second order phase transitions.

Totally asymmetric states can arise spontaneously in two ways:
(a) when the chirally symmetric state itself becomes *unstable* (due to the non-equilibrium conditions of system) and the system makes a transition to a chirally asymmetric state.
(b) an initially achiral state is *kinetically* driven to a chirally asymmetric state, though the corresponding chirally symmetric state is *not unstable*; the state remains asymmetric because the free energy barrier to the symmetric state is very large.

We shall refer to (a) as a *spontaneous symmetry breaking process,* to (b) as a *kinetic symmetry breaking processes* . Examples of either case are given below.

Following the terminology of chemistry, we shall refer to a chiral molecule and its mirror-image molecule as *enantiomers*. The two forms of the molecule are referred to as L- and D-enantiomers. As for the mechanism, symmetry breaking can occur only when both *chiral autocatalysis* and some form of *competition* between the two enantiomers is present. By competition we mean a process by which the growth of one enantiomer, directly or indirectly, prevents the opposite enantiomer from growing. Autocatalysis alone is not sufficient to cause symmetry breaking. This is because, if one enantiomer can proliferate through autocatalysis so can the other, and the result will be a symmetric state. Only when the growth of one somehow suppresses the growth of the other can we expect a highly asymmetric state to arise.

Systems that fall under the category (a) have been theoretically extensively studied[8-11]. Using the standard stability analysis and bifurcation theory [5, 6, 12], it can be seen that transitions in such non-equilibrium systems have a theoretical description that is very similar to the theory of second order phase transitions[7]. This type of symmetry breaking has not been realized experimentally.

On the other hand, a symmetry breaking process of type (b) was recently identified in crystallization in our laboratory[13]. In this case, chiral symmetry breaking occurs in a remarkably simple way. $NaClO_3$ is an achiral molecule, i.e. it does not have chirality. However, when $NaClO_3$ crystallizes, the unit cell of the crystal structure is chiral; hence the crystals are optically active, i.e. every crystal is either levo- or dextro-rotatory. If $NaClO_3$ is crystallized in a static unstirred solution, statistically equal number of levo and dextro- rotatotry crystals are formed. If the crystallization is performed in a stirred solution, however, more than 99% of the crystals formed in each crystallization all have the same handedness, either levo or dextro. In this case the simple act of stirring generates both autocatalysis and competition[14] that are needed for symmetry breaking. The sensitivity of such systems to small chiral asymmetries is not well understood at this time. The implication of such processes to the origin of biomolecular chiral asymmetry is not known. But the system gives us a clear example of a symmetry breaking process that occurs under very simple conditions.

A model of process (a) is the following reaction scheme[8, 9] which is a variation of a model proposed by Frank in 1953[15]:

S + T	$\Longleftrightarrow$	X_L or X_D	(A)
S + T + X_L	$\Longleftrightarrow$	$2X_L$	(B)
S + T + X_D	$\Longleftrightarrow$	$2X_D$	(C)
X_L + X_D	$\longrightarrow$	P	(D)

In the above, arrows on both sides denote that both forward and reverse reactions are considered. We shall denote the rate constants of the above reactions by K_{Af} for the forward reaction of A, K_{Ar} for the reverse reaction of A and so on. Thus, the rate of forward reaction (A), for example, is $K_{Af} [S] [T]$, in which [S] and [T] are the concentrations.

Under the non-equilibrium conditions of an inflow of S and T and an outflow of P, this system exhibits the phenomenon of spontaneous symmetry breaking: if the product of the concentrations [S] and [T] is below a certain critical value, the concentrations $[X_L]$ and $[X_D]$ of the chiral intermediates will be equal; but if [S][T] exceed this critical value, the *symmetric state* of equal $[X_L]$ and $[X_D]$ becomes unstable and the system *spontaneously* makes a transition to an *asymmetric state* in which $[X_L] \neq [X_D]$. When the transition to the asymmetric state has occurred, whether $[X_L] > [X_D]$ or $[X_L] < [X_D]$ is a matter of chance. A more detailed analysis of the above reaction scheme shows the close analogy between systems such as these and second order symmetry breaking phase transitions[8]. In fact, using this analogy and the methods of group representation theory one can develop a general theory of chiral symmetry breaking for these systems and derive the following equation for $\alpha = [X_L] - [X_D]$:

$$\frac{d\alpha}{dt} = - A \, \alpha^3 + B(\lambda - \lambda_c) \, \alpha \qquad (1)$$

in which A and B are constants that depend on the kinetic rate constants and $\lambda = $ [S][T]; λ_c is the critical value of λ beyond which the symmetry is broken. A term $\sqrt{\varepsilon}F(t)$ representing additive fluctuations can be added to eqn. (1):

$$\frac{d\alpha}{dt} = -A\alpha^3 + B(\lambda - \lambda_c)\alpha + \sqrt{\varepsilon}F(t) \qquad (2)$$

Here $F(t)$ may be assumed to be Gaussian white noise and $\sqrt{\varepsilon}$ the root-mean-square value of this noise. For such additive fluctuations, one can obtain the Fokker-Planck equation for the probability distribution $P(\alpha)$ of α. The solution of the Fokker-Planck equation shows that for $\lambda<\lambda_c$ the stationary probability $P(\alpha)$ has one peak and for $\lambda>\lambda_c$, it is a two-peak distribution. The similarity to Ginzburg-Landau mean field theory of symmetry breaking transitions is clear.

Let us now consider a situation in which there is a small chiral asymmetry in the chemical reaction rates, i.e., $K_{Bf} \neq K_{Cf}$. This asymmetry could be due to any parity violating interaction including WNC. Then eqn.(2) is modified to[8]:

$$\frac{d\alpha}{dt} = -A\alpha^3 + B(\lambda - \lambda_c)\alpha + Cg + \sqrt{\varepsilon}F(t) \qquad (3)$$

in which, C is a factor that depends on the rate constants and the concentrations and g is a dimensionless factor that is characteristic of the strength of the chiral interaction. For example, if due to WNC or other chiral interactions the reaction energy barrier of reactions (B) and (C) differ by an amount ΔE, then $g = \Delta E/kT$.

Using eqn.(3), we can formulate our question as follows: Let us assume that λ is a function of time such that, starting at a subcritical value $\lambda_i < \lambda_c$ it increases linearly at a constant rate γ, i.e., $\lambda=\lambda_i + \gamma t$, till it reaches a supercritical value, λ_f, at which it stops. Each time λ sweeps from λ_i to λ_f, the system will make a transition to a state of broken symmetry in which either $\alpha > 0$ or $\alpha < 0$. Due to the fluctuations, this transition will be random.

Let $P_\pm$ be the probabilities for the states $\alpha > 0$ and $\alpha < 0$ respectively. If g=0, then $P_+ = P_- = 0.5$. If $g \neq 0$, however, these probabilities will be different. At first, one might expect to see no significant difference in P_+ and P_- if $\sqrt{\varepsilon} > Cg$. But this turns out to be false. Even if $Cg << \sqrt{\varepsilon}$, P_+ can be very large if γ, the sweeping rate, is small, thus revealing the sensitivity of symmetry breaking transitions to small chiral asymmetries. A careful analysis[9] of equation (3) gives the following result for P_+:

$$P_+ = \frac{1}{\sqrt{2\pi}} \int_{-\infty}^{N} e^{-x^2/2}dx \qquad \text{in which} \qquad N = \frac{Cg}{\sqrt{\varepsilon/2}}\left(\frac{\pi}{B\gamma}\right)^{1/4} \qquad (4)$$

This is a general result valid for all transitions that break a two-fold symmetry. It says that if γ is sufficiently small, then N can be large. Numerical and electronic simulation of equation (3) and related equations that include fluctuations in the coefficients A and B show that P_+ predicted by (4) is an excellent approximation[16, 17].

We shall use the above theory to discuss plausible origins of biomolecular asymmetry and its relation to WNC or other asymmetries.

A PREBIOTIC SCENARIO FOR THE ORIGIN OF CHIRAL ASYMMETRY

It is generally asumed that simple organic molecules were formed in the atmosphere and in the ocean. The energy of the sun and lightning are assumed to be the driving forces for these reactions. More specifically, we assume the following:
• The prebiotic molecules created in the atmosphere enter the oceans.
• By some mechanism similar to the scheme (A)-(D), a chiral symmetry breaking process occurs in the oceans.
• The critical parameter is an increasing function of some concentrations (such as the product [S||T]). Due to the slow increase of the concentrations of reactants in the oceans, the system undergoes a symmetry breaking transition as described above.
This scenario is shown schematically in Fig.1.

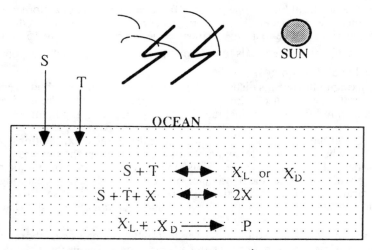

Fig.1. *A prebiotic scenario for the origin of biomolecular asymmetry. Due to chiral autocatalysis and competition between the enantiomers of X, symmetry breaking occurs when the concentration of S and T exceed a critical value. The system then makes a transition to a state dominated by either X_L or X_D. If the increase in concentration of S and T occurs slowly over a period of the order of*

10^4years, then even parity violation of one part in 10^{17} could have a very significant effect in selecting one enantiomer over the other.

With the above assumptions we can conceive of symmetry breaking occurring in the oceans. But this still leaves us with the following problem: if the assumed symmetry breaking process takes place in different regions on this planet, if there is no systematic bias favoring one enantiomer, then we will be left with a situation in which some regions of the oceans are dominated by one enantiomer while the other regions are dominated by the opposite enantiomer -- somewhat like the magnetic domains in a ferromagnet. We can have homochirality over the entire planet only when a systematic bias make the same enantiomer dominate wherever symmetry breaking occurs.

What might the source of such a systematic bias be ? There are several suggestions[2] whose estimated asymmetry ranges from one part in 10^{17} for WNC effects in amino acids[2, 18-20 21] to one part in 10^{12} -10^8 for certain processes involving β-decay[22-25]. The effects of WNC are intrinsic to all chiral molecules while the theories using β-decay make assumptions regarding the levels of β-radiation.

For our study[9] we used the WNC estimates, the smaller value. Then using reasonable values of concentration in the model given by the reaction scheme (A)-(D), and reasonable values for the random fluctuations[9] we considered the following prebiotic conditions in addition to those stated above:
• All concentrations were of the order of 10^{-3}M or less.
• A systematic chiral bias of one part on 10^{17} favoring the L-enantiomer exists.
• Slow increase (not necessarily uniform) in concentration that takes the system from a subcritical value to a super critical value in about 10,000 to 15,000 years. (In the model, this corresponds to the slow increase in λ with a corresponding small value of γ in eqn. (4))
• The mixing in the oceans disperses the reactants and maintains homogeneity on a scale of at least 30 km x 30 km x 10 m, on a time scale of hundreds of years.
• Random chiral fluctuations include circularly polarized components of sunlight.

Under these conditions, somewhat contrary to most expectations, we found that when symmetry breaking occurs the chance that the favored enantiomer will dominate is 98%. Thus almost every instance of symmetry breaking will give rise to the dominance of the *same enantiomer*, resulting in homochirality over the entire planet. The results are not model sensitive: inclusion of racemization and other such reactions do not change the basic conclusions[26].

From these considerations we see that, what is small on a laboratory volume and time scales, may not at all be "small" on an evolutionary scales of time and volume. We could indeed conceive of processes through which the observed homochirality of biochemistry is linked to chiral asymmetry of electroweak interactions, either WNC or β–decay. If systematic asymmetries as small as one part in 10^{17} could have a decisive role, what about chiral asymmetries other than those due to electroweak interactions ? It can be shown that static electric and magnetic fields could not produce chiral asymmetries[8, 27, 28]. Cosequently, a third static field is necessary. This field cluld be gravitational or centrifugal field. When the chiral asymmetries at a molecular scale due to three static fields are considered, it turns out that their strength is much smaller than that due to electroweak asymmetry[8]. When we consider dynamic fields such as circularly polarized light three fields are not necessary. However, for such cases the chiral asymmetries are not systematic and of constant sign but fluctuating and somewhat random. The best known estimates of fluctuating circularly polarized component of sun light was included in our analysis.

In the past, there have been suggestions as to how singular geological factors of a particular location might produce chiral asymmetries[2]. From this point of view, the observed chiral asymmetry is a result of some special geological conditions. I prefer to consider such theories as a last recourse, when all other systematic influences have been ruled out.

ACKNOWLEDGMENTS

Acknowledgment is made to the Donors of The Petroleum Research Fund, administered by the American Chemical Society, for support of this research.

REFERENCE

1.   Hegstrom, R. & Kondepudi, D.K. *Sci. Am.* <u>262</u>, 108-115 (1990).
2.   For an extensive recent review see: Bonner, W.A. *Origins of Life and Evol. Biosphere* <u>21</u>, 59-111 (1991).
3.   Avetisov, V.A., Goldanskii, V.I. & Kuz'min, V.V. *Phys. Today* <u>44</u>, 33-41 (1991).
4.   Field, R.J. & Burger, M. *Oscillations and Travelling Waves in Chemical Systems* (Wiley, New York, 1985).
5.   Nicolis, G. & Prigogine, I. *Self-Organization in Non-Equilibrium Systems* (Wiley, New York, 1977).
6.   Vidal, C. & Pacault, A. *Non-Linear Phenomena in Chemical Dynamics* (Springer, Berlin, 1981).
7.   Nitzan, A., Ortoleva, P. & Ross, J. *J. Chem. Phys.* <u>61</u>, 1056-1074 (1974).
8.   Kondepudi, D.K. & Nelson, G.W. *Physica* <u>125A</u>, 465-496 (1984).
9.   Kondepudi, D.K. & Nelson, G.W. *Nature* <u>314</u>, 438-441 (1985).
10.   Thieman(ed.), W. *Origins of life* <u>11</u>, 1-194 (1981).
11.   de Min, M., Levy, G. & Micheau, J.C. *J. de Chim. Phys.* <u>85</u>, 603-619 (1988).
12.   Haken, H. *Synergetics- An Introduction* (Springer, Heidelberg, 1977).

13. Kondepudi, D.K., Kaufman, R. & Singh, N. *Science* 250, 975-976 (1990).
14. Kondepudi, D.K., Bullock, K.L., Digits, J.A., Hall, J.K. & Miller, J.M. *J. Am. Chem. Soc.* (1993, to appear).
15. Frank, F.C. *Biochem. Biophys. Acta.* 11, 459 (1953).
16. Kondepudi, D.K., Prigogine, I. & Nelson, G.W. *Phys. Lett.* 114A, 29-32 (1985).
17. Kondepudi, D.K., Moss, F. & McClintock, P.V.E. *Physica* 21D, 296-306 (1986).
18. Mason, S.F. & Tranter, G.E. *Chem. Phys. Lett.* 94, 34-37 (1983).
19. Mason, S.F. *Nature* 311, 19-23 (1984).
20. Mason, S.F. & Tranter, G.E. *Molecular Physics* 53, 1091-1111 (1984).
21. Hegstrom, R.A., Rein, D.W. & Sandars, P.G.H. *J. Chem. Phys.* 73, 2329-2341 (1980).
22. Mann, A.K. & Primakoff, H. *Origins of Life* 11, 255 (1981).
23. Mann, A.K. & Primakoff, H. *Origins of Life* 13, 113-118 (1983).
24. Hegstrom, R. *Nature* 315, 749 (1985).
25. Hegstrom, R., Rich, A. & Van House, J. *Nature* 313, 391 (1985).
26. Kondepudi, D.K. *BioSystems* 20, 75-83 (1987).
27. Mead, C.A. & Moscowitz, A. *J. Am. Chem. Soc.* 102, 7301-7302 (1980).
28. Peres, A. *J. Am. Chem. Soc.* 102, 7389-7390 (1980).

Nuclear-spin-dependent P-odd energy difference of chiral radicals

I.B. Khriplovich[1]

Budker Institute of Nuclear Physics, 630090 Novosibirsk, Russia

Abstract

Parity nonconserving nuclear-spin-dependent weak interaction of an electron with the nucleus induces difference between hyperfine structure constants in optical isomers. For heavy radicals with $Z \sim 80$ the effect may exceed 10 Hz, so that its observation is not so far away from the possibilities of the modern experimental technics.

The parity-nonconserving weak interaction induces energy difference between left- and right-handed molecules. This fact is well established now, but theoretically only. The energy splitting is so tiny that its observation seems to be far beyond the modern experimental accuracy. In the present talk I wish to point out a possible exception to this pessimistic assertion [1].

The origin of the P-odd energy difference in chiral molecules can be explained intuitively as follows [2]. Parity-nonconserving weak interaction induces a spin helix of a sign determined by this interaction itself and by the molecular electronic wave function. It is only natural that the energy of the molecule is different depending on whether its own structural, cooordinate helix is of the same or opposite sign as the weak interaction one.

By the way, it is clear from this picture that the spin-orbit interaction is essential for the P-odd energy splitting. However, in heavy molecules it does not cause an extra suppression of the effect. This and other features of the phenomenon discussed can be conveniently demonstrated with a simple model, almost a cartoon, of a chiral molecule.

Let a heavy atom be surrounded by three other atoms that differ from it and from one another. It is easily seen that the simplest structure that could have optical isomers is a molecule consisting of four atoms not lying in the same plane. Suppose, further, that the external electron state in the heavy atom is $p_{3/2}$. Let us take account of its interaction with one of the neighbouring atoms, which we shall call atom 3, assuming that the levels with different projections μ of the angular momentum on the axis passing through the heavy

---

[1] E-mail: khriplovich@inp.nsk.su

atom and atom 3 are non-degenerate. We will assume the field of two other neighbbouring atoms (more precisely, ions), 1 and 2, to be Coulombic with the charges Z_1 and Z_2.

The calculation of the energy splitting δE is conveniently performed in the Cartesian basis and leads to the following result:

$$
\delta E = -\frac{4}{5} Z_1 Z_2 \left([\vec{n}_1 \times \vec{n}_2] \cdot \vec{n}_3 \right) \left((F_1 \vec{n}_1 - F_2 \vec{n}_2) \cdot \vec{n}_3 \right)
$$
$$
\cdot \frac{G m^2 \alpha^2 Z^2 R}{\sqrt{2}\, \pi \, (\nu_s \nu_p)^{3/2}} \frac{Ry^3}{E_s E_p} \left\{ \begin{array}{c} Q \\ \kappa \, (4/3) K \mu \nu / I(I+1) \end{array} \right. \tag{1}
$$

Here $\vec{n}_i$ is the unit radius-vector of the ith atom; F_i is a product of the dimensionless Coulomb interaction integrals; G is the Fermi weak interaction constant; m is the electron mass; $\alpha = e^2 = 1/137$; Z is the charge of the heavy nucleus; R is the relativistic enhancement factor which reaches the value ~ 10 at $Z \sim 80$; $\nu_{s,p}$ are the effective principal quantum numbers of the $s_{1/2}$, $p_{1/2}$ states admixed to the initial $p_{3/2}$ one by the Coulomb interaction; $E_{s,p}$ are their energies reckoned from the initial level; $Ry = m\alpha^2/2$ is the Rydberg constant.

The first entry in the last column of this formula refers to the nuclear-spin-independent energy splitting induced by the nuclear "weak" charge Q which is close numerically to $-N$, N being the neutron number.

But we are interested here in the second entry which describes the nuclear-spin-dependent (NSD) splitting. This effect is dominated by the electromagnetic interaction of the electron with the anapole moment of the nucleus, its P-odd multipole [3, 4]. This interaction is conveniently characterized by a dimensionless constant κ, its typical value for a heavy nucleus being about 0.3. Other notations here are as follows: I is the nuclear spin, ν is its projection on the axis 3, $|K| = (I + 1/2)$.

By the way, formula (1) demonstrates indeed that, though the P-odd splitting depends on the spin-orbit interaction, in heavy molecules where the fine structure is comparable with the electrostatic interaction, this dependence does not cause an additional suppression of the splitting.

Let us note that the NSD effect is just a pseudoscalar correction to the constant A of the molecular hyperfine structure (HFS), with its sign depending on whether the radical is a right- or left-handed isomer. A conservative numerical estimate for it constitutes $10 - 100$ Hz at $\kappa \sim 1$. Such an accuracy in the HFS measurements seems by itself quite realistic. It should be emphasized, however, that the above estimate pertains only to the case when the optical

isomer is a radical. So, the demands to a possible experiment are very strict:
1. Stable heavy chiral radical.
2. Unpaired electron close to both heavy atom and asymmetric group.
3. Sufficient population in a given molecular state. This is perhaps the most serious difficulty. But due to the modern experimental progress with low temperature magnetic traps, the situation with it does not look hopeless.

The experiment discussed would be extremely interesting not only from the point of view of the present Symposium subject. It would make it possible to detect nuclear anapole moments, a first rate contribution to nuclear physics. In principle, there is a possibility of comparing the anapoles of different isotopes, and this comparison obviously requires no molecular calculations.

References

[1] Khriplovich, I.B., Z.Phys. **A322**, 507 (1985).

[2] Khriplovich, I.B., Parity Nonconservation in Atomic Phenomena, Gordon and Breach Science Publishers, 1991.

[3] Flambaum, V.V., and Khriplovich, I.B., Zh.Eksp.Teor.Fiz. **79**, 1656 (1980) [Sov.Phys. JETP **52**, 835 (1980)].

[4] Flambaum, V.V., Khriplovich, I.B., and Sushkov, O.P., Phys.Lett. **B145**, 367 (1984).

High-frequency asymptotics of circular dichroism

M.E.Pospelov

Budker Institute of Nuclear Physics. 630090 Novosibirsk. Russia[1]

and

Institute for Nuclear Theory, University of Washington

Seattle. WA 98195

Abstract

Circular dichroism of optically active isotropic media of chiral molecules $Im(n_+ - n_-)/Im(n_+ + n_-)$ falls off as ω^{-2} at frequencies $Ry \ll \omega \ll Z^2 Ry$ and as ω^{-3} at $\omega \gg Z^2 Ry$, where Z is a typical nuclear charge of atoms in the chiral group. The contribution of the spin of electron to the circular dichroism appears in the second order in spin-orbit perturbation only. The polarization of photoelectrons in the absorption of unpolarized light is connected with the chirality of molecule and constitutes $Z^2\alpha^3$ from the degree of geometrical asymmetry.

PREPARED FOR THE U.S. DEPARTMENT OF ENERGY
UNDER GRANT DE-FG06-90ER40561

---

[1] permanent address

1 Introduction

Optical isomers are molecules and crystals that are mirror images of one another. The media. dominantly containing the isomer of a definite sign rotates the polarization plane of transmitted light and absorbs left and right quanta by different ways. This last phenomenon is called the circular dichroism (CD). We will investigate CD at frequencies $\omega \gg Ry$, where $Ry = m\alpha^2/2 = 13.6\,eV$ is the characteristic atomic energy.

Our interest in the high-frequency behaviour of CD was dictated by the following.

First. according to many speculations (see for ex. [1]), circular polarization of light might be the main cause of the origin of optical activity (OA) and the apparent left-right asymmetry of living matter. In the ultraviolet region prochiral photochemical reactions leading to OA were seen in a laboratory [2, 3, 4, 5].

Second. we are interested in the question of whether or not correlations of the momentum and spin of the molecular electron. originating from the spin-orbital interaction [6, 7], show up in the asymptotics of CD. It is known that such correlations determine the high-frequency behaviour of OA [8]. The rotation angle of the polarization plane at $\omega \sim Z^{-3/4}\alpha^{-1}$ allows the knowledge of P-odd energy difference of left- and right-handed molecules [8]. It is natural in this respect to investigate the contribution of the spin-orbital interaction to the high-frequency asymptotics of CD.

The refraction index $n(\omega)$ is related to the forward-scattering amplitude on the single molecule through the well known formula:

$$n = 1 + \frac{2\pi}{\omega^2}\frac{N}{V}f(\omega), \tag{1}$$

where N/V is the concentration. The imaginary part of the scattering amplitude in its turn can be rewritten using the optical theorem via the total cross section of the absorption by single molecule:

$$\operatorname{Im} n(\omega) = \frac{2\pi}{\omega^2}\frac{N}{V}\operatorname{Im} f(\omega) = \frac{1}{2\omega}\frac{N}{V}\sigma(\omega). \tag{2}$$

CD originates from the difference of the absopbtion cross sections of photons with opposite signs of the circular polarization λ:

$$\lambda = i([\bar{\epsilon}\bar{\epsilon}]\bar{n}) = \pm 1. \tag{3}$$

Here $\bar{\epsilon}$ is the vector of the polarization of photon; $\bar{n}$ is the unit vector of its momentum.

Quantatively, CD is characterized by the ratio of this difference to the cross section. independent of the spin of photon:

$$\eta = \frac{\sigma_+ - \sigma_-}{\sigma_+ + \sigma_-}. \tag{4}$$

As a first step to the asymptotics of $\eta(\omega)$ at large frequencies we rewrite (4) via the ratio of amplitudes. When the frequency of absorbing light is not very high ($\omega < m\alpha$), it is reasonable to operate only with two first terms in the multipole expansion:

$$< f|V_{int}|in > = < f|i\omega(\bar{\epsilon}\bar{d}) + i\omega([\bar{n}\bar{\epsilon}]\bar{\mu})|in > . \tag{5}$$

Here $\vec{d} = e\vec{r}$, $\vec{\mu} = e/2m(\vec{l} + \vec{\sigma})$ are operators of electric and magnetic moments. $E2$ - transition is omitted in (5) because its contribution to the CD vanishes under the average over random orientation of scatterer. The last procedure leads to the following formula for CD η in the transition between eigenstates in and f:

$$\eta = -\frac{\text{Im} < in|\vec{l} + \vec{\sigma}|f > < f|\vec{r}|in >}{m < in|\vec{r}|f > < f|\vec{r}|in >}.$$ (6)

At the frequencies $\omega \sim Ry$ the magnitude of CD is $\eta \sim \alpha\xi$. The fine structure constant originates from the ratio of M1 to E1 amplitudes. The factor $\xi \sim 10^{-2}$ reflects the degree of the molecular geometrical asymmetry. To take the next step we need information about the wave function of the electron inside a chiral molecule.

2   The wave functions of molecular electron

We shall deal with the electron, forced by the potential V of several static centers, placed at points $\vec{r_a}$:

$$V = \sum_{a=1}^{N} V_a(|\vec{r} - \vec{r_a}|)$$ (7)

To possess OA the molecule should be composed of four such centers at least ($N \geq 4$). If the nuclear charge Z of atoms in the chiral group is not very high, $Z^2\alpha^2 \ll 1$, the spin-orbital interaction

$$V_{SO}(\vec{r}) = -\frac{1}{4m^2}([\vec{\sigma}\vec{p}]\nabla V(\vec{r}))$$ (8)

can be treated as a small perturbation and we neglect it in the lowest order in α. It is convenient for us to place the origin at one of these centers:

$$V = U(r) + \sum_{a=1}^{N-1} V_a(|\vec{r} - \vec{r_a}|),$$ (9)

and use the basis of eigenfunctions, corresponding to the motion in the potential $U(r)$. The distortion of the spherical symmetry by V_a we shall describe using perturbation theory. To be concrete, we take $N = 4$ and $V_a(|\vec{r} - \vec{r_a}|) = -\alpha Z_a^{eff}/|\vec{r} - \vec{r_a}|$.

It is clear from the very beginning, that there is no degeneracy in the potential (9); all wave functions are real and do not possess definite parity. As a first step we take two levels of energy E_{1s} and E_{2s}, corresponding to different s-states with radial dependencies $s_1(r)$ and $s_2(r)$. From here and below the angular normalization $(4\pi)^{-1/2}$ is omitted, while all integrations over solid angles in matrix elements are taken as an average. The dipole part of the potential V

$$V_{dip} = -\sum_{a=1}^{3} \alpha Z_a^{eff}(\vec{n_a}\vec{n})[\frac{r}{r_a^2}\theta(r_a - r) + \frac{r_a}{r^2}\theta(r - r_a)] \equiv -\sum_{a=1}^{3} \alpha Z_a^{eff}(\vec{n_a}\vec{n})\kappa_a(r)$$ (10)

leads to the mixing of p-states to the given s. For our purposes it is sufficient to consider the admixture of two p-waves with different radial dependencies $p_1(r)$ and $p_2(r)$ to lower level and one of them (nearest) to the upper level:

$$\psi_{in}(\vec{r}) = s_1(r) + (\vec{A}\vec{n})p_1(r) + (\vec{B}\vec{n})p_2(r) \qquad (11)$$
$$\upsilon_f(\vec{r}) = s_2(r) + (\vec{C}\vec{n})p_2(r).$$

Vectors $\vec{A}$, $\vec{B}$ and $\vec{C}$ are determined by the geometry of the molecule. Explicit formulae for them in our case read as follows:

$$\vec{A} = -\sum_{a=1}^{3} \vec{n}_a \left[\frac{\alpha Z_a}{E_{s1} - E_{p1}} \int_0^{\infty} s_1 p_1 r^2 \kappa_a(r) dr \right]$$

$$\vec{B} = -\sum_{a=1}^{3} \vec{n}_a \left[\frac{\alpha Z_a}{E_{s1} - E_{p2}} \int_0^{\infty} s_1 p_2 r^2 \kappa_a(r) dr \right]$$

$$\vec{C} = -\sum_{a=1}^{3} \vec{n}_a \left[\frac{\alpha Z_a}{E_{s2} - E_{p2}} \int_0^{\infty} s_2 p_2 r^2 \kappa_a(r) dr \right]. \qquad (12)$$

The calculation of $E1 - M1$ interference in the transition between wave functions (11) is rather trivial:

$$\mathrm{Im} < i|\vec{l}|f > < , f|\vec{r}|i > = -\frac{1}{9}(\vec{A}[\vec{B}, \vec{C}]) \int_0^{\infty} s_2 p_2 r^3 dr$$
$$= -\frac{1}{9}(\vec{n}_1[\vec{n}_2, \vec{n}_3])\epsilon_{abc}\rho_a^A \rho_b^B \rho_c^C \int_0^{\infty} s_2 p_2 r^3 dr \qquad (13)$$

Here ρ_a^A denotes dimensionless quantities, separated by squared brackets in (12). The effect is proportional to $(\vec{n}_1[\vec{n}_2, \vec{n}_3])$, as it should be, and vanishes in the case of planar molecule. We want to stress, also, that nontrivial radial dependence $\kappa_a(r) = \theta(r_a - r)r/r_a^2 + \theta(r - r_a)r_a/r^2$, presented in the expression for dipole perturbation is essential for the existence of OA in the transition between (11). Indeed, the use of V_{dip} in the form

$$V_{dip} = -\sum_{a=1}^{3} \alpha Z_a^{eff} \frac{(\vec{r}_a \vec{r})}{r_a^2} \equiv (\vec{r}\vec{D}) \qquad (14)$$

leads to the fact that all three vectors $\vec{A}$, $\vec{B}$ and $\vec{C}$ acquire the same direction and $(\vec{A}[\vec{B}, \vec{C}]) = 0$.

The model of a chiral molecule presented above duplicates in fact [9, 10] in the usual physical scale $Z^2\alpha^2 \ll 1$. Wave functions obtained here help us in the analysis of the high-frequencies behaviour of $\eta(\omega)$.

3 Asymtotics of $\eta(\omega)$

At frequencies $\omega \gg Ry$ the final state of electron lands in the continuum, far away from the threshold. Its energy $E_f = E_i + \omega = k^2/(2m)$ is close to ω. This means, that the velocity of photoelectrons is restricted by the following:

$$\frac{\alpha}{v} \ll 1. \qquad (15)$$

79

This condition allows use of Born approximation and consideration of the potential energy V in the final state as a small perturbation. This approximation gives in the leading order for $\sigma_+ \div \sigma_-$ the well-known asymptotic formula (see. for ex.. [11]):

$$\sigma_+ \div \sigma_- \sim \alpha (m\alpha)^{-2} Z_{eff}^5 (Ry/\omega)^{7/2}. \tag{16}$$

Here Z_{eff} depends on the number of the shell from which the ionization occurs. Now matrix elements of $E1$ and $M1$ transitions are defined as integrals with the function $v_f \sim \exp(i\vec{k}\vec{r})$, rapidly oscillating on atomic distances $1/(m\alpha)$. Therefore, the leading contribution to the asymptotics of CD comes from the lowest multipole components of the wave function v_{in}. Due to the plane wave taken as a final state. $E1$ and $M1$ transitions become orthogonal and lead to the vanishing of OA. The effect, which differs from zero, appears if we take into account admixtures to the final state induced by the potential V_{dip}. This means that the λ-dependent cross section at high frequencies is suppressed in comparison with $\sigma_+ + \sigma_-$ by additional powers of the Born parameter $\alpha/v = \sqrt{Ry/\omega}$. Because all radial integrals are determined now by small distances $r \sim 1/k$. it is reasonable to use V_{dip} in the form (14).

We choose the wave function of initial state in the form (11). The angular momentum operator in the matrix element of $M1$-transition separates in the final state p-wave admixtures with the same radial dependencies $p_1(r)$ and $p_2(r)$. All other admixtures give only small corrections to the $E1$ transition and can be omitted in our treatment. As a result. the wave function of the final state reads as follows:

$$v_f = s_k(r) + (\vec{D}\vec{r}) \left[\frac{p_1(r)}{\omega + E_i - E_{1p}} \int_0^\infty p_1 s_k r^3 dr + \frac{p_2(r)}{\omega + E_i - E_{2p}} \int_0^\infty p_2 s_k r^3 dr \right]. \tag{17}$$

At distances $r \sim 1/(m\alpha)$ function $s_k(r)$ coincides with the s-wave radial function of the free motion $\sin(kr)/r$. The calculation of $E1 - M1$ interference in the transition between (11) and (17) is quite straightforward now and leads to the following answer:

$$\operatorname{Im} < in|\vec{l}|f >< f|\vec{r}|in > = \tag{18}$$

$$= \frac{1}{9}([\vec{A}\vec{B}]\vec{D}) \int_0^\infty p_2 s_k r^3 dr \int_0^\infty p_1 s_k r^3 dr \left(\frac{1}{\omega + E_i - E_{2p}} - \frac{1}{\omega + E_i - E_{1p}} \right) \simeq$$

$$\simeq \frac{1}{9}([\vec{A}\vec{B}]\vec{D}) \int_0^\infty p_2 s_k r^3 dr \int_0^\infty p_1 s_k r^3 dr \frac{E_{2p} - E_{1p}}{\omega^2}.$$

Taking into account that in order of magnitude $|\vec{D}| \sim \alpha(m\alpha)^2$ we get for $\eta(\omega)$ the following estimation:

$$\eta(\omega) \sim \alpha\xi \frac{\alpha^2 m (E_{1p} - E_{2p})}{\omega^2} \frac{\int p_2 s_k r^3 dr \int p_1 s_k r^3 dr}{(\int s_1 p_k r^3 dr)^2}. \tag{19}$$

Here $p_k(r)$ is p-wave radial function, interpolated on the atomic scale by correspondent function of the free motion. The ratio of radial integrals in (19) cannot be calculated in a general case, without specifying potential energy $U(r)$ and wave functions p_1. p_2 and s_2. In the limit $\omega \gg Z^2 Ry$ all wave functions could be considered as rapidly oscillating even inside K-shell, at $r \leq 1/(Zm\alpha)$. Than this ratio is determined by the behaviour of $p_1(r)$. $p_2(r)$ and $s_1(r)$ near the origin and can be calculated up to the end:

$$\frac{\int p_2 s_k r^3 dr \int p_1 s_k r^3 dr}{(\int s_1 p_k r^3 dr)^2} = \frac{4}{9} \frac{1}{k^2} \frac{p_1''(0) p_2''(0)}{(s_1'(0))^2} \sim Z^2 \frac{(m\alpha)^2}{k^2} \sim Z^2 \frac{Ry}{\omega}. \tag{20}$$

Combining (19) and (20) we derive the asymtotics of $E1 - M1$ interference. normalized on the square of $E1$ amplitude:

$$\eta(\omega) \sim \alpha\xi \left(\frac{Ry}{\omega}\right)^2 \begin{cases} const & Ry \ll \omega \leq Z^2 Ry \\ Z^2 Ry/\omega & \omega \geq Z^2 Ry. \end{cases} \tag{21}$$

It should be mentioned. in conclusion. that the ratio of cross-sections $(\sigma_+ - \sigma_-)/(\sigma_+ + \sigma_-)$ at $\omega \gg Z^2 Ry$ is Z^2 times smaller than the correspondent value in (21) because in that case the cross-section independente of spin of the photon is determined by ionization of the internal electronic shell and contains additional factor Z^2.

4 Circular dichroism and spin-orbit interaction

The rapid drop of CD with growing frequency provides reason to believe that smoother behaviour could be a result of relativistic effects neglected in the previous consideration. Below we investigate the contribution to the asymptotics of CD caused by the spin of the electron.

In the zeroth order in the spin-orbit interaction for any eigenstate of the electron in the chiral molecule $|i:\sigma>$ spin and coordinate parts of the wave function are factorized:

$$w_i(\vec{r};\sigma) = w_i(\vec{r})\chi, \tag{22}$$

where χ is two-component spinor. All levels of the discrete spectrum have two-fold degeneracy. According to the Kramers theorem this degeneracy should exist even after introducing the spin-orbit interaction. First order in this perturbation mixes to the initial state $|in.\sigma>$ arbitrary intermediate states $|n.\sigma'>$:

$$|in.\sigma> + \sum_{n.\sigma'}' \frac{|n,\sigma'><n,\sigma'|V_{SO}|in,\sigma>}{E_{in} - E_n}. \tag{23}$$

Summation over σ' can be done using the completeness relation with the explicit formula for spin-orbit interaction (8) it allows transformation of (23) to the following form:

$$\left(1 - \sum_n' \frac{|n><n|[\vec{p}\nabla V(\vec{r})]\vec{\sigma})}{4m^2(E_{in} - E_n)}\right)|in> \chi_{in}. \tag{24}$$

Now we are able to write down the contribution of the spin part of the $M1$ transition to the $E1 - M1$ interference. Operator $\vec{\sigma}$ gives nonvanishing transition between states of the form (22) only with the same coordinate dependencies:

$$<n,\sigma|\vec{\sigma}|k,\sigma'> = \chi^\dagger \vec{\sigma}\chi'\delta_{nk}. \tag{25}$$

It significantly simplifies the calculation of OA in the first order in spin-orbit interaction removing the sum over intermediate states:

$$\text{Im} < in|\frac{V_{SO}\vec{\sigma}}{E_i - E_f} + \frac{\vec{\sigma}V_{SO}}{E_f - E_i}|f><f|\vec{r}|in>$$

$$= \frac{1}{\omega}\text{Im} < in|[\vec{\sigma}, V_{SO}]|f><f|\vec{r}|in> = \frac{1}{2m^2\omega}\epsilon_{ijk}\text{Re} < in|L_j\sigma_k|f><f|r_i|in>. \tag{26}$$

where $\vec{L} = [\vec{p}\nabla V(\vec{r})]$. Since the direction of initial spin in the chiral molecule is evidently random. the average over it should be performed. This immediately leads to the vanishing of (26). So. the first order in spin-orbit interaction is not sufficient to obtain a nonzeroth contribution of the electronic spin to the effect of interest. This conclusion is in agreement with the fact that the corresponding asymptotics of the real part of scattering amplitude arises only in the second order in this perturbation [8].

The whole calculation of relativistic corrections to the OA and CD in the second order is a very complicated problem. Nevertheless. it could be shown that significant Z-enhancement arises in the corrections originated from admixtures to the wave function of the electron caused by spin-orbit interaction and is not contained in contributions from modification of current. higher order terms in multipole expansion etc.

The consideration of the admixtures to the wave function in the second order in combination with square of first order corrections leads to the following expression for contribution of spin to the OA of transition:

$$\frac{1}{8m^4\omega}\epsilon_{ijk}\mathrm{Re}\sum_n[\frac{<in|L_i|n><n|L_j|f><f|r_k|in>\omega}{(E_{in}-E_n)(E_f-E_n)}$$
$$+\frac{2<in|L_i|n><n|r_j|f><f|L_k|in>}{(E_{in}-E_n)}$$
$$+\frac{2<in|r_i|n><n|L_j|f><f|L_k|in>}{(E_f-E_n)}].\tag{27}$$

The average over the initial orientation of spin is already performed. The relative magnitude of the effect of interest at frequencies $\omega \sim Ry$ can be estimated as follows:

$$\eta_{SO}(\omega) \sim (Z\alpha)^4\alpha\xi.\tag{28}$$

Significant Z-enhancement arises from the singular behaviour of the operator

$$\vec{L} = \frac{1}{r}\frac{dU}{dr}\vec{l} + \sum_{a}[p\nabla V_a(r)]\tag{29}$$

near the origin. i.e. from the first term in (29).

With the growth of frequency of absorbing light the expression (27) remains nonvanishing even in neglection of the influence of potential energy on the final state f of the photoelectron. Simultaneously. in the sum over intermediate states n. main contributions come from the levels for which the distortion of spherical symmetry of potential is significant . In other words. $E_{in} - E_k \sim Ry$ and $E_f - E_k \sim \omega$. Thus. the last term in (27) can be neglected and this expression transforms to the form:

$$\frac{Z^2a^2}{8m^4\omega}\epsilon_{ijk}\sum_n\frac{\mathrm{Re}}{(E_{in}-E_n)}<in|\frac{l_i}{r^3}|n>(<n|\frac{l_j}{r^3}|f><f|r_k|in>$$
$$-2<n|r_k|f><f|\frac{l_j}{r^3}|in>)\tag{30}$$

Let us restrict ourselves on the estimation of the last expression in the region of frequencies $Ry \ll \omega \ll Z^2Ry$. In this case all matrix elements of the spin-orbit interaction are determined by small distances $\sim (Zam^{-1})$, independent of the velocity of photoelectrons and constitute in order of magnitude

$$< V_{so} > \sim Z^2 \alpha^2 Ry. \tag{31}$$

Finally, the estimation of the high-frequencies behaviour of (30), normalized on the square of $E1$-amplitude is given by the following:

$$\eta_{SO}(\omega) \sim \frac{(Z^2\alpha^2 Ry)^2}{\omega m Ry < r >}\xi \sim (Z\alpha)^4 \frac{Ry}{\omega}\frac{k}{m}\xi \sim \alpha\xi(Z\alpha)^4 \sqrt{Ry/\omega}. \tag{32}$$

5 Polarization of photoelectrons

Intramolecular spin-orbit interaction causes a new interesting effect connected with the polarization of photoelectrons.

Let us consider the ionization of a chiral molecule by unpolarized light. The probability of geting the electron in its final state with the spin directed along or opposite to the momentum of the photon occurs to be differently. The similar effects in the transmission of electrons through an optically active media were discussed previously in the work [12].

The final state density matrix for the electron polarized along $\vec{n}$ has the form:

$$\rho_f = \frac{1}{2}(1 + \vec{\sigma}\vec{n}). \tag{33}$$

The effect discussed above is characterized by the ratio ζ:

$$\zeta = \frac{W(\vec{n}) - W(-\vec{n})}{W(\vec{n}) + W(-\vec{n})}, \tag{34}$$

where $W(\vec{n})$ is the probability of finding the electron in its final state with the spin directed along $\vec{n}$. Expectation values arising in (34) are proportional to the square of amplitude (5) averaged over the polarization of the photon and the random orientation of a molecule with the weight ρ_f:

$$W(\vec{n}) - W(-\vec{n}) \sim i\epsilon_{ijk}Im < in|\mu_i|f > \sigma_j < f|d_k|in > . \tag{35}$$

Here σ-matix originates from ρ_f. To obtain the nonvanishing result for ζ we have to consider the influence of spin-orbit interaction on the wave functions of electron. Now it is sufficient to consider first order corrections (24). The final formula for the degree of polarization can be written in the form:

$$\zeta = \frac{2\mathrm{Im} < in|\vec{r}|f > < f|\vec{L}|in >}{m^3\omega < in|\vec{r}|f > < f|\vec{r}|in >} \tag{36}$$

$$+\epsilon_{ijk}\sum_n [\frac{< in|L_i|n > < n|r_j|f > < f|l_k|in > + < in|L_i|n > < n|l_k f > < f|r_j|in >}{2(E_{in} - E_n)m^3\omega < in|\vec{r}|f > < f|\vec{r}|in >}$$

$$-\epsilon_{ijk}\sum_n \frac{< in|r_i|n > < n|L_j|f > < f|l_k in > + < in|r_i|f > < f|L_j|n > < n|l_k|in >}{2(E_f - E_n)m^3\omega < in|\vec{r}|f > < f|\vec{r}|in >}].$$

In atomic range of frequencies $\omega \sim Ry$ where velocities of photoelectrons are about α the ratio ζ constitutes in order of magnitude:

$$\zeta \sim \alpha \xi (Z\alpha)^2. \tag{37}$$

This result is α-times smaller than the estimation of the helicity [12] acquired by slow electrons ($v \sim \alpha$) scattering by a chiral molecule.

6 Discussion

The main asymptotics (21) is obtained in our approach using the model of a chiral molecule in which the distortion of spherical symmetry was treated as a small perturbation. Nevertheless, it is clear that the result does not change qualitatively outside this approximation. Indeed, the applicability of the perturbation theory for the derivation of wave functions (11) requires only that $|\bar{A}|, |\bar{B}|, |\bar{C}| \ll 1$. This fact was not used in the further consideration.

Let us discuss once more the contribution to the asymptotics of CD which came from multipole components of the wave functions of higher degrees than s and p. At $\omega \gg Z^2 Ry$ the asymptotics is determined by the behaviour of wave functions near the origin and these contributions are therefore suppressed in comparison with the $s - p$ effect by additional powers of Ry/ω. At $Ry \ll \omega \ll Z^2 Ry$ these components may induce the effect of the same ω-dependence that in $s - p$ one. We neglect it in our consideration simply because these contributions have another, more complicated than $(\bar{n}_1 [\bar{n}_2, \bar{n}_3])$ geometrical structure and a priori smaller factor ξ.

For organic compounds characteristic Z usually constitute 6-8. Therefore, the asymptotics of CD ω^{-3} shows up below the frequencies where multipole expansion (5) breaks down. Spin-orbit effects in that case are negligibly small. These effects could determine the asymptotics for chiral compounds containing elements with higher Z. The simple comparison of (21) and (32) shows that the spin-orbit contribution to CD began to dominate at $\omega \geq (Z\alpha)^{-8/3} Ry$. This frequency satisfies the relation $\omega \leq m\alpha$ when $Z > 22$. One should have in mind however that for exotic molecules with heavy atoms in the chiral group the spin-orbit interaction cannot be regarded as a small perturbation in comparison with the potential violating spherical symmetry. It is clear that in this case the separation of contributions to CD on the "spin" part and "usual" part is meanless.

The effect of polarization of photoelectrons is interesting for us because it may indicate the measure of spin-momentum correlations inside chiral molecules. Spin-orbit interaction arises in this effect in the first order which gives some chance to its experimental observation. For compounds with heavy atoms inside such effect could reach 10^{-2} from the degree of geometrical asymmetry.

I would like to thank I.B.Khriplovich for his interest to this work and critical remarks. P.G.Silvestrov and A.S.Yelkhovsky for helpful discussions, and A.Clinton for useful comments. I thank the Institute for Nuclear Theory at the University of Washington for its hospitality during completion of this work.

References

[1] S.F.Mason. Nature 311 (1984) 19.

[2] W.Kuhn. F.Braun. Naturwissenchaften. 17 (1929) 227.

[3] W.Kuhn. E.Knopf. Naturwissenchaften. 18 (1930) 183.

[4] W.J.Bernstein. M.Calvin. O.Buchardt. J. Am. Chem. Soc. 95 (1973) 527.

[5] H.Kagan et al.. Tetrahedron Lett. (1971) 2479.

[6] A.Garay, P.Hrasko. J. Mol. Evolut. 6 (1975) 77.

[7] R.Hegstrom, Nature 297 (1982) 643.

[8] I.B.Khriplovich. M.E.Pospelov. Phys. Lett.) A171 (1992) 349.

[9] I.B.Khriplovich. JETP 52 (1980) 177. Zh. Eksp. Teor. Fiz. 79 (1980) 354).

[10] I.B.Khriplovich. M.E.Pospelov. Z. Phys. 17 (1990) 81.

[11] V.B.Berestetskii. E.M.Lifshitz. L.P.Pitaevskii. Relativistic Quantum Theory (Pergamon Press. 1979)

[12] R.Hegstrom. A.Rich, J.Van House. Phys. Rev. Lett. 48 (1982) 1341.

Molecular Structures derived from Deterministic Theory of Atomic Structure

K. U. Lu

Department of Mathematics, California State University, Long Beach, CA 90840

Based on the deterministic thoery of the atomic structure, details of the electronic orbits and electronic configurations of H, He, Li, Be, B, C, N, O, and P atoms are calculated. The bond angles of water, benzene, Adenine, Thymine, Uracil, Guanine, and Cytosine are deduced.

INTRODUCTION

The Quantum Mechanics undeniablely provides a physical base for the shell structure of atoms, and in turn explanations of covalent bonds in molecules. But because of the probabilistic nature of the interpretation of the electronic distrbution in the shells, the picture of the electron in the shell is described as a "spherical electron cloud". This spherically symmetric electron cloud can not provide a physical base for the molecular structure. Take the molecular structure of water for instance, there is no physical rationale for the bond angle of 104°27′ based on the spherical electron cloud picture.

There is room for improvement of the standard Quantum Mechanics. A deterministic Theory of atomic structure was advanced. This theory avoids the probabilistic interpretation of the electronic orbit, yet retains,through the mathematical solution, most of the mathematical results of Schrödinger equation. A deterministic electronic orbits in the shell of the atom are provided in the mathematical solution of this deterministic theory, which in turn provides the physical base for the bond angles.

In Section 2, a brief introduction of the deterministic theory of the atomic structure and its mathematical solution are given. In Section 3, the numerical locations of the first few electronic orbits around the nucleus are calculated. In Section 4, the atomic structures of H, He, Li, Be, B, C, N, O and P are given. In Section 5, the molecular structures, in particluar the bond angels, of water, benzene, Adenine, Thymine, Uracil, Guanine, and Cytosine are explained as examples.

INTRODUCTION TO THE DETERMINISTIC THEORY
OF ATOMIC STRUCTURE

The Maxwell Equation works well for macroscopic electromagnetic wave propagation, but does not work well for electrodynamics in atoms. A system of equations that governs the electrodynamics in atoms and accounts for the stability of atoms is derived. They are as following:

$$\nabla \times E = -\frac{\mu}{c}\frac{\partial H}{\partial t} \tag{1}$$

$$\nabla \times H = 0 \tag{2}$$

$$\nabla \cdot E = 0 \tag{3}$$

$$\nabla \cdot H = 0 \tag{4}$$

Following the approach in (Heitler, pp.2-3), we set $H = \nabla \times A$ into (1) to obtain

$$E + \frac{\mu}{c}\frac{\partial A}{\partial t} = -\nabla \phi \tag{5}$$

where A is the vector potential and ϕ is the scalar potential. Substituting $H = \nabla \times A$ into (2), we obtain

$$\nabla(\nabla \cdot A) - \nabla^2 A = 0 \tag{6}$$

Taking divergence on (5) and making use of (3), we obtain

$$-\nabla^2 \phi - \frac{\mu}{c}\nabla \cdot \frac{\partial A}{\partial t} = 0 \tag{7}$$

Making use of the Lorentz Gauge, if A_0 and ϕ_0 represent certain possible values of A and ϕ, and χ is determined from

$$\nabla^2 \chi - \frac{\mu^2}{c^2}\frac{\partial^2 \chi}{\partial t^2} = \nabla \cdot A_0 + \frac{\mu}{c}\frac{\partial \phi_0}{\partial t} \tag{8}$$

and

$$A = A_0 - \nabla \chi$$

$$\phi = \phi_0 + \frac{\mu}{c}\frac{\partial \phi_0}{\partial t}$$

the Lorentz relation

$$\nabla \cdot A + \frac{\mu}{c}\frac{\partial \phi}{\partial t} = 0 \tag{9}$$

is obtained. Substituting (9) into (7), we obtain

$$-\nabla^2 \phi + \frac{\mu^2}{c^2}\frac{\partial^2 \phi}{\partial t^2} = 0 \tag{10}$$

Substituting (9) into (6), we obtain

$$\nabla\left(\frac{-\mu}{c}\frac{\partial \phi}{\partial t}\right) - \nabla^2 A = 0 \tag{11}$$

The task now is solving (10) to obtain ϕ; substituting into (11) and solving it to obtain A. The H and E are obtained from

$$H = \nabla \times A$$

and

$$E = -\nabla\phi - \frac{\mu}{c}\frac{\partial A}{\partial t}$$

Since the electron is constrained by the nucleus due to electric force, we need to use spherical coordinates. (10) is solved by separation of variables; setting

$$\phi = \psi(r, \theta, \varphi)T(t)$$

where (r, θ, φ) is the spherical coordinates, into (10), we obtain the solutions as following:

$$\phi_l^m(r, \theta, \varphi, t) = \cos C_1 t (or \sin C_1 t) Y_l^m(\theta, \varphi) r^{\frac{-1}{2}} J_{l+\frac{1}{2}}(C_1 \frac{\mu}{c} r)(or Y_{l+\frac{1}{2}}(C_1 \frac{\mu}{c} r))$$

where C_1 is a constant; $Y_m^l(\theta, \varphi)$ is the spherical harmonics; l is restricted to be a positive integer; $m = -l, -l+1, \cdots, l$; $J_{l+\frac{1}{2}}(C_1 \frac{\mu}{c} r)$ and $Y_{l+\frac{1}{2}}(C_1 \frac{\mu}{c} r)$ are Bessel Functions.

LOCATIONS OF ELECTRONIC ORBITS

We shall derive a collection of the first few possible electronic orbits around the nucleus from the mathematical solution at the end of last Section. We choose the Bessel function of the first kind as the solution to look for the electronic orbits, since there is quite a distance to the nucleus.

Since the electrons are expected to stay at the minima of the electric potential, ϕ; the minima of the Bessel function of the first kind of the order of

$l + 1/2$ are the locations of the electronic orbits in the radius coordinate, r, direction. From the mathematical table (pp. 154, Jahnke and Emde) for Bessel function of the first kind of order $l + 1/2$, the approximate values of the first minimum for

$$l = 1, J_{\frac{3}{2}}; l = 2, J_{\frac{5}{2}}; l = 3, J_{\frac{7}{2}}; l = 4, J_{\frac{9}{2}}$$

are respectively

$$6; 7; 9; 10$$

We denote them by

$$J_{\frac{3}{2}}^{(1)} = 6; J_{\frac{5}{2}}^{(1)} = 7; J_{\frac{7}{2}}^{(1)} = 9; J_{\frac{9}{2}}^{(1)} = 10.$$

These are their relative values. A normalization is needed, when they are applied to individual atoms. They are the relative positions of the first four shells.

The electronic orbits are the intersections of the orbital planes and the shells. It is well-known that $(l(l + 1)/2)^{1/2}\hbar$ is the total angular momentum ; and $m\hbar$ is the projection of the total angular momentum on z-axis. After some geometric consideration, we obtain

$$cos\bar{\theta} = \frac{m}{\sqrt{l(l + 1)}}$$

where $\bar{\theta}$ is the angle between the orbital plane and the x-axis, which is the pole axis, and y-z plane is the equatorial plane of the atom.

The following list are the possible orientations of the orbital planes:

$$l = 1 : m = 0; cos\bar{\theta} = 0; \bar{\theta} = 90°.$$

$$l = 1 : m = \pm 1; cos\bar{\theta} = \frac{\pm 1}{\sqrt{2}}; \bar{\theta} = 45° or 135°.$$

$$l = 2 : m = 0; cos\bar{\theta} = 0: \bar{\theta} = 90°.$$

$$l = 2 : m = \pm 1; cos\bar{\theta} = \frac{\pm 1}{\sqrt{6}}; \bar{\theta} = 65° or 114.1°.$$

$$l = 2 : m = \pm 2: cos\bar{\theta} = \frac{\pm 2}{\sqrt{6}}; \bar{\theta} = 35.2° or 144.8°.$$

$$l = 3 : m = 0; cos\bar{\theta} = 0: \bar{\theta} = 90°.$$

$$l = 3 : m = \pm 1; cos\tilde{\theta} = \frac{\pm 1}{\sqrt{12}}; \tilde{\theta} = 73.2° \, or \, 106.8°.$$

$$l = 3 : m = \pm 2; cos\tilde{\theta} = \frac{\pm 2}{\sqrt{12}}; \tilde{\theta} = 54.7° \, or \, 125.3°.$$

$$l = 3 : m = \pm 3; cos\tilde{\theta} = \frac{\pm 3}{\sqrt{12}} : \tilde{\theta} = 30° \, or \, 150°.$$

$$l = 4 : m = 0; cos\tilde{\theta} = 0 : \tilde{\theta} = 90°.$$

$$l = 4 : m = \pm 1; cos\tilde{\theta} = \frac{\pm 1}{\sqrt{20}} : \tilde{\theta} = 77° \, or \, 103°.$$

$$l = 4 : m = \pm 2; cos\tilde{\theta} = \frac{\pm 2}{\sqrt{20}}; \tilde{\theta} = 63.4° \, or \, 116.6°.$$

$$l = 4 : m = \pm 3; cos\tilde{\theta} = \frac{\pm 3}{\sqrt{20}}; \tilde{\theta} = 47.8° \, or \, 132.2°.$$

$$l = 4 : m = \pm 4; cos\tilde{\theta} = \frac{\pm 4}{\sqrt{20}}; \tilde{\theta} = 26.5° \, or \, 153.5°.$$

ATOMIC CONFIGURATIONS

We see from Section 3 that the first four shells of atoms in the deterministic theory are determined by

$$J^{(1)}_{\frac{3}{2}} = 6; J^{(1)}_{\frac{5}{2}} = 7; J^{(1)}_{\frac{7}{2}} = 9 : J^{(1)}_{\frac{9}{2}} = 10,$$

and the orientations of the orbital planes are determined by l and m. The intersections of the shells and the orbital planes are the electronic orbits around the nucleus.

We shall list a collection of the electronic orbits and the electronic configurations for H, He, Li, Be, B, C, N, O, and P.

$$H:$$

$$l = 1, m = 0, J^{(1)}_{\frac{3}{2}} : 1e$$

$$He:$$

$$l = 1, m = 0, J^{(1)}_{\frac{3}{2}} : 2e$$

$$Li:$$

$$l = 1, m = 0, J^{(1)}_{\frac{3}{2}} : 2e$$

$$l = 2, m = 0, J^{(1)}_{\frac{5}{2}} : 1e$$

$$Be:$$

$$l = 1, m = 0, J^{(1)}_{\frac{3}{2}} : 2e$$

$$l = 2, m = 0, J^{(1)}_{\frac{5}{2}} : 2e$$

$$B:$$

$$l = 1, m = 0, J^{(1)}_{\frac{3}{2}} : 2e$$

$$l = 2, m = 0, J^{(1)}_{\frac{5}{2}} : 2e$$

$$l = 3, m = 0, J^{(1)}_{\frac{7}{2}} : 1e$$

$$C : Gasio - Carbon$$

$$l = 1, m = 0, J^{(1)}_{\frac{3}{2}} : 2e$$

$$l = 2, m = 0, J_{\frac{5}{2}}^{(1)}; 2e$$

$$l = 3, m = 0, J_{\frac{7}{2}}^{(1)}; 2e$$

$$C : Bio - Carbon$$

$$l = 1, m = 0, J_{\frac{3}{2}}^{(1)}; 2e$$

$$l = 3, m = 0, J_{\frac{7}{2}}^{(1)}; 2e$$

$$l = 3, m = \pm 3, J_{\frac{7}{2}}^{(1)}; \bar{\theta} = 30° or 150°; 2e$$

$$N : Gasio - Nitrogen$$

$$l = 1, m = 0, J_{\frac{3}{2}}^{(1)}; 2e$$

$$l = 2, m = 0, J_{\frac{5}{2}}^{(1)}; 2e$$

$$l = 3, m = 0, J_{\frac{7}{2}}^{(1)}; 1 \text{ e or } 2 \text{ e}$$

$$l = 3, m = \pm 2, J_{\frac{7}{2}}^{(1)}; 1 \text{ e or } 2 \text{ e}$$

$$N : Bio - Nitrogen$$

$$l = 1, m = 0, J_{\frac{3}{2}}^{(1)}; 2e$$

$$l = 2, m = 0, J_{\frac{5}{2}}^{(1)}; 2e$$

$$l = 3, m = 0, J_{\frac{7}{2}}^{(1)}; 1 \text{ e or } 2 \text{ e}$$

$$l = 3, m = \pm 3, J_{\frac{7}{2}}^{(1)}; J_{\frac{7}{2}}^{(1)}; \bar{\theta} = 30° \, or \, 150°; 1 \text{ e or 2 e}$$

$$O : Gasio \, or \, Hydro - Oxygen$$

$$l = 1, m = 0, J_{\frac{3}{2}}^{(1)} : 2e$$

$$l = 2, m = 0, J_{\frac{5}{2}}^{(1)} : 2e$$

$$l = 3, m = 0, J_{\frac{7}{2}}^{(1)} : 2e$$

$$l = 3, m = \pm 2, J_{\frac{7}{2}}^{(1)}; \bar{\theta} = 54.7° \, or \, 125.3°; 2e$$

$$O : Bio - Oxygen$$

$$l = 1, m = 0, J_{\frac{3}{2}}^{(1)} : 2e$$

$$l = 2, m = 0, J_{\frac{5}{2}}^{(1)} : 2e$$

$$l = 3, m = 0, J_{\frac{7}{2}}^{(1)} : 2e$$

$$l = 3, m = \pm 3, J_{\frac{7}{2}}^{(1)}; \bar{\theta} = 30° \, or \, 150°; 2e$$

$$P :$$

$$l = 1, m = 0, J_{\frac{3}{2}}^{(1)} : 2e$$

$$l = 2, m = 0, J_{\frac{5}{2}}^{(1)} : 2e$$

$$l = 2, m = \pm 1, J_{\frac{5}{2}}^{(1)} : 2e$$

$$l = 3, m = 0, J_{\frac{7}{2}}^{(1)}; 2e$$

$$l = 3, m = \pm 1, J_{\frac{7}{2}}^{(1)}; 2e$$

$$l = 4, m = 0, J_{\frac{9}{2}}^{(1)}; 1e$$

$$l = 4, m = \pm 1, J_{\frac{9}{2}}^{(1)}; 2e$$

$$l = 4, m = \pm 2, J_{\frac{9}{2}}^{(1)}; 1e$$

MOLECULAR BOND ANGLES

In order for two atoms to share electrons, i.e. to form a covalent bond, their respective orbital planes where the electrons stay must be coplanar, because of the conservation of angular momentum. This should be the key for atoms to join together to form molecules.

If the electrons configuation of an Oxygen is

$$l = 1, m = 0, J_{\frac{3}{2}}^{(1)}; 2e$$

$$l = 2, m = 0, J_{\frac{5}{2}}^{(1)}; 2e$$

$$l = 3, m = 0, J_{\frac{7}{2}}^{(1)}; 2e$$

$$l = 3, m = \pm 2, J_{\frac{7}{2}}^{(1)}; \bar{\theta} = 54.7° \, or \, 125.3°; 2e$$

then two Hydrogen atoms can attach to the orbits

$$l = 3, m = \pm 2, J_{\frac{7}{2}}^{(1)}; \bar{\theta} = 54.7° \, or \, 125.3°;$$

to form two covalent bonds. Recall that the angles $\tilde{\theta} = 54.7° \, or \, 125.3°$, (i.e. $-54.7°$); are the angles between the two orbital planes and the polar axis. Hence the angle between the two covalent bonds is then 109.4°, which is a close approximation to the observed bond angle 104.45°.

If the electrons configuration of the Carbon is

94

$$l = 1, m = 0, J_{\frac{3}{2}}^{(1)} : 2e$$

$$l = 3, m = 0, J_{\frac{7}{2}}^{(1)} : 2e$$

$$l = 3, m = \pm 3, J_{\frac{7}{2}}^{(1)}; \bar{\theta} = 30° \, or \, 150°; 2e$$

i.e. the Bio-Carbon, then the angels between the the equatorial orbital plane

$$l = 3, m = 0, J_{\frac{7}{2}}^{(1)} : 2e$$

and the planes

$$l = 3, m = \pm 3, J_{\frac{7}{2}}^{(1)}; \bar{\theta} = 30° \, or \, 150°; 2e$$

are 60° and 120° respectively. Hence no.1 Carbon atom can join to no.2 Carbon at the equatorial orbits

$$l = 3, m = 0, J_{\frac{7}{2}}^{(1)} : 2e$$

to form a double covalent bond, and no.3 Carbon can join to no.2 Carbon at the orbit

$$l = 3, m = \pm 3, J_{\frac{7}{2}}^{(1)}; \bar{\theta} = 150°;$$

to form a single covalent bond, and the bond angle is 120°; and so on. Six of these Carbons join together to form the Kekule structure. Six Hydrogen atoms join respectively to the Carbon at each corner to form the benzene. All the bond angels are 120°.

Since the outer orbital planes of Bio-Carbon. Bio-Nitrogen and Bio-Oxygen are the same, except the number of electrons in the equatorial orbits; the Carbon, Nitrogen and Oxygen can be interchanged in the Kekule structure and its attachments. For the pentagon structure that attaches to the Kekule structure, the four angles near to the Kekule structure are 120°. and the furtherest angle away from the Kekule structure is 60°, which are allowable by the outer orbital configurations of Bio-Carbon, Bio-Nitrogen, and Bio-Oxygen. These explain the molecular bond angles of Adenine. Thynine, Uracil. Guanine ,and Cytosine.

95

REFERENCES

1. Emilio Segre. From X-rays to Quarks, W. H. Freeman and Co., New York, (1980).
2. J. S. Bell, Speakable and unspeakable in quantum mechanics. Cambridge University Press. Cambridge, (1987).
3. N. David Mermin, Phys. Today, April,(1989).
4. Richard Feynman. The Strange Theory of Light and Matter, Princeton University Press. Princeton,(1985).
5. P. A. M. Dirac. The Principles of Quantum Mechanics, Oxford University Press, New York,(1986).
6. Encyclopedia Britanica, Encyclopedia Britanica Inc.,(1979).
7. Harry Lass. Vector and Tensor Analysis, McGraw–Hill, New York, (1950).
8. W. Heitler, The Quantum Theory of Radiation, Dover, New York, (1954).
9. John L. Powell and Bernd Crasemann, Quantum Mechanics. Addison Wesley, Reading, (1961).
10. C. Ray Wylie. Advanced Engineering Mathematics, McGraw–Hill, New York, (1975).
11. Herbert Goldstein. Classical Mechanics, Addison Wesley, New York.(1981).
12. A. Einstein. The Principle of Relativity, Dover, New York, (1952).
13. Peter Bergmann. Introduction to the theory of Relativity, Dover, New York, (1952).
14. J. A. Yeazell and C. R. Stroud,Jr., Phys. Rev. A **35**, 2806 (1987).
15. J. A. Yeazell and C. R. Stroud,Jr., Phys. Rev. Lett. **60**, 1494 (1988).
16. K. U. Lu, Report. Calif. State University, Long Beach, (1991).
17. E. Jahnke and F. Emde, Tables of Functions, Dover, New York. (1945).
18. L. Pauling, General Chemistry, Dover, New York, (1988).
19. J. March, Advanced Organic Chemistry, Wiley– Interscience, New York, (1992).

The Origins of Homochirality
A Simulation by Electrical Circuit

J. Park, C.W. Cheng, D. Cline, Y. Liu, * H. So, and H. Wang

Department of Physics, University of California at Los Angeles,
*Physics Department, Occidental College

Abstract. Chiral asymmetry is apparent in almost all living organisms on earth. However, the origins of chirality are unknown. One proposed theory of chiral asymmetry is by a symmetry-breaking transition such as the explosion of a supernova emitting neutrinos into the earth's biosphere during the prebiotic era. The discovery of weak neutral currents (WNC) reinforces this theory. We present a simulation of this theory by an electrical circuit that models the bifurcation equation for the chiral symmetry breaking process. There are three components that constitute this circuit: chiral circuit, sawtooth wave generator, and noise generator. The solution to the chiral circuit, which is the core of the hardware, generates the bifurcation equation. The sawtooth wave generator controls the bifurcation point (critical point), and the noise generator gives the system randomness. We included an adjustable bias voltage and saw a tendency to favor one symmetry over another (D or L) with randomness generated by noise.

INTRODUCTION

Chiral symmetries in life organisms puzzled scientists for many years. There have been several theories proposed in predicting the origins of chiral symmetries. The theory proposed here is of a terrestrial event occurring near the earth during the prebiotic era such as the explosion of a supernova. After an explosion, intense emission of neutrinos disrupted the hydrocarbon systems on earth including single cell DNA and RNA contained in the earth's oceans. With the disturbance, nature dictated the symmetries of the structures randomly favoring one type of symmetry (D or L) over another in DNA and RNA that was established in all life forms.[1,9,10,11,12]

THEORY

In a model proposed by Frank (1953)[2] and expanded by Kondepudi (1994)[3] the origins of chiral asymmetry began in the prebiotic era. Simple organic molecules formed in the earth's atmosphere and oceans. Energy from the sun and lightning were assumed to be the driving forces of the reaction toward chiral symmetry. Three assumptions were made in developing this theory: (1) the prebiotic molecules created in the atmosphere descended to the oceans; (2) by some mechanism, a chiral symmetry breaking process occurred in the oceans; and (3) the critical parameter is an increasing function of concentrations. The system undergoes a symmetry breaking transition due to the slow increase in concentration of the reactants (see figure 1).[3]

The model is as follows:

$$S + T \quad \Leftrightarrow \quad X_L \quad or \quad X_D \qquad (A)$$

$$S + T + X_L \quad \Leftrightarrow \quad 2X_L \qquad (B)$$

$$S + T + X_D \quad \Leftrightarrow \quad 2X_D \qquad (C)$$

$$X_L + X_D \quad \Leftrightarrow \quad P \qquad (D)$$

S and T are organic molecules, (<=>) denote forward and reverse reactions, and where [S] denote the concentration of S, K_{AF}[S][T] is the rate of forward reaction in (A). Under non-equilibrium conditions, an inflow of S and T, and an outflow of P exhibits conditions of spontaneous symmetry breaking if: (1) the product [S][T] is below a certain critical value; (2) the concentrations of [XL] and [XD] are equal; and (3) [S][T] exceed this critical value, the symmetric state of equal [XL] and [XD] becomes unstable and the system spontaneously makes a transition to an asymmetric state where [XL] ≠ [XD].

A general theory of chiral symmetry system derives the following equation:[3]

$$\frac{d\alpha}{dt} = -A\alpha^3 + B(\lambda - \lambda_c)\alpha \qquad (1)$$

A and B are positive constants that depend on the kinetic rate constants, λ = [S][T], and λ_c is the critical value of λ beyond which the symmetry is broken (critical point of bifurcation). If $\lambda < \lambda_c$ there is only one equilibrium solution at $\alpha = 0$. If $\lambda > \lambda_c$ then there are three equilibrium solutions:

$$\alpha = 0$$

$$\alpha = \pm\sqrt{\frac{B(\lambda - \lambda_c)}{A}}$$

Under these conditions, the solution splits into two trajectories symmetrically in the positive and negative α-directions at point where $\lambda = \lambda_c$ called the bifurcation point[4] (see Figure 2).

A term $\varepsilon^{1/2}F(t)$ can be added to equation (1) representing randomness or fluctuations (noise):

$$\frac{d\alpha}{dt} = -A\alpha^3 + B(\lambda - \lambda_c)\alpha + \varepsilon^{\frac{1}{2}}F(t) \qquad (2)$$

where F(t) is Gaussian white noise and $\varepsilon^{1/2}$ is the root-mean-square of this noise. Fluctuations make both states equally probable (see Figure 3).

If we consider a situation where there is a small chiral asymmetry in the chemical reaction rates of (B) and (C), where $K_{BF} \neq K_{CF}$ we obtain the following expression:

$$\frac{d\alpha}{dt} = -A\alpha^3 + B(\lambda - \lambda_c)\alpha + Cg + \varepsilon^{\frac{1}{2}}F(t) \qquad (3)$$

where C is a positive constant that depends on the rate constants and the concentrations, and g is a dimensionless factor that is characteristic of the strength of the chiral interaction, or bias (see Figure 4). This asymmetry could be due to disturbances such as WNC regulating the bias factor. The reaction barrier of (B) and (C) would then differ by ΔE, and $g = \Delta E/kT$, where k is the Boltzmann constant and T is temperature.[5,6,7]

Assume λ is a function of time that increases linearly at a constant rate γ, such that it starts at a subcritical value $\lambda_i < \lambda_c$, i.e. $(\lambda = \lambda_i + \gamma t)$ until it reaches a supercritical value λ_f at which it stops. Each time λ sweeps from λ_i to λ_f, the system will make a transition to a state of broken symmetry in which either $\alpha > 0$ or $\alpha < 0$. Due to fluctuations, this transition will be random.

APPARATUS

The theory above was simulated by electrical circuit. There are three components that constitute this circuit: chiral circuit, sawtooth wave generator, and noise generator. This configuration revolved around the chiral circuit by which we derived an analogous expression to equation (3).

Chiral Circuit

The main component of the chiral circuit is the AD534 multiplier chip. By the property of the AD534 chip, the output voltage is as follows:

$$V = \frac{(X_1 - X_2)(Y_1 - Y_2)}{V_0} - (Z_1 - Z_2) \qquad (4)$$

(see figure 5 for pin assignment). V_0 is a scalar factor, $V_0 = 10.0$ V if supply voltages of ± 15 V are assumed. The solution to the chiral circuit generates an expression analogous to equation (3). To solve for the first AD534 chip, we substituted voltages into equation (4), (see figure 6):

$$V_1 = \frac{(V_{out} - 0)(V_{out} - 0)}{V_0} - V_\lambda \quad = \quad \frac{V_{out}^2}{V_0} - V_\lambda \qquad (5)$$

and solving for the second AD534 chip, we obtain:

$$V_2 = \frac{(V_{out} - 0)(V_1 - 0)}{V_0} + V_g \quad = \quad \frac{V_{out}V_1}{V_0} + V_g \qquad (6)$$

substituting V_1 into V_2 we obtain the final result:

$$V_2 = \frac{V_{out}}{V_0}\left(\frac{V_{out}^2}{V_0} - V_\lambda\right) + V_g \qquad (7)$$

$$V_2 = \frac{V_{out}^3}{V_0^2} - \frac{V_{out}V_\lambda}{V_0} + V_g \qquad (8)$$

By the property of the op-amp, the net sum of all currents at point P is zero:

$$\frac{V_n}{R_n} + \frac{V_2}{R} + C\frac{dV_{out}}{dt} = 0$$

$$\frac{V_n}{R_n} + \frac{1}{R}\left(\frac{V_{out}^3}{V_0^2} - \frac{V_\lambda}{V_0}V_{out} + V_g\right) + C\frac{dV_{out}}{dt} = 0 \qquad (9)$$

$$\frac{dV_{out}}{dt} = -\frac{V_n}{R_n C} - \frac{1}{RC}\left(\frac{V_{out}^3}{V_0^2} - \frac{V_{out}}{V_0}V_\lambda + V_g\right)$$

Therefore we obtained the final expression:

$$\frac{dV_{out}}{dt} = -\frac{1}{RCV_0^2}V_{out}^3 + \frac{V_\lambda}{RCV_0}V_{out} - \frac{V_g}{R_n C} - \frac{V_n}{R_n C} \qquad (10)$$

which is analogous to equation (3)

$$\frac{d\alpha}{dt} = -A\alpha^3 + B(\lambda - \lambda_c)\alpha + Cg + \varepsilon^{\frac{1}{2}}F(t).$$

V_n is the input voltage from the noise generator that provides randomness (noise). V_λ is the input voltage from the sawtooth wave generator that regulates the bifurcation point or the point of symmetry breaking transition. V_g is the bias voltage. Small bias voltages generate symmetric results, 50% left handed and 50% right handed. Large bias voltages generate one-directional results, left or right handed.

Sawtooth Wave Generator

The sawtooth wave generator is comprised of five major components: transistor, comparator (A), gain 1 amplifier (G1), timer, and switch (see figure 7).

The sawtooth wave is generated by using a current source to charge a timing capacitor.

The transistor is used in a common emitter configuration. The collector current (I_C) is very small while the base current (I_B) is negligible. The voltage at point (1) is $V_1 = 6.5$ V acting as a voltage divider. The voltage across the base is approximately $V_{BE} = 0.7$ V where as the voltage at point 2 is $V_2 = 7.2$ V which is the emitter voltage V_E. This configuration acts as a current generator. The output charges a capacitor (C_T) and feeds into an op-amp with high impedance and negative feedback acting as a buffer. The sawtooth waveform appearing on the capacitor terminal must be buffered with an op-amp since it is at high impedance.

The combination of the NE555 and the RCA623 acts as a timer. It regulates the signal received from the comparator producing 1μs timing intervals.

The AD7511 is the switch that activates the sawtooth wave signal. When it receives a signal ($V_+ > V_{\lambda max}$ on comparator (A)) a switch closes creating a short circuit allowing the capacitor to discharge. When no signal is received ($V_+ < V_{\lambda max}$ on comparator (A)) the switch opens producing infinite impedance, increasing the charge and voltage on the capacitor.

The output of the buffer feeds into the positive input of comparator (A), a high-gain differential amplifier. The negative input voltage is at a constant value $V_{\lambda max}$. If $V_+ > V_{\lambda max}$ the capacitor discharges and a square wave signal is produced. If $V_+ < V_{\lambda max}$ the capacitor continues to charge.

The output of the buffer feeds into the negative inputs of a gain 1 amplifier (G1) and the negative input of comparator (B). Another quantity given by the power supply $V_{\lambda c}$ feeds into the positive input the gain 1 amplifier and the positive input of comparator (B). The output of the gain 1 amplifier produces the voltage that regulates the bifurcation point (V_λ). The output of the comparator produces a square-wave signal for triggering.

Noise Generator

The noise generator used in our simulation was borrowed from Horowitz and Hill with a few slight modifications (see figure 8).[8] We used a 1 MHz oscillator instead of 2 MHz. Unfortunately we were unable to match exact values on all resistors, but we approximated using 1% or 5% resistors singularly or two in series. Instead of an exclusive NOR (XNOR) gate, we used an exclusive OR (XOR) gate. These changes had no effect on the results because a random noise signal was expected, and slight modification also produced noise. We divided the noise generator by sections to simplify our discussion.

In section 1 we used a 1 MHz oscillator instead of a 2 MHz oscillator. This change has no effect on the quality of the output. It merely slows the frequency of input voltage of the circuit.

Section 2 consists of a 24-stage divider which acts as a counter for the circuit. It turns on after a certain number of pulses is received from the oscillator (section 1). It turns off after it receives the same number of pulses. The number of pulses can be adjusted by connecting a combination of binary switches.

Section 3 includes a 4093 NAND gate used as a reset. It activates the shift registers in section 4 when all components are operating. It ensures that the shift registers are synchronized with the incoming pulses.

Section 4 consists of four shift registers aligned adjacently creating a finite number of states. These states are determined by the number of bits on the shift registers. Values arrive in cycles. A particular value cannot be repeated until the cycle repeats. A 32-bit shift register, like one presented here, creates roughly two-billion states. Each cycle takes about an hour to complete. Each pulse from the 24-stage divider (section 2) is sent sequentially to the Q's of each shift register turning that bit on. Defining Q_1 of the first shift register as 1 to Q_8 of the fourth shift register as 32 (left to right), each bit is activated when its D (data) is turned on. Each bit is deactivated when its D is turned off. Clocks in each shift register are synchronized with one another moving all signals at the same rate. Data in bits 18 and 31 are sent through an exclusive OR gate feeding back to bit 1. The XOR gate creates randomness by looping shifting sequences back through the shift registers.

Section 5 generates digital noise output. However, the output signal is reduced by a factor of 1/6 to cover a 2 V amplitude range. The inverter is another 4093 NAND gate. However this output was not used in the experiment.

Section 6 is the analog noise output used in the experiment. Random signals from the shift registers (section 4) create random voltages feeding to the differential amplifier (LF412). This creates random amplification ratios. The outputs of the first paired shift registers (1st & 2nd) and the outputs of the second paired shift registers (3rd & 4th) feed into the op-amps of the LF412 chip. Signals from each shift register are random enabling an analog noise output generated from digital input. Outputs of the two amplifiers feed into a second amplifier with an amplification ratio of 1/1. Random changes in the voltages across the inputs combine two signals into a single, random, analog noise output.

EXPERIMENT

We applied an 8 V DC power source to the input terminals and obtained the expected results (see figure 9 for block diagram). The results are presented in figure 10. In 10.a we have symmetric results due to small biasing, 10.b we have biasing favoring upper V_{out} values, and 10.c we have V_{out} favoring lower V_{out} values. The bias voltages were regulated using variable resistors.

CONCLUSION

The aim of this simulation was to reproduce a transitional phase occurring millions of years ago. By using analog noise, "true" randomness that occurs in nature was produced. Digital noise created by random number sequences generated by computer are insufficient because digital electronics produce reoccurring, predictable signals. A computer simulation was devised and published by the authors.[1] Extremely small step-sizes were taken when numerically analyzing the bifurcation point. This process was extremely time consuming while "true" randomness was not achieved. With the circuit described above we can analyze the bifurcation point, sensitivity to initial conditions, and the favoring of one type of symmetry over another by varying the bias voltages. With an analog-to-digital converter we can interface the circuit to a computer and utilize step-sizing to take a deeper analysis of the critical point with the aid of analog noise.

REFERENCES

1.   Liu, Y., Wang, H., Cline, D. "Simulation of a Weak Interaction Induced Chiral Transition in a Pre-Biotic Medium," *AIP Conference Proceedings 300*, American Institute of Physics, 1994, pp. 499-505.

2.   Frank, F.C. *Biochem. Biophys. Acta.* 11, 459 (1953).

3.   Kondepudi, D.K. "Selection of Handedness in Prebiotic Chemical Processes," *AIP Conference Proceedings 300*, American Institute of Physics, 1994, pp. 491-498.

4.   Boyce, W.E., DiPrima, R.C. *Elementary Differential Equations and Boundary Value Problems;* 5th Edition, John Wiley & Sons, 1992, pp. 66-67.

5.   Kondepudi, D. K., & Nelson, G. W. "Weak neutral currents and the origin of biomolecular chirality," *Nature,* 314, pp. 438-441, 4 April 1985.

6.   Moss, F., Kondepudi, D.K., McClintock, P.V.E. "Branch Selectivity at the Bifurcation of a Bistable System with External Noise," *Physics Letters* , 112A, N. 6,7, 4 November 1985.

7.   Kondepudi, D.K., Moss, F., McClintock, P.V.E. "Observation of Symmetry Breaking, State Selection and Sensitivity in a Noisy Electronic System," *Physica* 21D (1986) pp. 296-306.

8.   Horowitz, P., and Hill, W. *The Are of Electronics*, Second Edition, Cambridge University Press, 1989 pp. 662-663.

9.   Jacques, J. *The Molecule and its Double,* Mcgraw-Hill, 1993.

10.   Avetisov, V. A., Goldanskii, V. I., and Kuz'min, V. V. "Handedness, Origin of Life and Evolution," *Physics Today,* July 1991, pp. 33-41.

11.   Kondepudi, D. K. "Nature's Chirality: Why Not Evenhanded?" Physics Today, April 1992, pp. 119-120.

12.   Bonner, W. A. "The Origin and Amplification of Biomolecular Chirality," Origins of Life and Evolution of the Biosphere 21: 59-111. 1991.

Figure 1

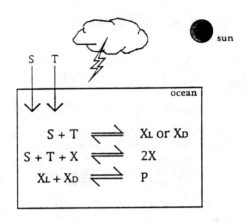

Figure 2

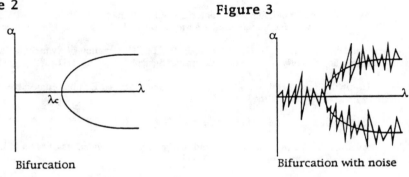

Bifurcation

Figure 3

Bifurcation with noise

Figure 4

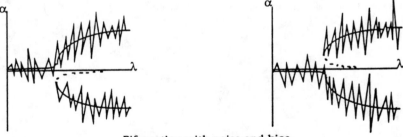

Bifurcation with noise and bias

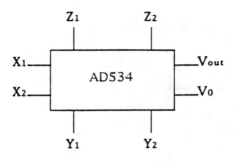

Figure 5

Chiral Circuit

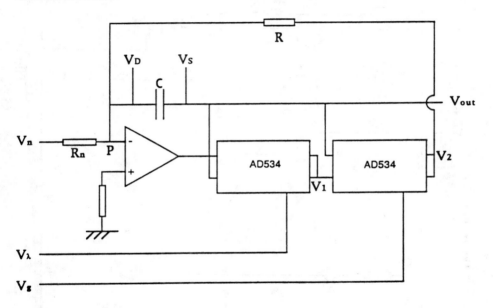

Figure 6

105

Sawtooth Wave Generator

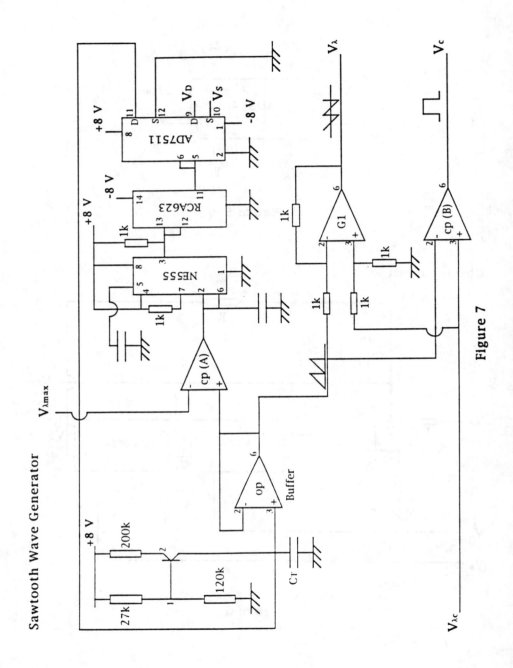

Figure 7

106

Noise Generator

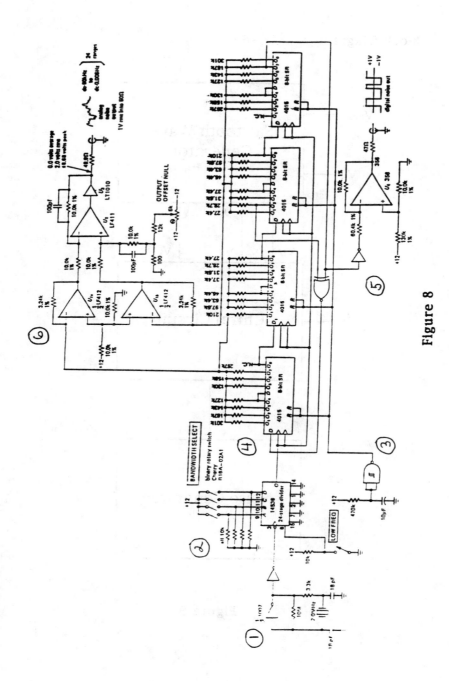

Figure 8

Block Diagram

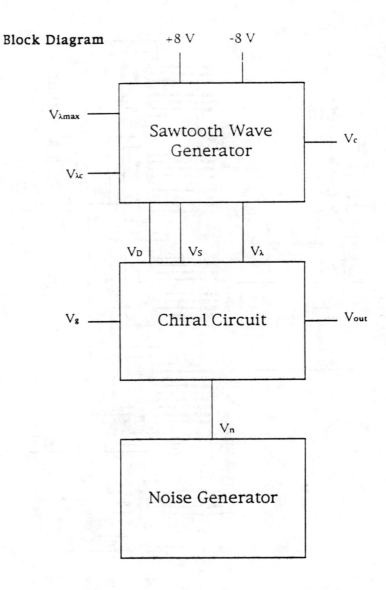

Figure 9

Simmulation Results

(a)

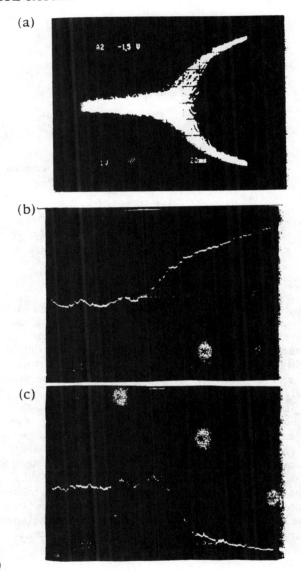

(b)

(c)

Figure 10

109

WEAK NEUTRAL CURRENT CHEMISTRY

R. Mohan

Max Planck Institute for Biophysical Chemistry
Am Fassberg 11
D-37077 Gottingen, Germany

"If it is not easy, it is not prebiotic"

S. L Miller

"Any theory of early evolution should have as it's main purpose the problem of explaining biochemistry"

G. Wächtershäuser

"It is asymmetry that creates the phenomenon"

P. Curie

ABSTRACT

Metal cluster organic complexes, neither atomic nor solid but in analogy to atomic nuclei and to mesoscopic systems, have unusual dynamics and catalytic properties. Organo-metal clusters as quintessence prebiotic enzymes could have originated the homochirality of the molecules from achiral precursors, controlled from the atomic nucleus, with the initial product itself serving subsequently as chiral auxiliary, transferring and amplifying the chirality in the autocatalytic process now. High resolution spectroscopic studies of diatomic molecules beginning now may lead to upper estimates of the interaction strength of weak neutral currents (WNC) with valence electrons of metal clusters and suggest kinetic pathways to dynamic symmetry breaking in the asymmetric synthesis of chiral molecules. An estimate of 10^{-5} kT (thousand times larger than for radiolysis) for the parity violating energy (PVE) could be sufficient to run an entropy driven spin-catalyzed asymmetric synthesis. Expect then, wherever there are metal clusters in interstellar dust or under the sea, chiral molecular production.

1. INTRODUCTION

1.1 Initialization

After the cessation of massive external bombardment [1], the terrestrial atmospheric ion-molecular plasma chemistry begun by lightening with CH_4, NH_3, H_2 and H_2O by the Urey-Miller synthetic process, or after degassing by the Fischer-Tropisch process producing the "nature's vast arsenal of chemicals", the soup of the relevant biochemicals of life, all within a period of 0.3 Gy, there were doubts. The assumed short time and the geochemical evidence against the early reducing atmosphere have not only forced F. Crick to look for outer space for frozen accidents at the macromolecular level, but also the prebiotic scientists to look for the interstellar gas and dust grains for the organic chemicals — the cold ion-molecular chemistry being different from the terrestrial one. Recently, attention is being paid to ion-molecular solution chemistry in near deep hydrothermal vents with H_2S as the reducing agent, and surface oxidation-reduction on FeS-pyrites for the carbon fixation. Carbon compounds seldom reach equilibrium (especially at high temperatures) and charged surface chemistry seldom promotes polymerization. Miller et al., have invoked thermodynamic criteria to show that excepting the atmospheric scenario, neither the hydrothermal vents [2] (with a supercritical phase under high pressure and high temperature) nor the FeS-pyrite catalytic surfaces (two dimensional world [3]) is plausible site for chemical evolution. Fox [4] considers an entirely phosphate driven uroboros, nonlinear dynamic world, while de Duve [5] presents an intervening thioester world.

While conventional stellar or cosmic chemistry involves cosmic radiation and radiative associative reactions [6] in the gas and surface reactions on passive grains, Greenberg [7] presents an u.v photochemistry in the grains (although the light is attenuated by the dust) and his laboratory simulation experiments of cold photochemistry with CO, NH_4, N_2, CH_4 and H_2O at 230 nm u.v in the grains (0.1 μ) yields an yellow stuff with a 3.4 μ C-H stretch spectra that led F. Hoyle to suggest his virus hypothesis. There were frozen free radicals in such grains and by collision or by slight raise in temperature to 27^0 K, the grains explode and the material mix with the gas. Although no quantitative data given, his experiments with circularly polarized light seem to selectively enrich chiral molecules. Bonner [8], who was earlier involved in various chiral experiments, has along with Greenberg is suggesting that neutron star renements of a supernova might have been the source for the chiral synchrotron radiation and that the chiral molecules produced in the interstellar medium gradually accumulated on the earth both by accretion and through the meteorites and comets. Goldanskii [9] has recently settled for formaldehyde as the primal chemical which trimerized (formose reaction) by tunneling in the cold universe, and delivered to the earth had a spontaneous chiral symmetry breaking through a Frank type of kinetic scheme, but without any advantage factor and the D-sugars produced went directly into a Kidrowoski type of self-replication with nucelic acids attached, as small chiral molecules may not be stable for long. Kidrowoski [10] has shown experimentally the feasibility of self-replication starting with trinucleotides.

Eschenmoser with Arrhenius [11] has recently shown that using positively charged metal hydroxide mineral sheets, selective diasteromeric sugars can be formed from glycoaldehyde phosphates, unlike the complex mixture of products formed in the formose reaction. Although the products are racemic here, connection has to be made to recent applied chemistry for chiral synthesis, transfer and amplification using metal or organo-metal catalysts. In addition to simple acid-base catalysis as in esther hydrolysis and self-replicating lipids, enantioselective reaction products using a variety of homo- and hetero- organo-metal complexes are known, although the kinetics and mechanisms are not fully elucidated. For instance, Titanium mediated epoxidation and aldehyde to enol production are known. While various routes to amino acid synthesis are used, very little is known about synthesis of chiral amino acids, and probably nothing is known of de novo synthesis of any chiral amino acid, although the reaction may simply involve the C-H bond breaking at the α-carbon and reforming selectively. Wynberg [12] quotes Betti's reaction and his attempts to produce chiral amino acids similarly by interacting imines with HCN turned out to be racemic. It is impossible to produce de novo chiral molecules by conventional chemistry. Could the handed Z^o bosons extend out to the valence electrons to influence a chemical reaction?

1.2 Continuity

From chiral elementary particles and atomic nuclei to achiral molecules dressed in strong forces to chirality back again in the biological world, which Goldanskii calls the biological bing bang, it is hard to discern any useful continuity, de facto or de lege. But most agree that the order produced in homochirality is essential for the self-organization of living things. The easy way out would be something like the D-sugar-nucleic acid system or the RNA temporarily produced an extra amino acid thereby asserting itself over the mirror counterpart, or even simpler, it was just spontaneous chiral symmetry breaking and it happened to be D-sugar and L-ammino acid.

Understanding the origin of homochirality of life, how, when and why would put some order in our speculations about the sequence of events in the prebiotic 0.3 Gy chemical to biological evolution period. Was it primitive heterotrophic or autotrophic metabolism (archists vs. ventists), was it metabolism, amino acids, lipids and then the nucleotides or was it the other way around (metabolists vs geneticists), was evolution structural or functional, continuos (with crises) or meandered (with some surprises), was it local or global? Similar to J. Wald's chance hypothesis, Miller and Orgel [13], while recognizing the necessity for chirality, argue that it was a matter of chance that we are with D-sugars and L-amino acids, and that the origin of chirality is origin of life, i.e., it is part of the replicator dynamics. Since evolution experiments in a test tube are becoming increasingly an experimental tool, this idea could be tested by allowing L-and and D-oligonucleotides to compete with each other for the resources in their oligomerization process. However, Goldanskii [9] emphasizes the need for chirally pure monomers without which life can not be started. It is possible that homochiral L- and D- oligonucleotides can coexist at the expense of the racemic form. Evans et al., [14] have shown by field molecular dynamic simulations that the L-, and D- amino acids differ from the LD- racemic compound in their rotational correlation functions in the far-infrared region. Goldanskii's [9] is a chance emergent mechanism that sets once for all the D-sugar selection. It is known that in spontaneous crystallization (what Norden [15] calls local selection and Bonner calls SRURC - spontaneous resolution under racemic conditions) crystals of single handed molecules can be obtained by chance mechanism due to temporary advantage in the secondary nucelii formed. In Goldanskii's mechanism, unless the chiral symmetry breaking occurred already in the cold, where racemization by tunneling is negligible, and with no deterministic but chance mechanism, the emergence of terrestrial chirality would have been local. and summing up over different sites would be needed. If it was amino acids, as Mann and Primkoff [16] have shown, hardly any sensible protein can be formed. Assuming that the sign and the values for PVED (parity violating energy difference) given by Mason et al., [17] is correct, the 10^{-14} kT advantage factor for sugars and amino acids could not have surmounted various difficulties so long, even though a well stirred rather than a placid pond is needed - it is not like the electron gaining spin polarization circle after circle in the cyclotron. Mechanisms such as Kondepudi's [18] should be tested in a large flask or pond with the PVED properly scaled for the smaller size and shorter time. Besides in Kondepudi's scheme there is no term for racemization. In Bada's list [19] isoleucine has the longest half-life time.

Although not prebiotic, chemical replication systems by host-guest mechanisms, self-assembly and self-organization of inclusion compounds and metallo-organic helices and knots and molecular motors attest to the capabilities of carbon compounds. Chemical reaction networks modeled after neural networks could handle logic and recognize patterns [20]; and it is further shown by simulation that mass action alone is sufficient to produce self-maintenance and self-organization [21]. It is also shown that carbon compounds are capable of ecological evolution and that the theories of heterochrony or self-similarity could span the entire evolution, including development and morphogenesis, starting from amino acids onwards. In all these models, homochirality is taken for granted. A computation based on plus, minus logic could easily reach a dead-end.

The chiral molecules due to electromagnetic chirality, i.e., first order in the spatial dispersion of the electromagnetic field (m.μ, where m is magnetic moment and μ is electric dipole moment) prefer to interact with the same kind; and induced chirality in intermolecular forces is what keeps the structure of the genetic code stable by 100 cal/mole per dimer compared to racemic mixtures, and in large molecules by more than a factor of ten exceeding thermal motion. The symmetry of the starting substance predetermines the symmetry of the more complicated compounds formed.

This is also true in modern day asymmetric synthesis where chiral ligands acting as templates influence chiral production. Permanent dipole moment as in amino acids is also favorable. Magnetic fields can increase chiral production than racemization. Higher order spatial dispersions (due to gradients in m and μ) are symmetric, not chiral, weaker in energy and play lesser role. The electromagnetic field also leads to collective excitations and cooperativity. The special role of chiral molecules is not in energetics but in ordering, coding, self-instruction, information storage and transfer at a cost of smaller energy requirements - the fundamental concept of information and dynamics. Strangely, the sugars are noted only for their energy supply and not for their informational ability. Being multifunctional, sugars may be more ancient than the amino acids in the chemical evolution chain. With loss of homochirality, the polynucleotides and proteins can not fold and the excitonic coupling is lost; even a single defect (opposite chiral nucleotide or amino acid) could disrupt the order, although the DNA could be an aperiodic crystal and protein a random medium. The symmetric second order spatial dispersions do not have these characteristics. Electromagnetic chirality endows this unique character to life. But it is a symmetrical parity force and for the origin of homochirality, we have to search elsewhere.

Against the continuity in evolution, from chemical to biological stage, denied by Miller and against his thermodynamic criteria for two dimensional living, Wächtershäuser [22] invokes K. Popper for retrodiction. His FeS-pyrites is an evolving machinery, a catalogue of possible chemical pathways, leading inexorably to higher levels of biological evolution. It is a positively charged surface chemistry with local dissipation and recycling than surface nonlinear dynamics. Unlike the heterotrophic prebiotic soup, it is chemolithoautotrophic and continuity of evolution is maintained. The soup, static, is probably gone, but the hydrothermal vents and the FeS -pyrites are dynamic and evolution could be followed as it is, probably, still happening now. Wächtershäuser's scheme is palimpset in principle, i.e. present day metabolism is as if the prebiotic metabolism is erased and written on. Understanding evolution - continuos, emergent, punctuated or recapitulating - could aid in finding when homochirality appeared.

1.3 Chirality

Any theory of continuity of evolution, to be useful, should indicate when the door was open for the homochiral molecules. The cosmological event could have continued through our geological period with a strong PNC force that was subsequently withdrawn, but during which time, living molecules feeding on negentropy could have started life as we know it today as a fossilized remainder of such an event. Such a cosmological intrusion in the early period of the earth is unlikely. Bada [19] asserts that neither the carbonaceous chondrites (meteorites) nor the oceanic sediments show any enantiomeric excess (e.e). Analytical chemistry is well advanced for precise quantitative measurements. This removes extra-terrestrial origin for chirality - although Bonner [23] claims radioracemization can occur during the voyage of such meteorites due to cosmic radiation and suggests that during one of its three voyages across the galaxy in 0.3 Gy, the solar system might have encountered a supernova neutron star. The elimination of redundancy by chance of one of the two equal antipodal systems in the biological stage is implausible as there is no universal criterion for this, and local events could have been of either handedness. The Goldanskii mechanism can be assumed to support such a suggestion. Equally impossible would be the Frank-Kondepudi nonlinear fluctuation model that requires enormous amplification of the PVED from 10^{-14} kT to reach a decent e.e above the fluctuation of a racemic mixture ($\sim 10^{-8}$ kT). Interestingly enough, no measurement of concentration fluctuations in a racemic mixture has been measured either by light scattering or by other means. It is not clear whether a stochastic resonance mechanism due to colored noise which, organized biological systems are supposed to use in their signal amplification mechanisms could have been effective. Such subtleties are not expected of a racemic pond. Kondepudi and Nelson [24] have considered Rhodium compounds with a PVED of 10^{-11} kT in their autocatalytic scheme but have not pursued it further. Radiation damage on the other hand, which is also related to the universal Z bosons, could give a better ratio of 10^{-8} kT (not more), while photolysis with circularly polarized light gives even higher yields (Bonner claims 10^{-4} e.e with alanine). Cline et al., [25] have recently given an interesting twist to the Kondipudi general first order cubic chemical reaction mechanism (derived from group theoretic method) by boot-strapping the advantage factor in the equation:·

$$\dot{x} = -Ax^3 + B(\lambda - \lambda_c)x + \sqrt{\varepsilon}f(t) + Cg + d\delta_{\lambda\lambda'} \qquad (1)$$

hundred times, d/g= 100, with positron pulses from neutrino showers ($\overline{V}_e + p = e^+ + n$). with an increase in the control parameter λ that would arise due to the selective decomposition of say D-amino acids upto a depth of 100 meters. The amplification occurs at near the bifurcation point and not above or below. This is understandable as disturbances at the critical regime are very effective. The neutrino showers could have been from a prominent supernova flare that the planet met on its voyage through the galaxy. It would be interesting if Cline et al., could indicate a definite time in the 0.3 Gy biochemical evolution period like in the time series analysis of the supernova γ-ray bursts, thereby setting a time when the chiral event occurred.

In addition to direct radiolysis and photolysis experiments, since chirality involves a pseudoscalar (chiefly as superposition of an axial vector and a polar vector), attempts were made to shift a chemical equilibrium or extend a reaction by external fields. Thus Dougherty [26] tried to promote the chiral epoxidation reaction, earlier studied only with a magnetic field, by imposing colinearly magnetic and electric fields on the reaction thereby influencing the magnet dipole in the transition state. The yields were low, as is to be expected, as even high voltages and fields at the molecular level in solutions corresponds to only about 10^{-17} kT. Similar experiments under the influence of a coriolis force and the geomagnetic field showed no improvement, perhaps because, additionally, the inhomogenity (rather than the well-stirred medium of Kondepudi) produced by the centrifugal force was a counter-influence. Magnetic field alone could give an e.e (enantiomeric excess), but under the influence of the magnetic field of an unpolarized light beam as in magnetochiral circular dichorism. No attempt has been made to study lightening as an 'in situ' experiment for e.e, as this phenomenon might have been intense in the prebiotic age with more than hundreds of ampers surging through the atmosphere with the ground as the positive pole. and with the electric and magnetic fields orthogonal.

Norden [15] has classified the various mechanisms proposed for the origin of chirality into global or local (similar to Bonner's classification into deterministic or SRURCs) phenomenon. Radiolysis and photochemical are global, while crystallization, desiccation - hydration, polymerization, quartz crystals and spontaneous symmetry breaking with no global advantage factor are local.

Metals play an important role in biochemistry. Ribozymes are metal catalysts. Nature uses a variety of enzymes with heterometal atoms in it, probably a carryover from the physics of metal clusters. Wächtershäuser's FeS system is metallo-organic chemistry. Metal clusters with heavy atoms have unusal catalytic properties and the study of fullerenes is of recent origin. Some rare earth or actinide rather than transition metal with high PVE could have induced a chiral product from achiral precursor. However slow this could have been, the chiral product could have takenover further chiral production by chiral transfer and amplification in an autocatalytic process attached to the metal clusters.

2. SEARCH FOR ORIGIN OF HOMOCHIRALITY

Unlike the Frank-Kondepudi-Goldanskii processes which rely on nonlinear reaction rate fluctuations, with or without an advantage factor, but is self-contained, the radiolysis method could yield 10^6 larger e.e than the starting material and has to be further amplified by other mechanisms. to high homochiral purity without which, according to Goldanskii's criterion, life can not start. Radiation damage with electrons, positrons or protons is a wasteful process, while photolysis could be more specific and give larger e.e. Bonner [8] has given an informative guide through the various experimental attempts in search of homochirality. With β- electrons and parity violation (even before the discovery of the Z^0 vector bosons which mediate interactions between e-e, e-p and e-n), the psuedoscalar handed forces appearing on the scientific horizon, it was natural to seize upon such an universal force for deterministic causality and radiolysis experiments were begun in earnest, and the experiments need only the end-products to be analyzed after certain irradiation time. The experimental accuracy for the asymmetric factor A_d was less than 10^{-4}, while the enantiomeric absorption coefficient differences were smaller and the e.e in the decomposed products even smaller. Although the yields were larger in photodecomposition experiments, the sun light is not known to have much circularly polarized light, and with a broad spectrum the asymmetry would cancel itself.

From Miller to Eigen to Luisi to ecosystems, from chemistry to molecular biology to cell to ecology, evolution experiments have tempered grandiose schemes about evolution. Hemmed in between an achiral world and the RNA $\rightarrow$ DNA$\rightarrow$ stromatolite mats that inhabited this world, there was hardly 0.1 Gy for chiral production. Chemistry at its best than biological big bang.

2.1 Radiolysis

The Vester-Ulbricht hypothesis suggests that the parity violation in nuclear emission might have determined the biological asymmetry we see now, as there are carbon and potassium isotopes in nature. Initial experiments started with radionucleides, moving on to artificial sources of electrons, positrons and protons. Spin polarized electrons with longitudinal spin opposite to the direction of motion or positrons with spin parallel to motion are supposed to absorb and destroy the D- component better than the L. Many compounds, but mainly D- and L- amino acids, in solution, sublimed, powdered or crystalline were used, A typical experiment is that of Bonner with 25 years irradiation by 100 Kev electrons. It was soon realized that they were not easy experiments. Radiolysis of solvent complicates, so the radio isotope was imbedded in the powdered sample. The original idea was that the bremsstrahlung created by the stopping power of the target was responsible for the damage or for the excited state ionization. Soon, it was realized that the low energy region of the γ-rays which is more responsible for the damage had much less helicity than the original electrons. Some experiments showed that it was the β-particles themselves that were more effective. Crystal sizes were controlled so that the entire secondary radiations were confined within one crystal. Also 15 kev $\gamma-$ray (low energy bremsstrahlung) absorption showed no difference between D- and L- amino acids. Bonner also showed that for about 30% material decomposition, there was also about 5% racemization. In solids, it was soon realized that defects may complicate the results. Proton experiments with 100 Mev, though of lower velocity, but with larger ionization density in the tracks, also showed negative results.

High energy or short wavelength particles produce secondary electrons, which carry little of the polarization of the primary electrons and is polydisperse in the wavelengths. Such electrons flying past the molecule excite it to higher excited states including ionization.Fluorescence decay measurements (cathodoluminescence) could have aided to follow the experiments, but was not done.

Zeldovich and Saakyan [27] showed by theoretical studies that even relativistic electron (with higher optical activity than slower electrons) by direct interaction could only decompose a fraction of that by a photon or by synchrotron radiation. Direct electronic interaction (due to its electromagnetic field) is less by $vh\omega/cE$ where $h\omega$ is the activation energy of the reaction, v the velocity and E the energy of the electron. as shown in equation (2) below, where A_d and A_0 are the asymmetry factors for e.e by decomposition by radiolysis and photolysis respectively.

$$A_d = \frac{vh\omega}{cE} \cdot A_0 \qquad (2)$$

Early results tried to show that the β-particles radiolysed the D-component more than the L-enantiomer. As of then and even now the exact mechanism involved, either theoretically or experimentally, is not understood. It is likely that the AP (antiparallel) electrons that have preference to interact with the protons gave negative parity violation to this real world.

Soon interest developed to study the mechanism for the interaction involved, rather than just estimate the relative end-products. Muon or positron can capture an electron from the molecule to form a singlet (μ^+ or e^+ parallel to e^-, APs) or a triplet (OPs) with antiparellel spins on slowing down. The triplet OPs have longer lifetimes than the singlet, decaying to γ-photons. From the amount of γ- emission, the relative amounts of triplets formed by the two enantiomers can be compared. Although the results were inconclusive. Garay and Hraskos' [28] innovative simple benzenoid-like theory of helical electron gas around helically fixed charges was later quantitatively developed as bound helical density by Hegstrom [29], Reich et al., [30] and others, correlating the bound positronium life-time measurements to ionization potentials, which is a measure of radiation damage and relating in general to

119

scattering and absorption of spin polarized electrons by optically active molecules. The equation relating asymmetry in positronium binding to the D- and L- forms to the relative radiation damage is given as:

$$A_d = c \frac{A_{ps}}{h(e^-)z^2} \leftarrow \left\{ \begin{array}{l} A_d = \eta_{e^-} \dfrac{(\alpha z)^2 \cdot h(e^-)}{2 E_p \ln(E_p)} \\[2ex] A_{ps} = h(e^+)(\alpha z)^2 \eta_{ps} \end{array} \right\} \quad (3)$$

where α is the fine structure constant, Z, the atomic number of the heavy atom, while η_{ε^-} and η_{ps} are molecular asymmetric factors for electron induced radiation damage and for positronium binding to the molecular electrons respectively; h (e^-) is the degree of helicity of the particle and E_p the energy of the β particles in 100 Kevs. Since asymmetry increases as z^2, Reich et al [30] studied Sb-cystine compound but the accuracy was yet insufficient.

An improved theory was given by Blum et al., [31] for absorption and scattering by optically active molecules. The spin polarization is the result of spin-orbit interaction and is of the order of $(\alpha Z)^2$, acquiring helicity as $\eta (\alpha Z)^2$. Asymmetry in triplet positronium formation in L- vs D- for positrons is of the same order. as also, the differences in electron ionization rates of L- vs D- for incident electron of unit helicity. The difference in the damage is of the order $10^{-5} \eta (\alpha Z)^2$, i.e., about $2 \times 10^{-5} \eta$ for 100 Kev of β electrons. A survey of spin polarized scattering (Rayleigh-like scattering) and absorption is given by Kessler[32]. Using improved instrumentation, Kessler has recently displayed [33] the first electron optic dichorism spectra of right handed and left handed Ybcamphor - yetribium-camphor compound-in the 0 - 5 ev range of spin electrons in the vapor phase. Contrary to earlier claims, camphor itself did not show any optical activity upto the resolution of the instrument at 10^{-4} for A_e, the CD asymmetry. In Faragas et al., [34] experiment with camphor showing an A_e of 10^{-4} could have arisen from molecular orientational effects due to ion formation. Resonant temporary ion lengthens the time in which the spin-orbit interaction takes place and hence enhancing the polarization effects. Such effects are pronounced in photoelectron spin polarization spectroscopy. No experiment of spin polarized photoionization spectroscopy is available (Cherepekov [35].). Spin-orbit coupling, inspite being weak, plays a conspicuous role in spin dependent scattering than spin exchange flip. As for the positron, due to repulsion by the nucleus, it does not reach near enough the nucleus to observe the spin effects by the spin-orbit interaction. Kessler estimates the polarization effects for positrons to be about 0.015% to that for electrons. Naturally, the earlier theory of Reich and others should include this. Kessler's instrumentation and the high resolution he means to achieve could be used along with Reich et al., type of experiments (with Sb-cystine) to explore the effects of weak neutral current (WNC) interaction with the spin-orbit coupling. Along with optical experiments, the avenue with spin polarized electrons and positrons could be another way to measure the PVED in organo-metal compounds and extend to organo metal clusters. Khriplovich and Popslov [36] have recently given the asymptotic refractive index equation for molecules due to spin helicity. Extrapolation to higher electron volts of the electron optical dichorism can yield the spin helicity and hence the PVED. It should also be possible to extract by subtracting the positive from the negative spectra of Kessler's Ybcamphor, with improved accuracy, a value corresponding to spin helicity. Similar experiments on Ybcamphor with positron, by comparison, could help interpret properly the positronium binding experiments. Since the steroselective radiation damage limit is known as at 10^{-8} kT and Bada's evidence against any earthly chiral excess, these experiments should be directed to metal clusters and their organic complexes as possible source for homochirality.

2.2 Photochemistry

While Bonner and Greenberg chose the cold photochemistry in interstellar cloud, Norden [37] chose terrestrial photochemistry with sun light as the photo source. The concern was on direct photodestruction, photoderacemization (partial resolution) and photoasymmetric fixation rather than chiral sensitized catalytic asymmetric syntheses for high yield (for a review on asymmetric photoreactions, see Inoue [38]). The optical activity of aromatic compounds is due to m.μ, while the optical activity of amino acids and sugars is due to the 1-electron mechanism and have a weaker. asymmetric factor, $A_O = \Delta\varepsilon/\varepsilon$ (where $\Delta\varepsilon$ is the absorption difference between the two enantiomers) of the order of $1o^{-4}$ or less., and the relative destruction ratio of the enantiomers would depend on this and is typically about 0.1%, but larger than what can be expected from radiolysis· Typically a high photolytically selective yield for DL- leucine was 2%. A two photon experiment with A_O values two orders of magnitude larger was reported [39] probably due to a larger $\Delta\varepsilon$ and smaller ε for the first excited state rather than as an artifact due to orientational effects that accompany photoionization. Photoinversion with temperature or at the excited state can occur with change from enthalpy to entropy dependent inversion rate.

Magnetochiral dichorism does not require polarized light, the sign of the e.e depends on the direction of the magnetic field being parallel or antiparellel to the light beam. This is essentially a partial resolution process and very weak for the consideration of the chiral origin phenomenon. An interesting theoretical suggestion for chiral production from achiral precursor is Norrish type II reaction by laser control of chemical reactions by interference either with two simultaneous laser beams with different wavelengths or between two laser pulses with precise time and phase differences. [40]

2.3 External Fields

Symmetry is invariance and indiscernability, indeterminancy is a manifestation of symmetry, and every conserved quantity is associated with impossible experiments. It may fail prercise testing leading to new physics. The degeneracy may be lifted by external fields as magnetic and electric fields leading to new phenomenon. Discrete symmetry violations intrinsic or in external fields, within the CPT theorem, help to decide the feasibility of an experiment. Selection rules for linear, nonlinear coherent and incoherent optics is given by Stedman [41]. Barron [42] has given a more intuitive approach covering a range of topics. Barron insists on viewing the molecular processes in the same light as with high energy particles.. Parity violation alone will give an energy or rest mass difference as seen in the natural optical activity of molecules between real world enantiomers, but not with their CP counter parts involving antimatter, and also not in particles or systems that T violate or CP violate as with neutral K mesons or as in Faraday effect. While Barron assumes that in CP violation, charge conjugation counter balances, Quack [43] distinguishes between de lege and de facto, between discrete and continuos symmetry violation and symmetry breaking in P and T, and argues that the CP mirror systems also could show an energy difference. Since parity violating spectroscopy for the H atom in the 2s, 1p bands are being contemplated, Barron suggests that comparison of H^+ and H^- spectra could verify his stand. But equally, the recent contemplated experiments to check charge equivalence of proton and antiproton would also be of equal interest.

While the magnetic and electric field effects are small in solution as evidenced in Dougherty's experiments, superposition of external fields as pointed out by Curie, increases the asymmetry and have important effects in oriented or ordered systems as in the phase transitions of ferromagnets, high temperature superconductors and magnetoelectric systems, the last of which is closely related to metal clusters. Most of them involve macroscopic spin helicity and PoddTodd effects (with PT invariant), and Khriplovich [44] has searched for the PNC Podd effects in ferromagnetic phase transitions and suggests that it could be contributing by higher than 10% to the Podd effects in these materials. Symmetry alone does not describe the new effects. Besides, it may be that in the finer detail CPT theorem itself, based on relativistic quantum mechanics, may fail. Additionally, the vacuum could be broken by colinear electric and magnetic fields. Stedman and Bilger [45] plan to use their ring laser with an accuracy of $\Delta\omega/\omega$ of 10^{-15} in frequency to measure the vacuum spin helicity by a Farady measurement. The parallel E and M field is expected to show much larger optical activity than M field alone. E -field induced magnetization and inverse Cotton - Mutton effects are known.

Barron molds the phenomenological K_L fission into e^- and e^+ products inspite of a meager difference in their rate constants of 1.004 as a closed 2-state enzymic system, far from equilibrium, and as catalyst reduces the equilibrium transition state activation energy, while the chemical potential or potential between the reactants $(e^- + \pi^+ + v^-_\tau)$ and the products $(e^+ + \pi^- + v_\tau)$ are the same, because of T violation or CP violation. He thus concedes that T violation with kinetic control could be better than P violation for chiral imbalance. Additionally, his arguments that the Onsager reciprocal relations (ORR) are invalid could be subjected to criticism, First, Onsager derived his relations only for near equilibrium systems. Second, the S-matrix based implications to kinetic foundations may depend on the choice of bases. The derivation of microscopic reversibility, microscopic inversion or detailed balance are based on different bases in the S-matrix. The ORR can be derived by other methods.

An interesting discussion of T violation and T- symmetry breaking (de lege and de facto) - irreversibility in general - is given by Quack [43]. Finally, magnetoresistance in mesoscopic systems was analyzed by Buttiker [46] in terms of Onsager symmetric and antisymmetric parts. Gelikonov et al., [47] showed that an earlier experiment with magnetic optical activity observed in nonmagnetic material as due to lack of ORR was artefactual.

In Dzyaloshinski's theory [48] of unconventional super-conductors, it is a magnetoelectric effect with linear relation between the internally produced (orthogonal) magnetic and electric fields (waves) inside the crystal. It is PoddTodd. The Todd arises from the two adjacent AFM ordered planes having opposite signs with respect to the gauge (chiral) fields and a magnetoelectric effect should be observable as a macroscopic spin helix, implying that the stacking layers do not introduce an inversion center. The Jahn-Teller vibronic effect in dipole and multipole moments and in polarizabilities of molecules due to degeneracy in the electronic states leading to instability and anisotropy would also lower the symmetry. They may also involve time-dependent (Barry's) geometric phases. It is suggested that in high temperature superconductors, it is $d_{x^2-y^2}$ asymmetry. The PNC P odd effect that Khriplovich searches for could be s symmetry.

Magnetochiral dichorism (m.m.μ and hence odd) which is axial birefringent with the light's magnetic field interacting with the external static magnetic field, as mentioned before, is a very weak phenomenon for chiral origin. Farady effect (m.μ.μ) is parity even and hence can be seen in all materials. Inverse Faraday effect which is nonlinear has recently been shown theoretically by Harris and Tinoco [49] as insensitive even to distinguish the left from the right enantiomer in spite of recent experiment claiming 1 HZ NMR chemical shift for methoxyphenylimminocamphor with 10W. laser source. And hence inverse Faraday effect is too weak to induce e.e.

3. HIGH RESOLUTION SPECTROSCOPY

High resolution spectroscopy experiments in atomic vapors to measure the WNC weak signals by circular dichorism (CD) and by circularly polarized luminescence (CPL) in Stark intrerference are beginning to be applied to diatomic molecules with a heavy atom in it. These studies are not yet extended to metal clusters or to organo-metal compounds as Ybcamphor. The interest in diatoms is mainly to anapole moment (nuclear spin-dependent and interacting with electrons electromagnetically) and electric dipole moment (EDM) of electron and nucleus. However, we can expect with advances in nonlinear optics (NLO) and molecular two or multislit interference experiments, new approaches to WNC spectroscopy, although Bouchiat's [50] NLO suggestion is not yet implemented. Drukarev and Moskalev had earlier suggested a two photon experiment [51] that would enhance the optical signal to 10^{-4} from 10^{-8} for hydrogen atoms.

3.1 Atoms

124

Sandars [52] has given a short classified list of expected P T even and odd effects in interactions between e-e, e-p and e-n electroweak manifestations in atoms. Stacey [53] summarizes the experiments achieved with atoms. The spin-independent (scaling as Z^5) and the spin-dependent (anapole moment, scaling as Z) and the elusive EDM are searched for in the experiments. While in ordinary molecules, the spin-dependent and independent strengths are nearly the same, with heavy atoms the spin-independent term is much larger and they set a limit to PNC in metal-organic compounds. The WNC aligns the spin (cf. s and p electrons) more in the direction of orbital motion and through the spin-orbit coupling mixes the even (E1) and odd (M1) parity spectral lines. The more forbidden is M1 and much nearer to E1 it is, the signal enhancement occurs, in addition to the Z^5 enhancement. The asymmetry factor, A_0 is given as:

$$A_0 = \frac{(M1 + \beta \cdot E1)^2 - (M1 - \beta \cdot E1)^2}{M1^2} \approx \beta \cdot \frac{E1}{M1} \qquad (4)$$

where $\beta = <V_p> / \Delta E$ gives the strength of the Z^0 interaction admixing E1 and M1 by the parity violating potential V_p, while ΔE is the energy difference between E1 and M1. A_0 can also be written as $2\beta \sqrt{\omega_1 / \omega_0}$,where ω_0 is the transition probability for the basic M1 transition, and ω_1 for E1 after the mixing. Thus there is enhancement with increasing Z, decreasing ΔE and ω_0 (or increasing forbiddenness). The experiments measure either circular dichorism or circular polarized emission in a Stark interference.

A list of recent experimental result and suggested experiments are :   Rotation in atomic Thallium at 1285 nm, $6p_{1/2} \rightarrow 6s_{3/2}$ (M1), $A_0 = -15.68 \times 10^{-8}$ and anapole spin-dependent $A_{os} = 0.04 \times 10^{-8}$ (N.A. Edwards et al., Phys. Rev. Lett., 74, 2654 (1995)): Francium, E1 amplitude, $A_0 = 10^{-7}$ (V.V. Flambaum, Phys. Rev. A51, 3454 (1994)): Uranium, $A_0 = 10^{-4}$ at 1O KV/cm (V. V. Karasiev et al., Phys. Lett. A172, 62 (1992)) and Yettribium at 408 nm (D. DeMille, Phys. Rev. Lett. 74, 4165 (1995)). Earlier experiments involve Cs, Bi and Pb.

Chaotic spectrum in Ce [54}, argon clusters etc., are known and possible dynamic enhancements of weak interactions in condensed matter physics, like in nuceli, are discussed by Flambaum [55]. Nachtman et al. [56] have an interesting suggestion of atomic interference with rotational mixing of polarized H atoms through high voltage capacitance for anapole moment. Polarized molecules are better suited as less external voltages would be needed.

Bouchiat and Mezard [57] have studied the Podd effects in metals and quote a A_0 of 10^{-6} due to delocalised conducting electrons.

3.2 Molecules

The earliest brave attempt to measure CD in O_2 by Bradely III and Wills in the 760 nm band with E1/M1 = 36 was shown by Bouchiat[58] as too large a value for good mixing. Kozlov and Labzowskys' [59] recent review of diatomic molecules with a heavy atom is mainly devoted to anapole moment and EDM and hence with highly polarizable molecules such as PbF, so that a lesser external voltage can be used for the spin-rotation mixing enhancement. Flambaum et al., [60] have suggested to use an inverse Faraday effect with a laser beam of 10W or more, in lieu of extrernal electric fields, and measure the NMR signal with respect to the direction of the beam. of the nuclear spin. The EDM (PoddTodd) is too broad for the PoddTeven laser beam to interfere. Barra and Weisenfeld [61] have calculated the nuclear spin-dependent contribution for $Pt(C_2H_2)_2$ for which the L-D difference in NMR is only of few mHz. Other methods suggested include beating between two frequency stabilized lasers with their references being two corresponding lines in the vibrational-rotational spectra; observation in the inverted Lamb dip and observation of oscillations of optical activity around the non-zero values in the chiral molecule. Recently, Letokhov [62] has suggested sympathetic cooling to measure optically the PNC effects in molecular ions. Quack [43] quotes Fox for the existence of PNC as well as for the violation of Pauli principle in pyramidal OsO_4.

Shape plays an important role in molecular and biological chemistry leading to potential wells and to reaction paths. It also plays an important role in atomic nuceli. Balachandran et al. [63] could deduce from a gauge theory of quantum shapes, P and T violation in molecules. However, the question is asked how shape interacting holistically with other molecular properties can be defined unambiguously. Wooley has given an example of allene's photoelectron spectroscopy and attests that more and more photoelectron spectroscopy and laser spectroscopy would show the failure of the Born-Oppenheimer approximation. He also describes his non-adiabatic generator coordination model [64]. There have been attempts to formulate optical activity by nonlinear Schròdinger equation with columb interaction thereby dispensing with potential wells for the L- and R-handed optically active compounds [65]. Bialynicki-Birula's nonlinear Schrödinger equation can also be used. Hund's metastability formulation is reformulated [70], while Harris and Stodlasky [71] among others, have made more detailed study of the "watched kettle does not boil' effect in terms of collision effects stablizing the optically active molecule against the tunneling effect. Cina and Harris [67] have recently suggested a two slit interference type of experiment using two sets of phase separated pulses for preparing, maintaining and detecting by time resolved fluorescence [68] the superposition of two wavepackets. Molecules like sulfoxides which invert in the excited state could be used in a proper matrix. The first pulse lifts the L-enantiomer to the excited state, while the second pulse brings the L and R to the ground state, after which two probe pulses timed half and full inversion period measure to verify the superposition equation (6). It is a parity-insensitive experiment.

$$\Psi_\pm(x) = \frac{1}{\sqrt{2}}\{\Psi_L(x) \pm \Psi_R(x)\} \qquad (5)$$

$$\Psi(x) = c_L\Psi_L(x) + c_R e^{i\varphi}\Psi_R(x) \qquad (6)$$

Quack [69] intends to study the mixed parity equation (5), which can be derived from equation (6), by measuring the inter-conversion times for the L and R molecules due to parity violation energy. Harris et al., had earlier suggested a parity sensitive all-or-none experiment separating the PNC effect term from the electromagnetic natural optical activity term by polarized light scattering in the presence of an external electric field [70]. Harris had also studied the electron-electron Podd effect in triplet state of a molecule with the effect being calculated to be of the same order as spin-orbit coupling [71].

Khrioplovich [44] had suggested an interesting experiment involving anapole moment by observing the electrophoretic mobility of free radicals which is Todd due to dissipation. His quoted calculations are high. Typically, he estimates for 10V/cm electric field with ionic conductance of solution of $0.1\Omega^{-1}cm^{-1}$, a mobility of 10^{-2} cm/sec. However, using long lived metallo-organic free radicals, it should be easily testable by using laser Doppler spectroscopy. Khriplovich and Poposlov [46] have recently given the asymptotic refractive index formula for the spin helicity due to the spin-orbit coupling PNC term:

$$n_+ - n_- = 1 + \frac{2\pi\alpha}{\omega^2} \cdot \frac{N}{r} \cdot f(\omega) \qquad (7)$$

127

The asymptoticity is ω^{-5}, although perturbation methods are not valid at high Z atomic numbers, and relativistic effects are not included, a rough $A_0 = 10^{-8}$ for $\omega = 100$ ev is quoted. It should revive to quantify the spin helical density that Garay and Reich et al., looked for, extend Kessler's type of precision experiments to higher electron energies, and probably suggest new optical methods. For one thing, properly scaled back scattering experiment for helices is possible. This could be extended to metal clusters, which may show a higher refractive index than free atoms. (In addition to other methods, by two slit interferometry, refractive index of atoms that are beginning to be measured routinely could be extended to small clusters of known size) If metal or their organic complexes are going to be involved in asymmetric synthesis then refractive index, specific heat and volumes would be useful physico chemical properties that, in addition to spectroscopy and cluster sizes, could be useful to evaluate chemical reaction possibilities involving them.

The P-odd van der Wall forces [44] can be used to verify and estimate the additivity of PNC forces in aggregates, crystals etc. The Yamagata formulae [72] for polymers and Tranter's claim [73] of PVED of 10^{-2} for quartz crystals with PNC per unit cell of 10^{-14} has not been verified either experimentally or on valid theoretical calculations. Equally, the FeS crystals which are purported to be chirally active has not been experimentally tested nor quantitatively calculated. While FeS crystals are hardly chiral, Wächtershäuser claims the existence of special low temperature triclinic crystals that are optically active. He also assumes that the intimate concomitant growth and cooperative effects lead to accumulative PVED as in Tranter's case. The macroscopic spiral helicity of quartz crystals unlike in those of high temperature superconductors but more similar to in liquid crystals and in monolayers are structural. Further, the evidences are that the two enantiomeric crystals are equally balanced, unlike in Holbium crystals which show a 0.03% excess in left-handed crystals [44]. Metals in metal clusters are van der Wall coupled and could show a higher PNC than in quartz or in Holmium metal crystals.

4, SPIN CHEMISTRY

Spin plays an important role in ferromagnetic, unconventional superconductors to in magnetoelectrics, and the metal cluster compounds could be closely related. Spin chemistry in metal -organic clusters from crossover from high spin (HS) to low spin (LS) states and spin control in electron transfer are beginning to be studied in detail. Relativistic and WNC effects on spin-orbit coupling due to heavy atoms leading to stabilization of the sign of spin polarization and spin polarization itself which could select the structure of the material could lead to specific spin catalysed reactions. Spin to influence or catalyze a reaction should be directly involved. These are close ranged but with possible long range effects.

4.1 Spin-Orbit Coupling

H_2 is the smallest molecule where spin-orbit coupling plays a role in the ortho- to para- conversion catalyzed by oxygen. Open shell heavy atoms, with fast electrons (cf. in the s orbitals) due to the heavy atomic charge, contract the orbitals and increase the spin-orbit coupling leading to larger ionization potential, changes in bond length ~0.08 A, and the bond energy and bond angle between the heavy atom and the organic molecule attached to it due to the electron density being larger near the nucleus [74]. The color of gold is due to the relativisitc effect. The relativistic spin chemistry of SbI in the infra-red region has been studied in detail [75]. Such studies in compounds containing Tl, Bi, and Cs would be useful in comparing the relativistic effect in relation to the WNC effect, and extend it to metal clusters. Miller et al., [76] studied the binding of cis Pt.$(NH3)^{+2}$ to intrastrand guanine in DNA. Molecular mechanics does not account for the ligand binding. It is apparent such relativistic effects should be seen in conformal changes and in reactions.

The spin-orbit coupling enters essentially in the calculation of scattering and absorption of spin polarized electrons [32] and in the photoelectron spectroscopy of optically active compounds [35].

A variety of chemical reactions exist that specifically invoke or is influenced by spin-orbit coupling, viz., in the hydrogen abstraction by ketones from hydrocarbons, phenols and ammonia and are similar to Norrish type II photoprocess

$$\text{>C=O} + HR \longrightarrow \text{>C-OH} + R$$

Manifestations of spin-orbit coupling in chemical reactions are well categorized [77] including calculations for reaction paths using spin-orbit estimations in the vicinity of potential energy crossings.

The spin density can be measured by neuron scattering and XMCD (X-ray magnetic circular dichorism) is increasingly employed to analyze spin and orbital momenta. MOKE (magnetic optical Kerr effect) can be used for spin polarization measurements in films.

129

4.2 Spin Catalysis

Although other types of electronic interactions can also be envisaged, spin controlled electronic interactions would be the most suitable, involving spin polarization effects, to search for WNC effects in chemistry. because of radical formation or close contact. A spin-orbit influence in the achiral precursor molecule could also be an advantage for mutual influence between the metal cluster compound as the catalyst and the reactants. Kinetic selection may occur in spin catalysis [78]. Electron spin may play a crucial role in control of reaction channels in the region of an activated complex. Catalysts can change the rate of a reaction by a spin switch with the substrate in the complex by supplying electrons to non-totally quenched angular momentum to the adsorbed complex of the common system (spin-orbit couple induced spin catalysis); or supply non-paired electrons to the activated complex of the common system (paramagnetic spin catalysis). These could be augmented by the magnons of the magnetic surface. With spin control, like in pentacene biradical [79[an organo-magnetic molecule, spin polarization and localization can facilitate the reaction. Some such related molecule could have been the precursor to the production of sugar-like chiral molecules. It has to be catalytic because it would reduce the transition state free energy for the feasibility of a possible weak reaction in the face of thermal fluctuations. - the cluster environs also can help insulate the reaction. Changes that occur on chiral inversion as in cyclooctane from enthalpy driven to entropy driven above the inversion point may also be useful. The metal cluster compound yet would be multifunctional like in an enzyme to hold the substrate in an induced fit - an effect assumed here as due to the far-off Z^0 interaction with electrons.

A typical example for spin catalysis is in the caged radical recombination in a magnetic field or in the presence of heavy atom. where the spin-orbit coupling influences inter system crossing, thereby controlling the rate constant for triplet to singlet with the products being formed through the triplet state. However, the known prebiotic routes for the sugars or for the amino acids are far removed. The reaction is sensitive to isotope effects in the heavy metal atoms used. The calculation by Harris [71] for the e-e spin interaction in a triplet state with influence of NWC could profitably be extended to include heavy atoms.

5. METAL CLUSTERS

5.1 Properties

A variety of metal clusters and organo-metal clusters from Na, Li, Zn, Fe, Mn,Co,Mo,Pt, PD,Rh, Ru,Ti,Hl,Yb, Os,U etc., are being studied. In addition to vapor phase, the metal clusters can be solublized with organic carbonyl groups etc., or protected by surfactants and other chemicals or in pores with sizes ranging from 1-10 nm. Different aspects of the physics and chemistry of clusters (molecular to metal) are reviewed [80]. Properties of gold colloids as semiconductor quantum dots with changes in spectroscopic and oxidation reduction chemical properties depending on size have been detailed by K. Weller [81]. The catalytic properties but mainly for industrial chemistry purposes is detailed by Lewis [82]. The close analogy to atomic nuclei from magic numbers, shape and deformations and the mean field density functional theory - beyond the jellium and closed shell models are being exploited. In addition, there are also closely related to nanometric mesoscopic systems as confined many body system. There are similarities, for instance, like two atomic quantum dots showing molecular-like properties, chaos in the spectra etc., and like in molecular clusters to defining the thermodynmaic properties of the metal clusters. Pump-probe ionization spectra of small (Na_2, Na_3) clusters to coherent evolution of compounds as $Mn_2(CO)_{10}$ are being studied. From atoms to clusters wave interferometry through multiple slits are used to study the cluster properties. In addition probe pulses are used for studies on dissociation energy that may include coulomb explosion, FTMS , FTICR etc.

5.2 Asymmetric Synthesis

Perhaps the accidental assembly of autocatalytic or self-replicating molecules could have been the origin of life. The small part of the electromagnetic interaction due to the handed forces could have been responsible for the initial asymmetric synthesis of chiral molecules which later became autocatalytic. At present, a variety of organo-metal complexes involving Li, Zn,Mn,Co,Mo,Ru, Rh, Pt, Pd. Ti, Hl, Yb, U etc. for asymmetric synthesis are known. Since they involve the same kind of metal clusters but usually of smaller aggregates, we may eventually understand the direct induction of chirality from the atomic center without the benifit of chiral ligand influence. However, at present, there is no connection established between those who study the physical properties of metal clusters and those who deliver the chiral products, and hence the mechanisms or spectroscopy relating to kinetics is not yet done, although it is known that there exists synergy between metal center autoinduction and the auxiliary effects of the chiral ligands, the later mainly providing the sterochemical environs. Asymmetric synthesis of reactant to product usually involves many steps and nonlinear and the initial products may influence both the subsequent reaction pathway and the aggregation of the catalysts. There is a mutual influence between the catalyst and the reactants. There are also cases where chiral auxiliary is more efficient when not completely enantiomerically pure. The information content of the cluster aggregates in terms of the reactions they promote are not known.

Kunz has recently [83] detailed a variety of chemical reactions that sugars as chiral auxiliaries can promote. It was previously thought that sugars as multifunctional chemicals may not be suitable as chiral promoters until Kuntz has shown that sugars can be selective and specific in these catalytic processes. Sugars were thought as energy sources and their information capacity as receptors are ignored. Eschenmoser et al., have recently shown [13] that, unlike the earlier formose reaction, metal hydroxide sheets could convert glycoldehyde phosphates into regio and stereo selective but racemic pentoses. It would be interesting to see if these sugar phosphates could be put in one of Kuntz's reaction schemes (there is a variety of different reaction schemes giving entirely different products) to convert into chiral sugars by a further selective process or used as chiral auxiliaries in some other reaction. Kunz's system of sugar chiral systems can also do Strecker synthesis, as he has shown, in the production of D-amino acids with titanium complexes. It is often suggested that the prebiotic amino acid production route was from immines interacting with HCN giving aminonitriles. from which amino acids can be obtained. Alternatively, in aldolization process to form sugar phosphate chiral products, the spin-orbit coupling influence is known, and one has to search for a selective metal cluster complex that could induce enantioselectivity in the reaction. Similarly, for amino acids we have to conceive of a ratchet mechanism that directionally breaks the asymmetric carbon α- hydrogen in C^*- H bond in the right direction to form, say. L-amino acid only, somewhat similar to the ratchet in flagellar rotation. Molecular pistons based on light driven donor-acceptor substances are known. Also, catalysts are known in polymerization that can be made to oscillate switching from syntactic to atactic polymer. Although no experiment is specifically known to produce chiral molecules from achiral ones, without aid from chiral influences, a judicious extrapolation to zero chiral ligand concentration of the kinetic rate may give an estimate of the chiral unaided reaction rate.

Laser guidance or teaching lasers how to control a chemical reaction as a dynamic symmetry breaking process is analyzed as a feedback control system for which system details need not be known. Genetic algorithm is also used to steer the chemical reaction course. Being selective interference phenomenon, a high intensity laser is not needed. An often used chemical formulae is of the form ABC which can be dissociated into A and BC or as AB and C by properly phased and pulsed two laser pulses. Brumer and Shapiro [46] have presented an interesting possibility involving Norrish Type II reaction : $D(CH_2)_3 . CO. (CH_2)_3. \rightarrow D.CH.CH_2 + L(CH_2)_3.CO. CH_3$ or $L.CH.CH_2 + D(CH_2)_3. CO. CH_3$ in which D and L are right handed and left handed amino acids. Before dissociation the compound is a flat molecule with reflection symmetry across the plane and on dissociation, a free D-aminoacid or L-amino acid can be obtained. It would be worthwhile to see if a metal cluster or its organic complex could be able to selectively and/or with a large yield of one of the two enantiomers put the WNC chemistry on a firm footing. Norrish Type II reaction is a spin-orbit promoted reaction and matching with the metal atom with its spin-orbit coupling, a specific reaction is possible. Besides a Norrish Type II reaction can be introduced into Kunz's schemes involving asymmetric synthesis by chiral sugar assisted metal catalysts

Although light guidance may not have been within its reach, nature might have had other devices such as the electron transfer control in photosynthetic systems through series of consecutive steps, a typical example of kinetic selectivity. Besides, it choses one of two equal routes. Or, as in hypercycle involving viral nucleotides directing enzyme synthesis and which in turn directs nucleotide synthesis, the metal cluster compounds might have played the role of prebiotic enzymes forming the D-sugar phiosphates, which in turn could become autocatalytic and probably could have also condensed into oligonuclotides on the metal surfaces. The metal clusters like enzymes could have been polyfunctional and the reaction steps in the cycle could have involved hetro- metal atoms as well as aggregates of different sizes. The ligands themselves play a certain role in controlling the catalytic property of the metal complexes and the products themselves could also steer the reaction course. Kagan [84] has a simplified equation for asymmetric synthesis but only for the chiral assisted ones, while Eigen and Schuster [85] deal with hypercycle. Norden [15] had attempted to apply the Eigen Darwinian selection equation to set of D and L compounds before the appearance of Kondepudi's equation.

Unusual catalytic properties are ascribed to metal clusters and their organo- complexes including bond-braking and stabilizing as carbo- and oxo- radicals in addition to oxidation-reduction properties. The metal cluster complex reaction for chiral molecule production from an achiral molecule could be entropic. In fact there are cases (like in cyclooctane) where one chiral form is enthalpic and on inversion entropic. The shape, size and the type of metal-

metal bonds depend sensitively to the organic molecules they bind. The shape and multifunctionality could be favorable, like in an enzyme, for a close fit to induce chiral production in the presence of fluctuations. And probably collective effects as in a confined many body system and giant dipole resonances as in the atomic nuceli can occur. It would be wothwhile to look for collective PNC effcts. Chaotic behavior can also occur in the reaction path of molecular clusters. Magnetic ordering, asymetry and directionality and shape dependent orientational effects and surface dependent effects may influence the reactions involving especially the rare earth and actinide atomic clusters. via. spin-orbit coupling, high magnetic moments and spin polarization leading to efficient catalysis. Jahn-Teller effect due to deformations may also play a dominant role [86]

For chemical reactions relating to structure, especially involving metal organic complexes, a variety of simplified approaches are taken based on Lewis acids, hardness and softness, electronegtivity and ionization potential and connect it to the density functional theory. The structure-correlation method is the prominent. It freezes the molecule at some point along the reaction coordinate arising from the energy minimization (or crystallographic data). An approach similar to it is the moment of inertia method to get the shape of the metal organic clusters. The reaction path is ascertained by conservation of symmetry along the reaction path, or by principle of least motion, or by maximal symmetry, or by the frontier orbital, or by the use of genetic algorithm, artificial intelligence or by information theory approach. Fukui's frontier orbital method can be interpreted in terms of spin density to highest occupied orbital. Fukui's function [87] as simplified in equation (8) involves N, the number of electrons and ρ, the charg density:

$$f(r) = \partial \rho(r)/\partial N$$

$$\rho = \downarrow \rho(r) - \uparrow \rho(r)$$

$$f(r) = \frac{\downarrow \rho(r) - \uparrow \rho(r)}{N \uparrow - N \downarrow} \tag{8}$$

High spin density may indicate high catalysis. For some ligands, soft metals form LS (low spin) and hard metals HS (high spin). Free radicals have large spin gradients and hence high reactivity. Negative charged clusters also have large spin gradients. The applications include variety of experiments including the inner mechanism of rare earth metals elucidated by photoelectron spectroscopy, redox behavior of carbonyl clusters, hetro- nuclear cluster compounds, and ligand stabilized metal clusters. Additionally, in understanding the structure and dynamics of metal clusters, wavepacket interference (for example as in the dynamics and coherent evolution of $Mn_2(CO)_{10}$) including fluorescence detected time resolved spectroscopy, photoelectron spectroscopy, multiphoton ionization, Fourier transform mass spectroscopy etc., are being employed. Further extension of these experimental studies connecting to reaction dynamics should pave the way for measuring weak forces in these compounds and connecting to their reactions.

DISCUSSION

Bada's observation that the meteorites and the ocean sediments are racemic and show no kind of any chiral excess is a crucial one..This strictly removes all extra-terrestrial origins for chirality, in spite of 100 tons/day of materials dumped on this planet - chiral molecules must have been made on this earth. It also eliminates the Vester- Ulbricht hypothesis of radiolytic origin of chirality. The weak point of terrestrial photolytic origin of chirality promoted by Norden is the feeble circularity in sun's light and alteration in the sign between the poles and between morning and evening, while the magnetochiral dichorism mechansim is weak. A process in which two simultaneous light beams but with two different frequencies selectively dissociates a Norrish type II reaction (in a magnetic field) is a possibility that needs further investigation.

The question of chiral origin has intrigued cosmologists, physicists, chemists and biologists. In the quest for the origin of chirality physical methods as radiolysis and photolysis were tried. They were too weak requiring further amplification by other means. It is akin to a ship with little propelling power in relation to its load. But the ship was an intelligent one, self-organizing and self-maintaining, increasing its load as well as its drive without external help. But how did it start?

With no clues coming from physics and chemistry, probably, there was no continuity - K.Popper or not - life emerged with two equivalent chiral systems doing the same job. Such redundancy is common feature of biological evolution meant mostly for future use. At some stage by chance, one of the two systems acquired more information than the other, i.e., a dynamic advantage than fluctuation which eventually took over. The soup was a reality. It is not a matter of belief but hard to understand. Similar problem exists in the left-right asymmetry in embryonic development. The assumption of Miller and Orgel should be tested in an evolution experiment (and in comparison to FeS experiments in a flask) consisting of mixed D- and L -ribooligonucleotides. Even so as Mann and Primkoff have pointed out, statistical averaging would lead to nowhere; or is life an extremely rare phenomenon as Monod theorized and also chirality as Prolog speculated? The neutrino hypothesis of Cline et al., in boosting the Kondepudi kinetic scheme is intriguing and attractive. It escapes Bada's strictum and if true would be the first case of punctuated equilibrium or historic contingency whereby one whole species went extinct. It is more biological than Goldanskii's big bang.

Little attention is paid to chemical catalytic possibilities but that invariably must include the influence of WNC interaction. Assuming such a possibility, how weak it could be and yet survive fluctuations? Wächtershäuser has not yet showed us how his FeS-pyrites could promote chirality in its recycling products and when.. Tranter assumes, in a quartz crystal with 10^{13} atoms, the PVED would be $10^{-15} \times 10^{13} = 10^{-2}$. for which there is little evidence. Such additivity is assumed in polymerization also. In Fe atom with d electrons, the spin-orbit coupling would be weaker than with s,p electrons but with a larger PVED than in Si. Again in FeS crystal aggregates, WNC effects are not shown. which are experimentally possible.

The prebiotic chemists seldom evinced much interest in metal organic chemistry while it's importance in biological world is known, and also there are evidences that enzymes with heteronuclear metal atoms might be a crossover from the physics of metal clusters to biochemistry. Thus Miller and Orgel in their text-book [13] on the Origins of Life on the Earth hardly mention any metal chemistry, although Orgel was trained as an inorganic spectroscopist.

The hydrovents according to their proponents are an active metabolic production factory. In fact evolution is taking in these places even today. as in Titanic atmosphere now. There are also simulation experiments in metallic autoclaves under high pressure and temperatures. But no interest is shown where and when chirality is introduced and used. It may well be the influence of some unknown metal colloids. Are we witnessing fresh evolution in these hydrovents and FeS pyrites and different from the one that happened 4 Gy ago? Would they all show L-amino acids and D-sugars?

There is much interest in metal clusters and organo-metal clusters both for their interesting spectroscopy and physical properties from atomic to solid state and for their immediate applications for novel technology and for their chemical behavior to new reaction processes. But the atomic physicists are more interested in testing the standard model and for GUT than measure and search for large PVE in metal clusters. Equally, the sterochemists are more interested in increasing the yield of chiral products than searching for a de novo synthesis of a chiral product from an achirai precursor, however small the yield may be or however slow the rate may be. The type of organo-metal systems that Kuntz uses with sugar (chiral) ligands to influence a variety of chiral chemical reactions could be a place to look for de novo synthesis. If successful, it may also show whether chiral sugar production and chiral amino acid productions are closely related or are different episodes.

CONCLUSION

It is likely that metal clusters and their organic complexes of sub-colloidal dimensions might exhibit a PVED of 10^{-6} kT or an asymmetric ratio of about 10^{-7} which is measurable by improvements in existing CD instruments fitted for flourescence detection. Earlier experiments of positronium triplet asymmetry should be continued. Improvements in Kessler's type of instrumentation for electron optical CD and starting of photoelectron spectroscopy of optically active compounds aimed at measuring PVED could help other optical measurements. Aggregates that could show enhanced or collective asymmetry factor should be targeted for study. Emphasis should be on rare earth elements and actinides. Methods of calculation through density functional theory should be implemented to facilitate such studies. Use should be made of phenomenological asymptotic refractive index equation for spin helicity like the one given by Khriplovich and Poposlov. Khriplovich's suggestive experiment of electrophoresis of free radicals in solution should be tested. Back scattering experiments to measure directly helical spin density should be explored. Assured of sufficient PNC force, catalytic properties can then follow. The role and use of chaos, if any, should not be ignored.

The search for chirality is shifting from impact of physical particles and nonlinear fluctuation kinetics to chemistry of FeS compounds and metal clusters. The chemical origin of chirality should be fully explored before yielding to chance. The implications would be profound than the alternative chance mechanism. Biology after all is essentially chemistry. At present, I do not know of any specific mechanism or metal cluster complexes that one could start with, but advances in organo-metal cluster physics and chemistry could lead to it.

REFERENCES

1. C. Chyba and C. Sagan Nature 355, 125 (1992)
2. J. L. Bada, S. L. Miller and M. Zhao. Origins Life. 25, 111 (1995)
3. C. De Duve and S. L. Miller. Proc. Nat. Acad. Sci. 88, 10014 (1991)
4. R, F. Fox Biological Energy Transducrtion, The Uroboros (Wiley-Interscience, 1982).
5. C. De Duve Blueprint for a cell (Neil Patterson, Burlington, NC.(1991)
6. G. Winnewisser and E. Herbert in Topics in Curr, Chem. v. 139 Springer. Verlag, 1987.
7. J. M. Greenberg and C.-X,. Mendoza- Gómez in The Chemistry of Life's origin. Eds. J. M. Greenberg et al. NATO C416 (Kluwer. Dordrecht 1993)
8. W. A. Bonner in Topics in Sterochemisrtry. v.18.(Wiley 1988).p 1
9. V. I. Goldanskii. This Conference. 1995
10 T. Achilles and G. von Kiedrowski. Angew. Chem Int. ed. 32, 1198 (1993)
11. S. Pitsch, A. Eschenmoser, B. Gedulin, S. Hui and G. Arrhenius. Origins Life 25, 297 (1995)
12. H. Wynberg. J. Marcomol. Sci. A26, 1033 (1989).
13. S. L. Miller and L. E. Orgel. The Origins of Life on the Earth. (Prentice-Hall, New Jersy. 1974).
14. M. W. Evans, S. Wozniak and G. Wagniere. Phys. Lett. A171, 355 (1992)
15. B. Norden. J. Mol. Evoln. 11, 313 (1978).
16. A. K. Mann and H. Primakoff Origins Life. 13, 113 (1983)
17. S. F. Mason and G. E. Tranter Proc. R. Soc. Lond. A397, 45 (1985)
18. D. K. Kondepudi and G. W. Nelson. Nature 314, 433 (1985).
19. J. L. Bada Phil. Trans. B333, 349 (1991).
20. J-P. Laplante, M. Pemberton, A. Hjelmfelt and J. Ross, J, Phys, Chem. 99, 10063 (1995).
21. W. Fontana and L. M. Buss. Proc. Nat. Acad. Sci. 91, 757 (1994).
22. G. Wächtershäuser Proc. Nat. Acad, Sci. 91, 4293 (1994).
23. W. A. Bonner Origins Life. 25, 175 (1995).
24. D. K. Kondepudi and G. W. Nelson. Physica A125, 465 (1984).
25. D. Cline, Y.Liu and H. Wang. Origins Life. 25, 201 (1995).
26. R. C. Dougherty Origins Life. 11, 71 (1981).

27. Ya. B. Zel'dovich and D. b. Saakyan. Sov. Phys. JETP. 51, 1118 (1980).
28. A. S. Garay and P. Hrasko. J. Mol. Evoln. 6, 77 (1975)
29. R. A. Hegstrom. Nature. 297, 643 (1982).
30. J. van House, A. Rich and P. Zitzewitz. Origins Life. 16, 81 (1985).
31. R. Fandreyer, D. Thompson and K. Blum. J. Phys. B23, 3031(1990)
32. J. Kessler.Electron Polarized Phenomena in Electron- Atom Collisions in Adv. At. Mol.Opt. Phys. 27, 81 (1991).
33. S. Mayer and J. Kessler. Phys. Rev. Lett. 74, 4803 (1995).
34. D. M. Campell and P. S. Farago. J. Phys. B20, 5133 (1987).
35. N. A. Cherepekov. Polarization and Orientation Phenomena in Photoionization of Molecules in At. Mol. Opt. Phys. 34, 207 (1994).
36. I. B. Khriplovich and M. E. Pospelov. Phys. Lett. A171,, 349 (1992).
37. B. Norden Nature 266, 567 (1977).
38. Y. Inoue. Chem, Rev. 92, 741 (1992).
39. D. N. Nikogosyan, Y. A. Repetev, E.V. Khoroshikova, I. V. Kryukov, E. V. Khoroshikov and A. V. Sharkov. Chem. Phys. 147, 437 (1990).
40. M. Shapiro and P. Brumer. J. Chem. Phys. 95, 8658 (1991).
41. G. E. Stedman. Adv. Chem. Phys. 85, Pt.2. p.489 (1993).
42. L. D. Barron. Chem. Soc. Rev. 15, 189 (1986).
43. M. Quack. J. Mol. Struct. 292, 171 (1993).
44. I. B. Khriplovich. Parity Nonconservation in Atomic Phenomena. Gordon & Breach. 1991
45. G. E. Stedman and H. R. Bilger. Phys. Lett. A122, 289 (1987).
46. M. Buttiker. Phys. Rev. Lett. 56, 1761 (1986).
47. G. V. Gelikonov and M. A. Novikov Opt. Spectr. 74, 663 (1993).
48. I. E. Dzyaloshinski. Phys. Lett. A155, 62 (1991).
49. R. A. Harris and I. Tinoco. J. Chem. Phys.101, 9289 (1994).
50. M. A. Bouchiat, J. Gueria, ph. Jacquier, M. Lintz and L. Pottier. Opt. Comm. 77, 374 (1990).
51.E. G. Drukarev and A. N. Moskalev. Sov. Phys. JETP. 46, 1073 (1977).
52. P. G. H. Sandars. Physica. Scr. T46, 16 (1993).
53. D. N. Stacey. Physica. Scr. T40, 15 (1992).
54. V. V. Flambaum Phys. Rev. A50, 267 (1994).
55. V. V. Flambaum. Phys. Scr. T46, 198 (1993).
56. G. W. Botz, D. Bruss and O. Nachtmann. Ann. Phys. 240, 107 (1995).
57. C. Bouchiat and M. Mezard J. physique 45, 1583 (1984).
58. M. C. Bouchiat and C. Bouchiat. J. physique 35, 899 (1974)
59. M. G. Kozlov and L. N. Labzowsky. J. Phys. B28, 1933 (1995).
60. V. V. Flambaum and O. Sushkov. Phys. Rev. A47, R751 (1993).

61. A. L. Barra, J. B. Robert and L. Wiesenfeld. Europhys. Lett. 5, 217 1988).

62. V. A. Alekseev, D. D. Krylova and V. S. Letokhov. Physica. Scr. 51, 368 (1995).

63. A. P. Balachandran, A. Simoni and D. M. Witt. Int. J. Mod. Phys. A7, 2087 (1992).

64. R. G. Wooley. J. Mol.Struct. 230, 17 (1990).

65. V. M. Perez-Garcia, I. Gonzalo and J. J. Perez-Diaz. Phys. Lett. A167, 377 (1992).

66. B. R. Fischer and P. Mittelstaedt. Phys. Lett. A147, 411 (1990).

67. J. A. Cina and R. A. Harris. Science. 267, 832 (1995).

68. N. F. Scherer, R. J. Carlson, A. Matro, M. Du, A. J. Ruggerio, V. Romero-Rochin, J. A. Cina, G. R. Fleming and S. A. Rice. J. Chem. Phys. 95, 1487 (1991).

69. M. Quack. Chem. Phys. Lett. 132, 147 (1986).

70. R. A. Harris. Chem. Phys. Lett. 223, 250 (1994).

71. R. A. Haris and L. Stodolsky. J. Chem. Phys. 73, 3862 (1980).

72. Y. Yamagata. J. theor. biol. 11, 495 (1966).

73. G. E. Tranter. Nature. 318, 172 (1985).

74. K. Balasubramanian and K. S. Pitzer. Adv.Chem. Phys. 67, Pt.1. p.287 (1987).

75. K. K. Das, A.B. Alekseyev, H.P. Liberman, G. Hirsch and R. J. Buenker. Chem. Phys. 196, 395 (1995).

76. K. J. Miller, E. R. Taylor, H. Bosch, M. Krauss and W. J. Stevens. J. Biomol. Str. Dynam. 2, 1157 (1985).

77. B. F. Minaev and S. Lunell. Z. Phys. Chem. 182, 263 (1993).

78. B. F. Minaev and H. Ågren. J. Phys. Chem. 99, 8936 (1995).

79. D. A. Dougherty. Acc. Chem. Res. 24, 88 (1991).

80. Clusters of Atoms and Metals. Vols 1 & 2. Ed. H. Haberland, Springer Series in Chem. Phys. v. 52 & 56, 1995 & 1994; W. A. de Heer. Rev. Mod. phys. 65, 611 (1993); M. Brack. Rev. Mod. Phys. 65, 677 (1993); Springer Series on Structure and Bonding. v. 79 (1992).

81. K. Weller. Angew. Chem. Int. Ed. 32, 41 (1993).

82. L. N. Lewis. Chem. Rev. 93, 2693 (1993).

83. H. Kunz and K. Ruck. Angew. Chem. Int. Edn. 32, 336 (1993).

84. D. Guillaneux, S. H. Zhao, O. Samuel, D. Rainford and H. B. Kagan. J. Am. Chem. Soc. 116, 9430 (1994).

85. M. Eigen and P. Schuster. The Hypercycle Springer 1979

86. A. Ceulemans and L. G. Vanquickenborne. Springer Series in Str & Bonding v. 71 (1989), p.125

87. J. L. Gàzquez, A. M. Vela and M. Galvàn. Springer Series in Str. & Bond. v. 66,(1987) p.79.

88. I. Bertini, S. Ciruli and C. Luchinat. Springer Series in Struct Bonding. v. 83 (1995).

89. G. Wächtershäuser. Prog. Biophys. Mol. Biol. 58 (1992) p. 85.

Spontaneous mirror symmetry breaking via enantioselective autocatalysis

Vladik A. Avetisov

N.N.Semenov Institute of Chemical Physics of the Russian Academy of Sciences.
ul.Kossygina 4, 117977 Moscow, Russia

Abstract. The conditions for spontaneous generation of optically active product under auto-catalytic type reactions is considered on the base of the general kinetic model of enantiose-lective autocatalytic stages. It is shown that the spontaneous generation of optical activity mostly depends on the enantioselectivity of catalytic transformations. The properties of auto-catalytic reactions, which are needed for an experiment, as well as the necessary chemical conditions, are discussed.

INTRODUCTION

In 1953 F.Frank suggested an idea that autocatalytic reactions can have the property for generation of the optically active product (1). One of the kinetic models he has considered, involves autocatalytic synthesis of **R** and **S** enantiomers from an achiral substrate **A**

$$A+R \xrightarrow{k_1} R+R \qquad A+S \xrightarrow{k_1} S+S \qquad (1a)$$

and the interaction between enantiomers resulting in an achiral product **P**

$$R+S \xrightarrow{k_2} P \qquad (1b)$$

Here k_1 and k_{-2} are the rate reaction parameters.

To demonstrate the main kinetic property of the model, let us assume that concentration of the achiral substrate x_A is constant ($x_A = c$) while concentrations of enantiomers x_R and x_S are variables. Then, the enantiomeric excess $\eta = (x_R - x_S)/(x_R + x_S)$ (so-called chiral polarization of the mixture) and the total concentration of chiral product $\theta = (x_R - x_S)/c$ obey kinetic equations

$$\frac{d\eta}{d\tau} = K\theta(-\eta^3 - \eta) \ , \quad \frac{d\theta}{d\tau} = \theta - K\theta^2(1 - \eta^2) \tag{2}$$

where $\tau = k_1 ct$ is dimensionless time and $K = k_{-2}/k_1$.

One can see that the values of chiral polarization at the steady states are the solutions of an equation

$$-\eta^3 + \eta = 0 \tag{3}$$

and so, there exist two types of attractors. namely the racemic state $(\eta = 0)$ and the chirally pure states $(\eta = \pm 1)$. It can be shown that the racemic state is unstable while the chirally pure states are stable for any positive values of K. Therefore a small enantiomeric excess. which can be induced, in particular, by random fluctuations of chiral polarization, will increase until the optically pure state will be formed. This process has a meaning of the spontaneous generation of optical activity, since the scheme (1a,b) is totally symmetric with respect to chiral conjugation $R \leftrightarrow S$.

Subsequently, this idea has been generalized (2-6) and the Frank's model became the most popular object in the theoretical investigations of so-called spontaneous mirror symmetry breaking (2,4,6), or bifurcation with mirror symmetry breaking (3,5) under chemical transformations. It have been initiated some attempts to find an experimental corroboration of the Frank's idea for last years (7-9), however these attempts cannot be recognized as decidable. Unambiguous demonstration of an autocatalytic reaction, which would be able to generate spontaneously optically active product, is still an open question.

It is usually believed, that the problem is that to find a chemical analog of the Frank's scheme. Whereas it is obvious, that in any case the spontaneous generation of optical activity may be realized under limited values of the reaction's kinetic parameters, there is not "controlling" parameter in the model (1a,b): for any (positive) value of K. the chirally pure states are a single type of stable attractors. If for no other reason than that, the scheme (1a,b), which of course is the best for demonstration of the idea, is too formal for chemical applications.

It should be noted that the numerical Frank's model modifications with controlling parameters have been suggested (10-17). These models were classified (6) and it was concluded that the existence of self-catalytic (with respect to enantiomers) stages is the common qualitative requirement for spontaneous generation of optical activity. However it seems to be an only recommendation, which might be exploited for chemical applications, because other theoretical estimations, in particular, expressions for the controlling parameters and the critical points, are closely associated with the concrete kinetic schemes of the considered models.

From our standpoint this is one of the causes why the spontaneous generation of optical activity is imaged as a kinetic property of some specific set of reactions and probably, such imagination creates an obstacle on the passage to an experiment. If this were the case, we would be in the face of a hard problem, because on the one hand, without total kinetic scheme of the chemical transformations, it would be impossible to define whether or not spontaneous generation of optical activity will be obtained. On the other hand, if there is a suitable kinetic model, it is practically impossible to design it's exact chemical analog.

In fact, the conditions for spontaneous generation of optical activity obey very simple quantitative requirement, which shows up not too specific for realization. This requirement can be formulated for autocatalytic reactions of arbitrary complexity, however, for definiteness, we will obtain it directly for bimolecular transformations. The conditions for spontaneous generation of optical activity, as well as the contribution of non-selective monomolecular transformations are also considered.

AUTOCATALYTIC STAGES

Since under autocatalysis, the chiral product acts as a catalyzer of it's own synthesis from an achiral substrate, there are stereochemical interactions between reactants to be consider. In the simplest case of the pair (bimolecular) interactions, a complete scheme of autocatalysis of **R** and **S** enantiomers is in the form:

$$A+R \underset{k_{-1}}{\overset{k_1}{\rightleftharpoons}} R+R \ , \qquad A+R \underset{k_{-2}}{\overset{k_2}{\rightleftharpoons}} S+R \ ;$$

$$A+S \underset{k_{-1}}{\overset{k_1}{\rightleftharpoons}} S+S \ , \qquad A+S \underset{k_{-2}}{\overset{k_2}{\rightleftharpoons}} R+S \ , \tag{4}$$

where k_1, k_{-1}, k_2, and k_{-2} are the rate parameters.

The set (4) reflects two basic features of stereochemical catalytic reactions: enantioselective ability of any chiral catalyzer is limited, so the catalytic action of each enantiomer can result in synthesis of the both enantiomers, and the catalytic stages can be reversible.

To take into account these features we introduce the enantioselectivities γ_+ and γ_-,

$$\gamma_+ = \frac{k_1 - k_2}{k_1 + k_2} \ ; \qquad \gamma_- = \frac{k_{-2} - k_{-1}}{k_{-2} + k_{-1}} \ , \tag{5}$$

for direct and reverse transformations correspondingly. These parameters can also be expressed by the energies of chiral discrimination:

$$\gamma_+ = \frac{\exp\{-E_+^{no}/kT\} - \exp\{-E_-^{ut} \ kT\}}{\exp\{-E_+^{no}/kT\} + \exp\{-E_-^{ut} . kT\}} = \text{th}\left(\frac{\Delta E_+}{2kT}\right)$$

$$\gamma_- = \frac{\exp\{-E_-^{it}/kT\} - \exp\{-E_-^{co} \ /kT\}}{\exp\{-E_-^{it}/kT\} + \exp\{-E_-^{co} \ kT\}} = \text{th}\left(\frac{\Delta E_-}{2kT}\right) \tag{6}$$

143

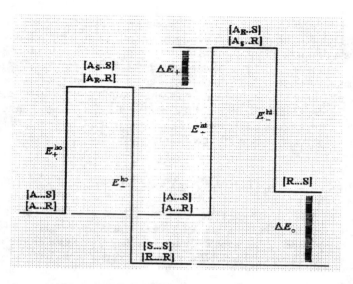

FIGURE 1. An example of the scheme of activation barriers for the direct [+] and the reverse [.] transformations: ($[A..R]$, $[A..S]$, $[R..R]$, $[S..S]$) are the basic states of interacting pairs of molecules and ($[A_R..R]$, $[A_S..R]$, $[A_R..S]$, $[A_S..S]$) are the excited transition states of the ones.

where E_+^{ho}, E_+^{ht}, E_-^{ho} and E_-^{ht} are the corresponding activation barriers (see Fig.1), $\Delta E_+ = (E_+^{ht} - E_+^{ho})$ is the energy of chiral discrimination of the direct transformations while $\Delta E_- = -(E_-^{ht} - E_-^{ho}) = (\Delta E_0 - \Delta E_+)$ is the discrimination energy of the reverse ones.

Note, that if $k_2 = k_{-1} = 0$, i.e. if, $\gamma_+ = \gamma_- = 1$, the set (4) is an exact kinetic analog of the scheme (1a,b) and therefore, the Frank's model corresponds to the extremely specific, unreal case of the absolute enantioselective interactions.

Consider the case of fast interchanging with "a big achiral reservoir", which maintains constantly the concentration of achiral substrate ($x_A = c$). By this technique, the chemical system can be kept far from the thermodynamic equilibrium. The controlling parameters of the model (4) can be obtained from analysis of the kinetic equations, which under "fast diffusion" are in the form:

$$\frac{d\eta}{d\tau} = -\frac{K}{2}\gamma_-\theta\eta^3 - \left[\frac{\bar{K}}{2}\gamma_-\theta + \gamma_+ - 1\right]\eta$$

$$\frac{d\theta}{d\tau} = \theta - \frac{K}{2}\left(1 - \gamma_-\eta^2\right)\theta^2$$

(7)

144

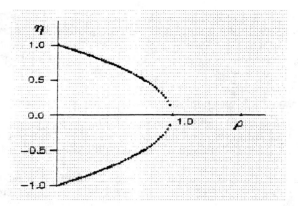

FIGURE 2. Bifurcation diagram for the equation (8): $\rho = (\gamma_+ + \gamma_- - 1)/\gamma_+ \gamma_-$ is the controlling parameter.

Here $\tau = c(k_1 + k_2)t$, is dimensionless time and $K = (k_{-1} + k_{-2})/(k_1 + k_2)$ is a parameter of reversibility of the autocatalytic stage.

For our aims it is enough to define the steady-state solutions of the equations (7) and then to examine theirs stability. It can be shown that the bifurcation equation of the model (4) is in the form

$$-\gamma_+ \gamma_- \eta^3 + (\gamma_+ - \gamma_- - 1)\eta = 0 \tag{8}$$

The real solutions of the equation (9) correspond to the values of chiral polarization of the steady states (Fig.2). It is easy to see that chiral polarization of the product depends only on the values of enantioselectivities γ_+ and γ_-. If $(\gamma_+ + \gamma_-) < 1$, there is a single physical solution $\eta = 0$, which corresponds to the stable racemic state. If

$$(\gamma_+ + \gamma_-) > 1 \tag{9}$$

there are three solutions: one of them is racemic, but it is unstable, and two others correspond to the stable optically active states. As soon as the value $(\gamma_+ + \gamma_-)$ reaches to unity, the racemic state losses it's stability and the transition to spontaneous generation of optical activity is realized. Thus, the critical point for such transition is defined by the simple expression

$$(\gamma_+ + \gamma_-) = 1 \tag{10}$$

It will enlarge on the physical meaning of this key correlation. First of all, by this is meant that there should be realized the conditions for the appearance of discrimination forces, which could be acting effectively under the formation of both the excited transition complexes ([A_R..R], [A_S..R], [A_S..S], [A_R..S]) and the "basic" diastereomeric complexes ([R..R], [S..S], [R..S]). These conditions play a key role for spontaneous generation of optical activity.

Next, the condition (9) may be satisfied only for positive values γ_- and γ_- (see the expressions (5), (6) and Figure 1). What this means is that:

a) the activation energy of the heterochiral transition states [A_S..R] and [A_R..S] should be higher then the activation energy of the homochiral states [A_R..R] and [A_S..S] ;

b) the dissociation energy of the heterochiral diastereomeric complexes [R..S] should be less then the dissociation energy of the homochiral complexes [R..R], [S..S] ;

c) the energy of chiral discrimination ΔE_0 of the basic diastereomeric complexes should be higher then the energy of chiral discrimination of the excited transition states ΔE_+ ($\Delta E_0 - \Delta E_+ = \Delta E_- > 0$)

Therefore the specific enantioselective interactions should be realized under the complexies formation.

Note lastly, that the thermodynamic equilibrium constant does not depend on the enantioselectivity of catalytic stages and so, the autocatalytic stage should be moved on right or on left from the thermodynamic equilibrium in order to spontaneous generation of optical activity would be possible. This is a general and important requirement.

Thus, if all these specific requirements are satisfied, it is left a few ways to control the transition to the spontaneous generation of optical activity. In reality, the critical point may be reached mostly by increasing of the corresponding discrimination energies or decreasing of the temperature.

To demonstrate the applicability of our result, consider the reaction of a liquid-phase autoxidation of tetralin, which was suggested in (9) as a reaction system for the Frank's model. In this reaction, the achiral tetralin (RH) reacts with oxygen to give chiral tetralin hydroperoxide (ROOH) via an autocatalytic pathway:

(i) $M^{(n+1)} + ROOH \longrightarrow M^{n+} + ROO^\bullet + H^+$

(ii) $ROO^\bullet + RH \longrightarrow ROOH + R^\bullet$

(iii) $R^\bullet + O_2 \longrightarrow ROO^\bullet$

The kinetic of transformations of the chiral peroxy radicals (ROO$^\bullet$), which become rate-limiting at high O_2 concentration, play the key role in that.

To examine whether or not the condition (9) may be satisfied under this reaction, consider the enantioselectivity of the autocatalytic pathway (i)$\Rightarrow$(iii) .

The stage (i) is enantioselective: the configuration of the chiral tetralin hydroperoxide (ROOH) gives the configuration of the chiral peroxy radical ROO$^\bullet$.

At the stage (*ii*), the configuration of the **ROO•** -radicals gives the **ROOH** configuration, but the configuration of the **R•** -radicals, which are formed from the achiral tetralin **RH**, is not controlled by the configuration of the peroxy radicals:

$$(R)\text{-}\mathbf{ROO^\bullet} + \mathbf{RH} \longrightarrow (R)\text{-}\mathbf{ROOH} + (R,S)\text{-}\mathbf{R^\bullet}$$
$$(S)\text{-}\mathbf{ROO^\bullet} + \mathbf{RH} \longrightarrow (S)\text{-}\mathbf{ROOH} + (S,R)\text{-}\mathbf{R^\bullet}$$

As a result, the stage (*ii*) is enantioselective with respect to the transformation of the peroxy radicals to the tetralin hydroperoxide, but it is not enantioselective with respect to the transformation of the achiral tetralin **RH** to the **R•**-radicals.

The stage (*iii*) is enantioselective similar to the stage (*i*), however at this stage, the configuration of the peroxy radicals **ROO•** is given by the configuration of **R•**-radicals rather than by the configuration of those **ROO•**-radicals, which were formed at the stage (*i*).

Therefore enantioselective control of the configurations of both the peroxy radicals and the tetralin hydroperoxide, is broken along the autocatalytic pathway from the stage (*i*) to the stage (*iii*):

$$(R)\text{-}\mathbf{ROOH} \longrightarrow (R)\text{-}\mathbf{ROOH} + (R,S)\text{-}\mathbf{ROOH}$$
$$(S)\text{-}\mathbf{ROOH} \longrightarrow (S)\text{-}\mathbf{ROOH} + (S,R)\text{-}\mathbf{ROOH}$$

Thus we can conclude that there is no enantioselective autocatalytic pathway in the autoxidation of tetralin. By this reason, the condition (9) cannot be satisfied and it seems to be impossible to reach the critical point and to obtain the spontaneous generation of optical activity.

NON-CATALYTIC STAGES

There are some arguments why it is important to estimate a contribution of non-catalytic, monomolecular stages to the critical condition for spontaneous genera-tion of optical activity. One of them is that if some chain of transformations con-sists of both the bimolecular stages and the monomolecular ones, then the chiral polarization of product can depend on the concentration of achiral substrate. This is well known fact in the technique of kinetic resolution. Note in addition, that in nu-merical publications dedicated to the problem of the origin of homochirality in Life (4,5,12,14-18), the concentration of an achiral substrate has been considered as a natural controlling parameter for passage through the critical point of spontaneous mirror symmetry breaking. It is clear however, that the critical condition should contain the kinetic limitations for the monomolecular stages, since they, as well known too (6), lead thyself to the racemization.

Let us add the set (4) by a scheme of reversible non-catalytic synthesis of the chiral product

$$A \underset{k_{-3}}{\overset{k_3}{\rightleftarrows}} R \ . \qquad A \underset{k_{-3}}{\overset{k_3}{\rightleftarrows}} S \ . \tag{11}$$

Here k_3 and k_{-3} are the rate parameters correspondingly.

The kinetic equations of the model (4)–(11) are in the form

$$\frac{d\eta}{d\tau} = -\frac{K}{2}\gamma_-\theta\eta^3 - \frac{K}{2}\gamma_-\theta + \gamma_+ - \left[1 - \frac{2\delta_+}{\theta}\right]\eta \tag{13}$$

$$\frac{d\theta}{d\tau} = 2\delta_+ + (1 - K\delta_-)\theta - \frac{K}{2}(1 - \gamma_-\eta^2)\theta^2$$

where $\delta_+ = k_3/(k_1 + k_2)c$, $\delta_- = k_{-3}(k_{-1} - k_{-2})c$ and the rest parameters were defined in the equations (7).

If the contributions of the monomolecular stages (11) to the equation (13) are small $(\{K\delta_+, K\delta_-\} \ll 1)$, then, to a first approximation with respect to the values of $K\delta_+$ and $K\delta_-$, the critical point for spontaneous generation of optical activity is defined by the expression

$$(1 - K\delta_+)\gamma_+ + (1 - K\delta_-)\gamma_- = 1 \tag{14}$$

The critical condition (14) involves now the parameters δ_+ and δ_- and it may be thought that there exist an independent, addition way to control the transition to the spontaneous generation of optical activity. To discuss the physical meaning of this result, let us assume that the chemical system is found in the subcritical region, so the racemic state $\{\eta = 0, \theta = 2K(1 + K\delta_+ - K\delta_-)\}$ is stable. By using the second of the equations (13), it can be estimated the part W_+ of those transformations $A \Rightarrow \{R, S\}$, which are catalytic:

$$W_+ = \frac{\theta}{\theta - 2\delta_+} = (1 - K\delta_+) \tag{15}$$

In a similar manner, the part W_- of the reverse catalytic transformations is given by the following expression:

$$W_- = \frac{\dfrac{K\theta^2}{2}}{\dfrac{K\theta^2}{2} - K\delta_-\theta} = \frac{\theta}{\theta + 2\delta_-} = (1 - K\delta_-) \ . \tag{16}$$

148

Therefore the factors $(1 - K\delta_+)$ and $(1 - K\delta_-)$ have a significance of "statistical weights" of the direct enantioselective (bimolecular) transformations and the reverse ones correspondingly. The physical meaning of the expression (14) becomes clear now: the transition to spontaneous generation of optical activity closely depends on the statistical weights and enantioselectivities of the catalytic stages. Thus the of catalytic transformations plays of a primary importance role

The part of the catalytic, bimolecular transformations increases when the concentration of the achiral substrate grows. In the limit case, this part is equal to unity and we obtain the expression (10). Therefore the critical condition (10) may be exploited for estimation of the lowest enantioselectivity necessary for spontaneous generation of optical activity. For example. if there is the case when the enantioselectivity γ_+ is of the order of enantioselectivity γ_-, then the value 0.5 will be an estimation of the lowest enantioselectivity needed for catalytic stages. Note, that in this case the energy of chiral discrimination should be of the order of the thermal energy. In reality, the more strong condition (9) should be satisfied..

GENERAL COMMENTARY

It is clear, that a set of nontrivial problems should be solved in order to find an experimental corroboration of the Frank's idea about spontaneous generation of optical activity.

The first problem is that to find an enantioselective autocatalytic reaction, under which the discrimination energy of the excited transition states would be of the order of the thermal energy, at least.

Next, it is necessary to provide the formation of diastereomeric complexes of a chiral product in the medium. In so doing, it should be borne in mind that the energy of chiral discrimination of the diastereomeric complexes in the media should be higher then the energy of chiral discrimination of the exited transition states are forming under autocatalysis.

Thus, sufficiently complex conditions should be provided. They can limit the search of the appropriate reactions. but if we are dealing with autocatalysis, it should be the high enantioselective, reversible autocatalytic reaction, which may be realized under strong interactions of enantiomers in condensed medium.

ACKNOWLEDGMENTS

The research described in this publication was made possible in part by Grant №M1W000 from the International Science Foundation.

REFERENCES

1. Frank. F., *Biochem. Biophys. Acta* 11, 459-463 (1953).
2. Morozov, L.L., *Origin of Life* 9, 187-218 (1979).
3. Nicolis. G. and Prigogine. I., *Proc. Nat. Acad. Sci. USA* 78, 659-663 (1981).
4. Morozov, L.L., Kuz'min. V.V. and Goldanskii. V.I., "Mathematical grounds and general aspects of the mirror symmetry breaking in prebiological evolution", in *Sov. Sci. Rev. D*, New York: Hoorwood Acad. Press, 1984, pp.357-405.
5. Kondepudi, D.K. and Nelson. G.W., *Physica A* 125, 465-496 (1984).
6. Goldanskii, V.I. and Kuz'min V.V., *Sov. Phys. Uspekhi* 32, 1-29 (1986).
7. Kondepudi, D.K., Kaufman. R.J. and Singh. N., *Science* 250, 975-976 (1990).
8. Avetisov, V.A., Goldanskii V.I., Grechukha. S.N. and Kuz'min V.V., *Chem. Phys. Lett.* 184, 526-530 (1991).
9. Buhse. T., Lavabre, D., Micheau. J.-C. and Thiemann. W., *Cirality* 5, 341-345 (1993).
10. Kondepudi, D.K. and Nelson. G.W., *Nature* 314, 438-441 (1985).
11. Klemm, A., *Z.Naturforsch.* 40a., 1231-1234 (1985).
12. Goldanskii, V.I., Anikin, S.A., Avetisov. V.A. and Kuz'min, V.V., *Comments Moll. Cell. Biophys.* 4, 79-98 (1987).
13. Gutman, I. and Klemm, A., *Z.Naturforsch.* 42a.. 899-900 (1987).
14. Avetisov, V.A. and Goldanskii, V.I., *BioSystems* 25, 141-149 (1991).
15. Avetisov, V.A., Goldanskii, V.I. and Kuz'min. V.V., *Physics Today* 44, 33-41 (1991).
16. Gutman, I. and Todorovic', D., *Chem.Phys.Lett.* 195, 62-66 (1992).
17. Gutman, I., Todorovic', D., and Vuckovic'. M.. *Chem.Phys.Lett.* 216, 447-456 (1993).
18. Cattani, M. and Tania Tome, *Origin of Life and Evolution of the Biosphere* 23, 125-136 (1993).

Symmetry Breaking by Autocatalysis

David Lippmann*† and Arjendu Pattanayak††

*Department of Chemistry, Southwest Texas State University
San Marcos, Texas 78666 U.S.A.
and
† Center for Studies in Statistical Mechanics and Complex Systems
The University of Texas, Austin, Texas 78712 U.S.A.
and
‡ Department of Chemistry, University of Toronto
Toronto, Ontario, Canada M5S1A1

The origin of homochirality in biochemical compounds has been a mystery since this homochirality was discovered. Various mechanisms that might favor one enantiomer over the other have been proposed, but they all have the defect of producing, at most, a tiny excess of one enantiomer rather than the nearly complete homochirality that exists. It is clear that some process that can convert an initial small excess of one enantiomer into a very large excess is required. In 1953, Frank proposed that a combination of autocatalysis and mutual destruction by a pair of enantiomers will produce an extremely large excess of one enantiomer if there is a small initial excess of that enantiomer. (1) The initial asymmetry could be arbitrarily small and could have any cause, including statistical fluctuations in the rates of production of the two enantiomers.

Frank's proposed set of reactions is shown in Figure 1.

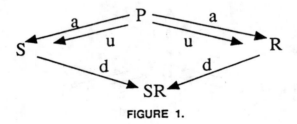

FIGURE 1.

P represents some achiral precursor or precursors of the S (sinister) and R (rectus) enantiomers, and SR is a dimer produced by reaction of one S molecule with one R molecule. The arrows labeled u represent the uncatalyzed production of S and R from P, the arrows labeled a represent autocatalytic production of S and R from P, and the arrows labeled d represent destruction of S and R by combination to produce SR. The differential equations for rates of production of S, R, and SR are

$$\frac{d[S]}{dt} = ku[P] + ka[S][P] - kd[S][R] \tag{1}$$

$$\frac{d[R]}{dt} = ku[P] + ka[R][P] - kd[S][R] \tag{2}$$

$$\frac{d[SR]}{dt} = kd[S][R] \tag{3}$$

Square brackets indicate molar concentration or-equivalently- number of molecules in a fixed volume, t is time, and the $k's$ are rate constants. Frank showed that any initial excess of either S or R would increase without limit. The mutual destruction step is required for the ratio $[S]/[R]$ or $[R]/[S]$ to increase. Without mutual destruction, any initial excess of either enantiomer will increase, but the ratio of the two concentrations will remain the same.

However, Frank made several simplifying assumptions which are, at best, only approximately valid. He assumed that the supply of P is unlimited, that there are no reverse reactions, and that the rate constant for destruction of S and R by reaction to produce the dimer SR is much larger than the rate constant for production of the possible dimers S_2 and R_2.

We modified Frank's set of reactions to eliminate some of his assumptions. This modification of Frank's set of reactions is shown in Figure 2.

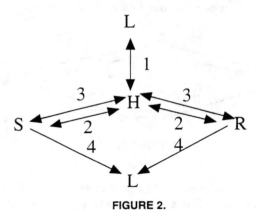

FIGURE 2.

In Figure 2, L represents some set of stable, low free energy compounds. H represents a set of high free energy compounds that are produced from

L. In order for H to be produced from L, free energy must be provided by some source such as sunlight, lightning, or a thermal gradient. S and R are enantiomers produced from H by reactions in which free energy decreases. S and R react with each other to reform L with another decrease of free energy. Double headed arrows indicate that the reactions can go in both directions. The rate equations for this set of reactions are

$$\frac{d[L]}{dt} = -f_1[H] + b_1[H] + 2f_4[S][R] \tag{4}$$

$$\frac{d[H]}{dt} = f_1[L] - b_1[H] - 2f_2[H] + b_2[S] + b_2[R] - f_3[S][H] + b_3[S]^2$$
$$\qquad - f_3[R][H] + b_3[R]^2 \tag{5}$$

$$\frac{d[S]}{dt} = f_2[H] - b_2[S] + f_3[S][H] - b_3[S]^2 - f_4[S][R] \tag{6}$$

$$\frac{d[R]}{dt} = f_2[H] - b_2[R] + f_3[R][H] - b_3[R]^2 - f_4[S][R] \tag{7}$$

The $f's$ and $b's$ are forward and reverse rate constants, respectively. The system is assumed to be closed, so

$$T = [L] + [H] + [S] + [R] \tag{8}$$

where T is the concentration (or number) of L molecules present before any $H, S,$ or R have been produced.

These equations could not be solved directly, but they were solved by numerical integration. If f_4 is sufficiently large relative to $f_1, f_2,$ and f_3, and if f_3, the forward rate constant for the autocatalytic reaction, is large relative to f_2, the rate constant for uncatalyzed formation of S and R, then any early excess of either S or R increases rapidly. A steady state in which there is a very large excess of the enantiomer that was initially present in slight excess is approached. If f_4 is not sufficiently large, racemization by way of reactions 2 and 3 eliminate any early excess of either enantiomer, and equal concentrations of S and R are present in the steady state.

A plausible hypothesis is that the catalytic species are not S and R themselves, but some oligomers of S and R. For simplicity we assumed that dimers, S_2 and R_2, are the catalysts. This assumption leads to the set of reactions shown in Figure 3.

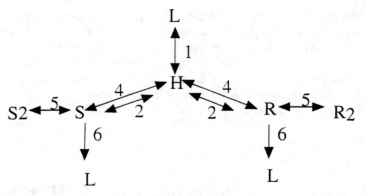

FIGURE 3.

A mutual destruction reaction is not required here. Reaction 6 is a first order decay of S and R to L. The differential equations are

$$\frac{d[L]}{dt} = -f_1[L] + b_1[H] + f_6[S] + f_6[R]$$

$$\begin{aligned}\frac{d[H]}{dt} &= f_1[L] - b_1[H] - 2f_2[H] + b_2[S] + b_2[R] - f_4[H][S_2] \\ &+ b_4[S][S_2] - f_4[H][R_2] + b_4[R][R_2]\end{aligned} \tag{9}$$

$$\begin{aligned}\frac{d[S]}{dt} &= f_2[H] - b_2[S] + f_4[H][S_2] - b_4[S][S_2] - f_5[S]^2 \\ &+ 2b_5[S_2] - f_6[S]\end{aligned} \tag{10}$$

$$\begin{aligned}\frac{d[R]}{dt} &= f_2[H] - b_2[R] + f_4[H][R_2] - b_4[R][R_2] - f_5[R]^2 \\ &+ 2b_5[R_2] - f_6[R]\end{aligned} \tag{11}$$

$$\frac{d[S_2]}{dt} = f_5[S]^2 - b_5[S_2] \tag{12}$$

$$\frac{d[R_2]}{dt} = f_5[R]^2 - b_5[R_2] \tag{13}$$

The conservation relation is

$$T = [L] + [H] + [S] + [R] + 2[S_2] + 2[R_2] \tag{14}$$

T was taken to be 100.001, and steady state solutions were obtained numerically. In each case it was assumed that S was present in slight excess over

154

R shortly after the reactions began. Calculations were carried out for a number of different sets of assumed values of the rate constants. The results of six sets of calculations with various values of the rate constants are shown in Table 1. The subscript ss indicates a steady state value.

See Table 1.

It is evident that very large ratios of $[S]_{ss}$ to $[R]_{ss}$ and $[S_2]_{ss}$ to $[R_2]_{ss}$ can be attained, that these ratios are sensitive to the values of the rate constants, and that, if f_6 is small, the steady state is a racemic mixture of S, R, S_2, and R_2.

ACKNOWLEDGEMENT

This work was partially supported by Department of Energy grant DE-FG03-94ER14465.

REFERENCES

1. Frank, F.C., *Biochimica and Biophysica Acta* **11**, 459-463 (1953).

TABLE 1. Steady State Numbers of Molecules in Fixed Volume With Various Values of Rate Constants

f_1	1×10^3	1×10^3	1×10^3	1×10^3	1×10^3	1×10^3
b_1	1×10^1	1×10^1	1×10^1	1×10^1	1×10^1	1×10^1
f_2	1×10^0	1×10^0	1×10^0	1×10^0	1×10^0	1×10^0
b_2	1×10^{-1}	1×10^{-1}	1×10^{-1}	1×10^{-1}	1×10^{-1}	1×10^{-1}
f_4	1×10^2	1×10^2	1×10^2	1×10^2	1×10^2	1×10^2
b_4	1×10^0	1×10^0	1×10^0	1×10^0	1×10^0	1×10^0
f_5	1×10^5	1×10^5	1×10^5	1×10^4	1×10^4	1×10^4
b_5	1×10^4	1×10^4	1×10^4	1×10^3	1×10^3	1×10^3
f_6	1×10^5	1×10^6	1×10^7	1×10^5	1×10^4	1×10^3
$[L]_{ss}$	2.211×10^3	2.000×10^4	8.540×10^4	2.211×10^3	2.234×10^2	3.163×10^1
$[H]_{ss}$	2.211×10^0	2.005×10^0	9.711×10^{-1}	2.215×10^0	2.234×10^0	1.581×10^0
$[S]_{ss}$	2.211×10^1	2.000×10^1	8.540×10^0	2.211×10^1	2.233×10^1	1.581×10^1
$[R]_{ss}$	2.222×10^{-5}	2.005×10^{-6}	9.711×10^{-8}	2.217×10^{-5}	2.358×10^{-4}	1.581×10^1
$[S_2]_{ss}$	4.888×10^4	3.999×10^4	7.294×10^3	4.888×10^4	4.988×10^4	2.498×10^4
$[R_2]_{ss}$	4.939×10^{-8}	4.019×10^{-10}	9.431×10^{-13}	4.913×10^{-8}	5.560×10^{-6}	2.498×10^4

P-odd energy splitting in helical antiferromagnets.

Can it influence the Néel transitions?

I.B. Khriplovich[1]

Budker Institute of Nuclear Physics, 630090 Novosibirsk, Russia

Abstract

Parity nonconserving energy difference between crystals with right-
and left-handed helical spin structure constitutes in rare earths 10^{-13} –
$10^{-12}\, eV$ per ion. We argue that there is a chance to detect experimentally its influence on the relative population of left and right domains
at the Néel transition

1 Introduction

Can weak interactions, tiny in the ordinary life, influence macroscopic phenomena? This question is actively discussed at this Symposium. My point
is that there are good reasons to expect that this influence can be detected
experimentally in the phase transitions in helical antiferromagnets [1].

A helical spin structure arises in some magnetics at low temperatures [2].
The origin of the effect is the oscillating coordinate dependence of the exchange
interaction between magnetic ions [3]. The coordinate dependence of the ion
spins in such structures is

$$\vec{S}(\vec{r}) = S\,(\,\cos(\theta z/c),\, \sin(\theta z/c),\, 0\,). \tag{1}$$

Here θ is the angle of spin rotation at the transition to the next crystal plane, c
is the distance between those planes. To simplify the further discussion we omit
here a possible ferromagnetic component which does not affect qualitatively
our conclusions. The handedness of the helix can be conveniently described
by the pseudoscalar [4]

$$P = \vec{S}\cdot[\vec{\nabla}\times\vec{S}] = -S^2\theta/c. \tag{2}$$

---

[1] E-mail: khriplovich@inp.nsk.su

As a complete order parameter of a spin helix one can take the pseudotensor

$$P_{\alpha\beta} = Pn_\alpha n_\beta \tag{3}$$

where n_α is directed along the helix axis.

Due to the invariance under reflections, in an ideal crystal in the absence of external fields nothing fixes the sign of the pseudoscalar P. In such a case at the transition to the helimagnetic phase, left and right domains should be equally populated.

A crystal with a helical spin structure is a kind of a giant chiral molecule. The parity-nonconserving weak interaction creates energy difference between left and right domains in it, as it does between left and right molecules. This energy difference can influence the formation of the helimagnetic phase making domains of one sign preferentially populated.

The origin of the P-odd energy difference both in chiral molecules and helical magnets can be explained intuitively as follows [5]. Parity-nonconserving weak interaction induces a spin helix of a sign determined by the interaction itself and by the electronic properties of the substance, molecule or crystal. It is only natural that the energy of a molecule or domain is different depending on whether its own structural helix, a cooordinate or spin one respectively, is of the same or opposite sign as the weak interaction helix.

By the way, does an energy difference between left and right molecules (or crystals) mean that the recording of scales with one pan filled with left-handed molecules and another one with the same number of right-handed, changes if the pans are viewed in a mirror? Of course, not. Under reflection not only does the P-odd interaction Hamiltonian change sign: $W \rightarrow -W$, but also right- and left-handed states are interchanged: $|l> \leftrightarrow |r>$. Therefore, the difference

$$< r|\,W\,|r> - < l|\,W\,|l >$$

remains the same.

2 Simple estimate

Let us now present a simple estimate of the P-odd energy difference in helimagnetic structures. We will confine here to metals (rare earths: Tb, Dy, Ho, Er) where one can expect the largest weak interaction effect. In metals the interior interaction responsible for the helimagnetic ordering, arises in second order in the spin-dependent exchange interaction of the conduction electrons with the electrons of the incomplete shells of the ions. The measure of this interion interaction is the Néel temperature T_N (i.e., the temperature of the phase transition of the type 'paramagnet - antiferromagnet').

It can be demonstrated that the leading contribution to the P-odd energy difference in metals arises also in second order in the same exchange interaction. But it should contain of course an additional small factor which characterizes the P-odd weak interaction on the atomic scale [5]

$$\eta \sim \frac{G}{\sqrt{2}} \frac{m^2 \alpha^2 Q Z^2 R}{\pi}. \tag{4}$$

Here G is the Fermi weak interaction constant; m is the electron mass; $\alpha = e^2 = 1/137$; Z is the charge of the nucleus; Q is its "weak" charge which is close numerically to $-N$, N being the neutron number; R is the relativistic enhancement factor which reaches the value 4.5 at $Z = 70$. The product $Z^2 R$ in expression (4) describes the ratio of the wave function squared of a conduction electron at a nucleus to its average value over the crystal.

Taking $T_N \sim 100\, K$, we find that the P-odd energy difference per ion, we are interested in, constitutes by an order of magnitude

$$\delta E \sim \eta\, T_N \sim 100\, Hz \sim 10^{-13} - 10^{-12}\, eV \tag{5}$$

in agreement with more detailed estimates made in Ref. [1]. Curiously enough, one of the contribution to the P-odd energy is of quite nontrivial dependence on the twisting angle θ, roughly

$$\frac{\theta}{|\theta|} - \frac{\theta}{\pi}, \quad -\pi < \theta < \pi. \tag{6}$$

3 External influence on the formation of helimagnetic structures

But can this tiny energy difference influence the relative population of left and right domains? We believe that it can, being strongly enhanced by the collective nature of the phase transition. Our argument is as follows.

It has been demonstrated experimentally [6] that applied external electric field $E = 2.5\, kV/cm$ allows one to control effectively the population of left and right domains in the spinel $Zn\, Cr_2\, Se_4$. Two more experiments [7, 8] have also demonstrated an external influence on the relative volume of helimagnetic domains. Unfortunately, in the last cases the origin of the effect observed is not clear [9].

Detailed theoretical investigation [4] has demonstrated that the effect discovered in experiment [6] is induced via spin-orbit interaction and is due in particular to specific properties of the crystal space symmetry group. The following estimate was obtained for the twisting energy per ion in the spinel under the conditions of experiment [6]:

$$\delta E_{el} \sim 10^{-10} \, eV. \tag{7}$$

The energy (7) is quite tiny as compared to the sample temperature. Then, how can it influence the sign of the spin helix? The explanation may look as follows. When a strong interaction (here the exchange one) has formed a long-range order, the whole ordered domain behaves as a huge molecule. In such a case the equilibrium function of the domains distribution depends on the twisting energy as $\exp(-N \, \delta E_{el}/T)$ where N is the typical number of ion spins in the domain.

It should be emphasized that the domain, which enters here, is not that of the final helimagnetic phase. It is the smallest non-dissipating, but growing, nucleus of this phase in the paramagnetic one. Still N is sufficiently large so that $N \, \delta E_{el}$ is at least comparable to the temperature.

Certainly, the collective nature of the helimagnetic state enhances strongly the role of a twisting energy.

However, the largest weak interaction effects can be expected in metals. The discussion of the possibility to influence the formation of the spin helices in them by external factors is pertinent as an estimate of possible backgrounds and of the feasibility of control experiments.

Electric field does not penetrate metals. but it can be in principle imitated by elastic mechanical deformations. The latter modulate microscopic electric fields which, via the spin-orbit interaction of conduction electrons, can influence the helical ordering of localized spins. To this end however, the magnetic ion should have an asymmetric environment. But in rare earths, which are apparently most interesting for the searches for weak interaction effects, the ion environment is symmetric. A deformation violates this symmetry as well. But then it gets obvious that in rare earths the influence discussed can be only of second order in deformations. So, it is negligible at any reasonable elastic deformation.

The very small influence of elastic deformations on the helimagnetic structure hampers control experiments in the searches for the weak interaction effects in metals. On the other hand, it diminishes the background problem in those searches.

Perhaps, much more serious background is due here to dislocations, but of special types only. Near the dislocation core the twisting energy per ion can reach $10^{-7}\,eV$.

Still, the experimental detection of parity nonconservation in the transition to the helimagnetic phase in rare earths looks feasible.

References

[1] Zhizhimov, O.L., and Khriplovich, I.B., Zh.Eksp.Teor.Fiz. **84**, 342 (1983) [Sov.Phys.JETP **57**, 197 (1983)].

[2] Hurd, C.M., Contemp.Phys. **23**, 469 1982.

[3] White, R.M., and Geballe, T.H., Long-Range Order in Solids, Academic Press, 1979.

[4] Khriplovich, I.B., and Kolokolov, I.V., Phys.Lett. **A160**, 204 (1991).

[5] Khriplovich, I.B., Parity Nonconservation in Atomic Phenomena, Gordon and Breach Science Publishers, 1991.

[6] Siratori, K., Akimitsu, J., Kita, E., and Nishi, M., J.Phys.Soc.Jap. **48**, 1111 (1980).

[7] Baruchel, J., Palmer, S.B., and Schlenker, M., J.de Phys. **42**, 1279 (1981).

[8] Patterson, C., Palmer, S.B., Baruchel. J., and Schlenker, M., Sol.State Commun. **55**, 81 (1985).

[9] Baruchel, J., and Schlenker, M., Physica **137B**, 389 (1986).

True and False Chirality, *CP* Violation, and the Breakdown of Microscopic Reversibility in Chiral Molecular and Elementary Particle Processes

Laurence. D. Barron

Chemistry Department, The University, Glasgow G12 8QQ, U.K.

Abstract. The concept of chirality is extended to cover systems that exhibit enantiomorphism on account of motion. This is achieved by applying time reversal in addition to space inversion and leads to a more precise definition of a chiral system. Although spatial enantiomorphism is sufficient to guarantee chirality in a stationary system such as a finite helix, enantiomorphous systems are not necessarily chiral when motion is involved, which leads to the concept of true and false chirality associated with time–invariant and time–noninvariant enantiomorphism, respectively. Only a truly chiral influence can induce an enantiomeric excess in a reaction that has reached true thermodynamic equilibrium (*i.e.* when all possible interconversion pathways have equilibrated); however, false chirality can suffice in a reaction under kinetic control due to a breakdown of microscopic reversibility analogous to that observed in particle-antiparticle processes involving the neutral K-meson as a result of *CP* violation, with the apparently contradictory kinetic and thermodynamic aspects being reconciled by an appeal to unitarity. This reveals that CP violation is analogous to chemical catalysis since it affects the rates of certain particle-antiparticle interconversion pathways without affecting the initial and final particle energies and hence the equilibrium thermodynamics. Consideration of falsely chiral influences, including the 'ratchet effect' arising from the associated breakdown in microscopic reversibility, greatly enlarges the range of possible chiral advantage factors in prebiotic chemical processes if far from equilibrium.

162

1. INTRODUCTION

A central theme in discussions of chemical evolution and the origin of life is the problem of the origin of biomolecular homochirality. Why is this homochirality based on L-amino acids and D-sugars rather than on the mirror-image versions? Any suspicion that, because of parity violation in the weak interactions, or something deeper, mirror-image life would not work at all appears to have been laid to rest by the recent synthesis of an enzyme (HIV-1 protease) exclusively from D-amino acids which was shown to have the same catalytic activity within experimental error limits as the original with respect to the corresponding mirror-image substrates (1). It has been argued that life, based on self-replication of organic homochiral polymers, could have originated only if the prebiotic organic medium had a high degree of enantiomeric purity, possibly achieved *via* a bifurcation-type transition to a chirally-pure state (2,3). Attention therefore becomes focused on *chiral advantage factors* which might influence the build up of an enantiomeric excess in the products of pre-biotic chemical reactions, and on whether or not any such factors could be sufficiently large to compete with the amplification of chance fluctuations at the bifurcation point.

The word *chiral* was first coined by Lord Kelvin (4), Professor of Natural Philosophy in the University of Glasgow, in his Baltimore Lectures delivered in 1884, to describe a figure "if its image in a plane mirror, ideally realized, cannot be brought to coincide with itself". This definition is essentially the same as that introduced earlier by Pasteur to describe objects he called *dissymmetric*. There is no disagreement when the term chiral is applied to a stationary handed object such as a finite cylindrical helix or an asymmetric molecule (Fig. 1); but when applied to less tangible systems the concept becomes less clear.

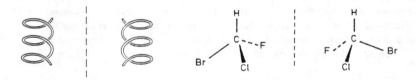

FIGURE 1. Dissymmetric (chiral) objects and their distinguishable mirror-image forms (enantiomorphs).

Pasteur, also in the last century, was the first to extend the concept of dissymmetry (chirality) to other aspects of the physical world in his search for universal forces which might be connected with optical activity in nature (5). For example, he thought that a magnetic field, since it can induce optical rotation (the Faraday effect), generates the same type of dissymmetry as that possessed by an optically active molecule. As we shall see, this idea is quite wrong and has been the source of much confusion. However, Pasteur was correct in thinking that the combination of linear motion with a rotation does generate the same type of dissymmetry as an optically active molecule.

The application of fundamental symmetry arguments provides a deeper description of chirality than that usually encountered in the literature of stereochemistry and facilitates a proper understanding of the structure and properties of chiral molecules and the factors involved in their synthesis and transformations. It is found that, *although dissymmetry is sufficient to guarantee chirality in a stationary object such as a finite helix, dissymmetric systems are not necessarily chiral when motion is involved.* Some years ago I introduced the concept of 'true' and 'false' chirality to draw attention to this distinction (6,7) and suggested that the word 'chiral' be reserved for systems that I call truly chiral. It will be appreciated from what follows that true and false chirality correspond to time-invariant and time-noninvariant enantiomorphism, respectively. This new classification has important implications for the discussion of chiral advantage factors.

2. TRUE CHIRALITY

2.1. Natural and Magnetic Optical Activity

A careful analysis of the natural and magnetic optical rotation experiments shows that the symmetry classifications of the two are quite different: the natural optical rotation observable associated with an isotropic collection of chiral molecules is a time-even pseudoscalar, whereas the magnetic optical rotation observable associated with a collection of achiral molecules is a time-odd axial vector (8). In fact Lord Kelvin was aware of this distinction, for his Baltimore Lectures (4) contain the following pronouncement:

"The magnetic rotation has neither left-handed nor right-handed quality (that is to say, no chirality). This was perfectly understood by Faraday and made clear in his writings, yet even to the present day we frequently find the chiral rotation and the magnetic rotation of the plane of polarized light classed together in a manner against which Faraday's original description of his discovery of the magnetic polarization contains ample warning."

This analysis tells us that the nature of the quantum states of molecules that can support natural optical rotation is quite different from that of the quantum states that can support magnetic optical rotation. Since natural optical rotation is an odd-parity, time-even observable, the quantum states must have mixed spatial parity but definite 'time parity'; conversely, since magnetic optical rotation is an even-parity, time-odd observable, the quantum states must have definite spatial parity but mixed 'time parity' (8).

2.2. A New Definition of Chirality

It should now be clear that the hallmark of a chiral system is that it can generate time-even pseudoscalar observables. This leads to the following definition (8,9) that enables true chirality to be distinguished from other types of dissymmetry:

> *True chirality is exhibited by systems existing in two distinct enantiomeric states that are interconverted by space inversion, but not by time reversal combined with any proper spatial rotation.*

This means that the enantiomorphism shown by truly chiral systems is time-invariant. Enantiomorphism that is time-noninvariant has different characteristics that I have called false chirality in order to emphasise the distinction. Notice that a static uniform magnetic field on its own is not even falsely chiral.

It is easy to see that a stationary object such as a finite helix that is chiral according to the traditional stereochemical definition is accommodated by the first part of this definition: space inversion, associated with the parity operator P, corresponds to inverting the positions of all the particles in a system through an arbitrary space-fixed origin, and is a more fundamental operation than the mirror reflection traditionally invoked but provides an equivalent result. Classical time reversal, associated with the operator T, corresponds to reversing the motions of all the particles in the system: although time reversal is irrelevant for a stationary object, the full definition is required to identify more subtle sources of chirality in which motion is an essential ingredient. A few examples will make this clear.

2.3. Translating Spinning Electrons, Photons and Cones

Consider an electron, which possesses a spin quantum number $s = 1/2$ with $m_s = \pm 1/2$ corresponding to the two opposite projections of the spin angular momentum onto a space-fixed axis. It is clear that a stationary spinning electron is not a chiral object because P does not reverse the spin sense and so does not generate a distinguishable P-enantiomer. However, an electron translating with its spin projection parallel or antiparallel to the propagation direction exhibits true chirality because P interconverts distinguishable right- and left-spin polarized versions propagating in opposite directions, but T does not. Fig. 2 illustrates the essential aspects of this behaviour.

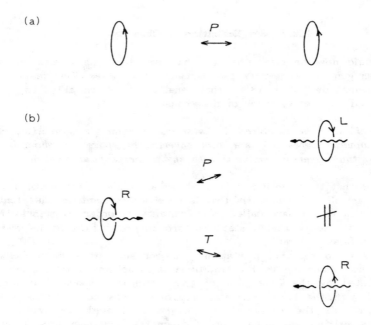

FIGURE 2. The effect of P and T on the motions of (a) a stationary spinning particle and (b) a translating spinning particle.

The photons in a circularly polarized light beam propagating as a plane wave are in spin angular momentum eigenstates characterized by $s = 1$ with $m_s = \pm 1$ corresponding to projections of the spin angular momentum vector parallel or antiparallel, respectively, to the propagation direction. The absence of states with $m_s = 0$ is connected with the fact that photons, being massless, have no rest frame and so always move at the velocity of light. Considerations the same as those in Fig. 2b above show that a circularly polarized photon exhibits true chirality.

Now consider a cone spinning about its symmetry axis. Since space inversion P generates a version that is not superposable on the original (Fig. 3a), it might be thought that this is a chiral system. However, the chirality is false because time reversal T followed by a rotation R_π through 180° about an axis perpendicular to the symmetry axis generates the same system as space inversion (Fig. 3a). But if the spinning cone is also translating along the axis of spin, time reversal followed by the 180° rotation now generates a different system to that generated by space inversion (Fig. 3b). Hence a *translating* spinning cone possesses true chirality.

(a)

(b)

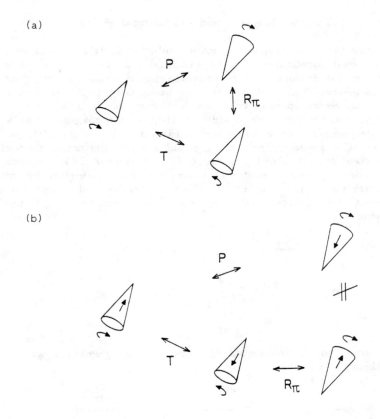

FIGURE 3. The effect of P, T and R_π on (a) a stationary spinning cone, and (b) a translating spinning cone.

167

These considerations expose a link between chirality and special relativity, because the chirality that an observer perceives in a spinning electron, for example, depends on the relative motions of the observer and the electron. Indeed, this relativistic aspect of chirality is a central feature of elementary particle theory, especially in relation to the weak interactions. Thus only left-handed extremely relativistic electrons (or right-handed extremely relativistic positrons) participate in charge-changing weak interactions. For nonrelativistic electron momenta, the weak interaction still violates parity but the amplitude of the violation is reduced to order v/c (10).

2.4. Electric, Magnetic and Gravitational Fields

From the foregoing, neither a static uniform electric or magnetic field on its own constitutes a chiral influence, true or false (a static electric field **E** transforms as a time-even polar vector so that its direction is reversed by P but not by T; a static magnetic field **B** is a time-odd axial vector so that its direction is reversed by T but not by P). Likewise for time-dependent uniform electric and magnetic fields. Furthermore, contrary to a suggestion first made by Curie (11), no combination of a static uniform electric and a static uniform magnetic field can constitute a chiral system. Collinear electric and magnetic fields do indeed generate enantiomorphism, but it is time–noninvariant and so corresponds to false chirality. Thus parallel and antiparallel arrangements are interconverted by space inversion and are not superposable:

$$
\begin{array}{ccc}
\mathbf{E} \longrightarrow & & \mathbf{E} \longleftarrow \\
& P \longleftrightarrow & \\
\mathbf{B} \longrightarrow & & \mathbf{B} \longrightarrow
\end{array}
$$

But they are also interconverted by time reversal combined with a rotation through 180°:

$$
\begin{array}{ccccc}
\mathbf{E} \longrightarrow & & \mathbf{E} \longrightarrow & & \mathbf{E} \longleftarrow \\
& T \longleftrightarrow & & R_\pi \longleftrightarrow & \\
\mathbf{B} \longrightarrow & & \mathbf{B} \longleftarrow & & \mathbf{B} \longrightarrow
\end{array}
$$

168

In fact the basic requirement for two collinear vectorial influences to generate chirality is that one transforms as a polar vector and the other as an axial vector, with both either time-even or time-odd. The second case is exemplified by *magneto-chiral* phenomena such as a birefringence and a dichroism induced in a chiral sample by a uniform magnetic field collinear with the propagation vector **k** of a light beam of arbitrary polarization (12,13). Thus parallel and antiparallel arrangements of **B** and **k** are true chiral enantiomers because they cannot be interconverted by time reversal since **k**, like **B**, is time-odd.

Analogous to collinear electric and magnetic fields is the case of a rapidly rotating vessel with the axis of rotation perpendicular to the earth's surface (14). Here we have the time-odd axial angular momentum vector of the spinning vessel either parallel or antiparallel to the earth's gravitational field, itself a time-even polar vector. The physical influence in this case therefore exhibits false chirality.

False Chirality and Anyons

Although not referred to as such, a version of false chirality arises in the anyon theory of high-temperature superconductivity. Anyons are spinning particles (or quasiparticle excitations) obeying quantum statistics that can vary continuously between those of fermions and bosons and which have been suggested to inhabit the two-dimensional world of copper oxide planes (15). Parity has a different effect in two dimensions than in three since it becomes simply a reflection in only one axis (16). This has the consequence that, unlike a spinning system in three dimensions, the sense of a spinning system in two dimensions is invariant under *PT* together but not under *P* and *T* separately. Hence time-noninvariant enantiomorphism (i.e. false chirality), albeit in two dimensions, is a central feature of the anyon theory of high-temperature superconductivity.

3. ABSOLUTE ASYMMETRIC SYNTHESIS

The use of an external physical influence to produce an enantiomeric excess in what would otherwise be a racemic product of a prochiral chemical reaction is known as an absolute asymmetric synthesis (17,18). The subject still attracts much interest and controversy (19,6,20), not least because it is an important ingredient in considerations of the prebiotic origins of biological molecules (3,21,22).

3.1. Truly Chiral Influences

If an influence can be classified as truly chiral we can be confident that it has the correct symmetry characteristics to induce absolute asymmetric synthesis, or some associated process such as preferential asymmetric decomposition, in any conceivable situation, although of course the influence might be too weak to produce an observable effect. Consider a unimolecular process in which an achiral molecule R generates a chiral molecule M or its enantiomer M^*:

$$M \underset{k_b}{\overset{k_f}{\rightleftharpoons}} R \underset{k_f^*}{\overset{k_b^*}{\rightleftharpoons}} M^* \tag{1}$$

In the absence of a chiral influence, M and M^* have the same energy so no enantiomeric excess can exist if the reaction is allowed to reach thermodynamic equilibrium. However, in the presence of such an influence M and M^* will have different energies so an enantiomeric excess can now exist at equilibrium. There are also kinetic effects because the enantiomeric transition states will have different energies.

Circularly polarized photons or longitudinal spin-polarized electrons are the obvious choice, and several examples in asymmetric synthesis or preferential asymmetric decomposition are known (3, 18, 22). Other things being equal, chiral effects induced by spin-polarized electrons should increase with increasing electron velocity because electron chirality is velocity-dependent.

Less obvious is the use of an *unpolarized* light beam collinear with a static magnetic field. This system exhibits true chirality so we can be confident it can induce asymmetric synthesis. The most favourable mechanism with this type of influence would appear to be one based on magneto-chiral dichroism (23).

3.2. Falsely Chiral Influences

When considering the possibility or otherwise of absolute asymmetric synthesis being induced by a falsely chiral influence, a distinction must be made between reactions that have been allowed to reach thermodynamic equilibrium (*thermodynamic control*) and reactions that have not attained equilibrium (*kinetic control*).

The case of thermodynamic control is quite clear. Because M and M^* are isoenergetic in the presence of, say, collinear electric and magnetic fields, or a spinning vessel with its axis perpendicular to the earth's surface (neglecting the very small difference due to parity violation), such falsely chiral influences cannot induce absolute asymmetric synthesis in a reaction mixture which is isotropic in the absence of the influence and which has been allowed to reach thermodynamic equilibrium (24). (The energy equivalence of M and M* follows from a consideration of the invariance of the Hamiltonian in the presence of the influence under the combined operations of space inversion and time reversal.)

The situation is less straightforward for reactions under kinetic control for, as discussed below, if microscopic reversibility and detailed balancing were to break down an enantiomeric excess could develop.

3.2.1. The Breakdown of Microscopic Reversibility induced by a Falsely Chiral Influence: Enantiomeric Detailed Balancing

I have suggested that conventional detailed balancing, and the associated kinetic principles, might not be valid for reactions involving chiral molecules in a time-noninvariant enantiomorphous influence (25,26). This can be seen from a quantum-mechanical description of the microscopic reaction event (27). The amplitude for a transition from some initial linear momentum state $\mathbf{p}$ to some final state $\mathbf{p}'$ is written $\langle \mathbf{p}' | \mathcal{T} | \mathbf{p} \rangle$, where $\mathcal{T}$ is the operator responsible for the transition. If $\mathcal{T}$ involves purely electromagnetic interactions it will be invariant under both parity and time reversal, which enables us to write

$$\begin{array}{ccccc} & \text{under } T & & \text{under } P & \\ \langle \mathbf{p}' | \mathcal{T} | \mathbf{p} \rangle & = & \langle -\mathbf{p} | \mathcal{T} | -\mathbf{p}' \rangle & = & \langle \mathbf{p}^* | \mathcal{T} | \mathbf{p}^{*\prime} \rangle \end{array} \qquad (2)$$

where we have allowed the particles to be chiral, the star denoting the P-enantiomer. The first equality in equation (2), obtained from time reversal alone, is the basis of the conventional principle of microscopic reversibility and, when averaged over the complete system of reacting particles at equilibrium, of the principle of detailed balancing (28). The second equality, obtained by applying space inversion to the time-reversed transition amplitude, describes the inverse process involving the *enantiomeric* particles.

Conventional detailed balancing is usually adequate for the kinetic analysis of reactions, even those involving chiral molecules, because conventional microscopic reversibility based on the assumption of T invariance is usually valid. This is conceptualized in terms of a potential energy profile that is the same in the forward and backward directions for a given reaction (Fig. 4).

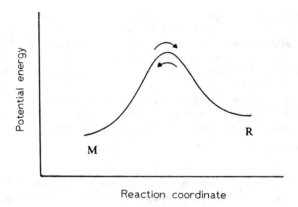

FIGURE 4. Microscopic reversibility conceptualized as the same potential energy barrier for the forward and backward processes.

However, in the presence of a time-noninvariant enantiomorphous influence such as collinear electric and magnetic fields, time reversal alone is not a symmetry operation since a different influence would be generated: space inversion must also be applied to recover the original relative orientations of **E** and **B**. The first equality in equation (2) is therefore no longer valid, and we must base any kinetic analysis on the relationship

$$\langle \mathbf{p'} | \mathcal{T} | \mathbf{p} \rangle \overset{\text{under } TP}{=} \langle \mathbf{p^*} | \mathcal{T} | \mathbf{p^{*'}} \rangle \tag{3}$$

which implies *enantiomeric* microscopic reversibility. Hence, as illustrated in Fig. 5, for reactions of chiral molecules in situations where only the combined *TP* invariance holds, the potential energy profiles for the forward and backward *enantiomeric* reactions are the same but the forward and backward profiles for the reaction of a given enantiomer are different in general. This can be modeled in terms of different *velocity-dependent* contributions to the potential energy profiles for the forward and reverse reactions involving a particular enantiomer induced by the falsely chiral influence.

Although conventional chemical kinetics is founded on the assumption of microscopic reversibility, the possibility of a breakdown in the presence of a time-odd influence such as a magnetic field does not conflict with any fundamental principles (29). Indeed, in his classic paper on irreversible processes, Onsager (30) recognized that

172

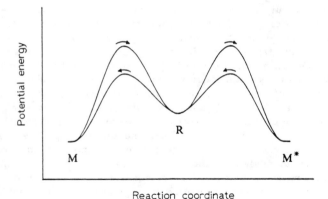

FIGURE 5. Potential energy profiles for enantiomeric reactions in the presence of a time-noninvariant enantiomorphous (*i.e.* falsely chiral) influence.

microscopic reversibility does not apply when external magnetic fields are present. However, Onsager's prescription of reversing **B** along with the motions of the interacting particles does not restore microscopic reversibility when **B** is a component of a falsely chiral influence; but even then, there will only be observable consequences if the particles are chiral. This is because, if the particles are achiral, the P-enantiomers are indistinguishable from the original so that $M = M^*$ and the barriers to the left and right of R in Fig. 5 must coalesce, which is only possible if the forward and reverse barriers shown for production of a particular enantiomer become identical; but if M is not identical to M^* they can, in general, remain distinct.

A breakdown in microscopic reversibility has recently been recognized to occur in reflective optical activity from chiral media (31). Although the context is different from that considered here, the basic ingredients are the same, namely a process involving chiral matter that breaks *T* and *P* separately but is *TP* invariant overall.

3.2.2. Unitarity and Thermodynamic Equilibrium

It therefore appears that enantiomeric microscopic reversibility and the associated enantiomeric detailed balancing allows $k_f \neq k_f^*$ and $k_b \neq k_b^*$ if the reaction (1) takes place in the presence of a time-noninvariant enantiomorphous influence such as collinear **E**, **B**.

However, this would generate an enantiomeric excess at thermodynamic equilibrium, which conflicts with the requirement that the concentrations of M and M^* must be the same. This conflict between the kinetic and thermodynamic requirements can be resolved by including other pathways by which M and M^* can be interconverted. Then at *true* thermodynamic equilibrium (i.e. when all of the possible infinite number of interconversion pathways have equilibrated), equal concentrations of M and M^* obtain because the different enantiomeric excesses associated with each separate pathway sum to zero. The proof of this assertion (25, 26) follows from the unitarity of the scattering matrix, which corresponds to the fact that the sum of the transition probabilities from a given initial state to all final states is unity. The argument is similar to that used to demonstrate the existence of equal numbers of particles and antiparticles at thermodynamic equilibrium, even when *CP* violation (see below) destroys the equality between the rates for specific particle → antiparticle and antiparticle → particle transitions, in the big bang model of the early universe (32). All this is consistent with the demonstration that it is *unitarity*, rather than microscopic reversibility, that lies behind the validity of Boltzmann's H-theorem and hence the second law (33, 34).

4. CHIRALITY AND SYMMETRY VIOLATION

A symmetry violation (often called symmetry nonconservation) arises when what was thought to be a 'nonobservable' is actually observed (35). The nonobservables of relevance here are absolute chirality (i.e. absolute left- or right-handedness), absolute sense of motion and absolute sign of electric charge, being associated with the symmetry operations of parity *P*, time reversal *T* and charge conjugation (particle-antiparticle exchange) *C*. However, even if one or more of these three symmetries is violated, the combined symmetry of *CPT* is always thought to be conserved (34, 35). A consideration of symmetry violation, and how it differs from spontaneous symmetry breaking, provides considerable insight into the phenomenon of molecular chirality.

4.1. Parity Violation and True Enantiomers

Following the discovery of the unification of the weak and electromagnetic interactions by Weinberg, Salam and Glashow in the late 1960s (10, 34, 35), it was shown that the absolute parity violation associated with the weak interaction infiltrates to a tiny extent into all electromagnetic processes by means of a *weak neutral current* interaction mediated by the massive neutral Z^o particle. One manifestation is a small energy difference between enantiomeric chiral molecules which is susceptible to quantum-mechanical calculations giving values of the order 10^{-20} a.u. (36-38).

It is easy to prove that only the space-inverted enantiomers of truly chiral systems show a parity-violating energy difference (7): this follows from the fact that the weak neutral current Hamiltonian transforms as a time-even pseudoscalar and so is odd under P but is invariant under T and R_π. Space-inverted enantiomers of falsely chiral systems such as a stationary rotating cone are strictly degenerate. Hence parity violation provides a cornerstone for the identification of true chirality.

Since the space-inverted enantiomers of a chiral molecule are not strictly degenerate, they are not true enantiomers (the concept of enantiomers implies the *exact* opposite). So where is the true enantiomer of a chiral molecule to be found? In the antiworld, of course! The molecule with the opposite absolute configuration but composed of antiparticles will have exactly the same energy as the original (8, 9, 39). So true enantiomers are interconverted by CP, which means that a chiral molecule is associated with two distinct pairs of true enantiomers (Fig. 6). Since P violation automatically implies C

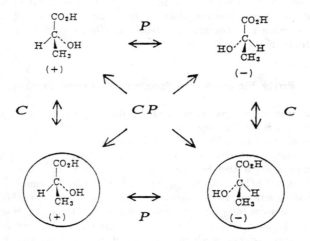

FIGURE 6. The two pairs of true enantiomers (i.e. strictly degenerate) of lactic acid that are interconverted by CP. The degeneracy is maintained even if CP is violated provided CPT is conserved (40).

175

violation here, it also follows that there is a small energy difference between a chiral molecule in the real world and the corresponding chiral molecule with the same absolute configuration in the antiworld. The original derivation (8, 9, 39) assumed that *CP* was not violated; but recently I have shown, from an extension of the proof that a particle and its antiparticle have identical rest mass, that the *CP* enantiomers of a chiral molecule remain strictly degenerate even in the presence of *CP* violation provided *CPT* invariance holds (40).

This more general definition of the enantiomers of a truly chiral system is consistent with the chirality that free atoms display on account of parity violation (41). The weak neutral current generates only one type of chiral atom in the real world: the conventional enantiomer of a chiral atom obtained by space inversion alone does not exist. Clearly the enantiomer of a chiral atom is generated by the combined *CP* operation. Thus the corresponding atom composed of antiparticles will of necessity have the opposite 'absolute configuration' and will show an opposite sense of parity-violating optical rotation (8).

The space-inverted enantiomers of objects such as translating spinning electrons or cones that only exhibit chirality on account of their motion also show parity-violating differences. One manifestion is that, as mentioned above, left- and right-handed particles (or antiparticles) have different weak interactions. Again, *true* enantiomers are interconverted by *CP*: for example, a left-handed electron and a right-handed positron. Notice that right- and left-handed circularly polarized photons are automatically true enantiomers since a photon is its own antiparticle.

4.2. Parity Violation and Spontaneous Parity Breaking

The appearance of parity-violating phenomena is interpreted in quantum mechanics by saying that, contrary to what had previously been supposed, the Hamiltonian lacks inversion symmetry (the weak interaction potential being a pseudoscalar). This means that *P* no longer commutes with the Hamiltonian, so the law of conservation of parity no longer holds. Such *symmetry violation* (nonconservation) must be distinguished from *spontaneous symmetry breaking* which arises when a system displays a lower symmetry than its Hamiltonian. Natural optical activity is therefore a phenomenon arising from spontaneous parity breaking since a resolved chiral molecule displays a lower symmetry than its associated Hamiltonian: if the small parity-violating term in the Hamiltonian is neglected, the symmetry operation that the Hamiltonian possesses but the chiral molecule lacks is parity, and it is this parity operation that interconverts the two enantiomeric parity-broken (mixed parity) states.

The conventional view is that the stability of chiral enantiomers arises from the existence of a high potential energy barrier separating the enantiomeric structures (8,26). Quantum-mechanical tunnelling through this barrier leads to a finite lifetime for a resolved enantiomer, so the corresponding quantum states are time-dependent. The optical activity therefore remains observable only so long as the observation time is short compared with the interconversion time between enantiomers, which is inversely proportional to the area of the barrier. The law of parity is saved in systems displaying spontaneous parity breaking because their pseudoscalar properties average to zero over a sufficiently long observation time, or, equivalently, the space-inverted experiment is realizable. In either case absolute chirality is not observable.

This exposes a crucial distinction between spontaneous parity-breaking and parity-violating natural optical activity phenomena. The former are time-dependent and average to zero; the latter are constant in time. It is still an open question as to whether the primary source of the high stability of resolved chiral molecules such as amino acids is parity violation (since the handed quantum states become the true stationary states if the parity-violating energy difference between the enantiomers is much larger than the tunnelling splitting), spontaneous parity breaking with very long tunnelling times, or spontaneous parity breaking stabilized by the environment. There has also been much discussion on the relationship between the microscopic and macroscopic aspects of molecular chirality (42,43) including the possibility of a phase transition to a chirally pure state induced by parity violation (44).

4.3. *CP* Violation

Violation of the combined *CP* symmetry was first observed by Cronin and Fitch in 1964 in certain decay modes of the neutral K-meson (10,34). The effects are very small (but nonetheless unequivocal) and have not been observed in any other system. Although it is *CP* violation that is observed directly, an accompanying *T* violation is implied from the *CPT* theorem.

For example, the decay mode of the long-lived neutral K-meson $K_L \rightarrow \pi^- e_r^+ \nu_l$ into negative antipions π^- plus right helical positrons e_r^+ plus left helical neutrinos ν_l occurs 1.00648 times faster than the decay mode $K_L \rightarrow \pi^+ e_l^- \nu_r$ into positive pions π^+ plus left helical electrons e_l^- plus right helical antineutrinos $\tilde{\nu}_r$. Since these two sets of decay products are interconverted by *CP*, this decay rate asymmetry indicates that *CP* is violated. If we naively represent this decay process in the form of 'chemical equilibria' as in equation (1),

177

$$\pi^+ + e_l^- + \tilde{\nu}_r \quad \underset{k_b}{\overset{k_f}{\rightleftharpoons}} \quad K_L \quad \underset{k_f^*}{\overset{k_b^*}{\rightleftharpoons}} \quad \pi^- + e_r^+ + \nu_l \qquad (4)$$

a parallel is established with absolute asymmetric synthesis associated with a breakdown of microscopic reversibility discussed above since in both cases $k_f \neq k_f^*$. Thus the K_L and the two sets of decay products are the equivalents, with respect to CP, of R, M and M* with respect to P. We can therefore conceptualize the decay rate asymmetry here as arising from a breakdown of microscopic reversibility due to a time-noninvariant CP enantiomorphism in the forces of nature (25): the CPT theorem guarantees that the two distinct CP-enantiomorphous influences are interconverted by T. The analogy is completed by the fact that, as mentioned above, the asymmetries cancel out when summed over all possible channels at true thermodynamic equilibrium.

4.3.1. *CP Violation is Analogous to Chemical Catalysis!*

In chemistry, a catalyst is defined as a substance that is not consumed in a chemical reaction and which increases the reaction rate at a given temperature by lowering the activation energy but without affecting the free energy change for the reaction. It follows that a falsely chiral influence acts as a special type of catalyst since it modifies potential energy barriers to change relative rates of formation of enantiomeric products without affecting the relative energies of reactants and products (remember that a falsely chiral influence does not lift the degeneracy of P-enantiomeric chiral molecules). *Since the CP-violating interaction responsible for the decay rate asymmetry of the K_L does not lift the degeneracy of the two sets of CP-enantiomeric products (a particle and its antiparticle have identical rest mass if CPT invariance holds), its action is analogous to that of a special type of chemical catalysis in that it affects the kinetics but not the thermodynamics of the reaction* (40).

5. CONCLUDING REMARKS

The use of a falsely chiral influence to induce an enantiomeric excess in a prochiral chemical reaction under kinetic control *via* a breakdown of microscopic reversibility has yet to be demonstrated experimentally. However, the close parallel with the neutral K-meson system where a breakdown of microscopic reversibility has been observed gives confidence in the prediction.

A basic requirement for the generation of the velocity-dependent contributions that must be added to the usual adiabatic potential energy surface of a molecule in the presence of a falsely chiral influence such as collinear electric and magnetic fields is a circular motion of charge in a plane perpendicular to the magnetic field direction as the chiral reaction intermediate evolves. The function of the electric field is to partially align the dipolar molecules in the fluid so that one sense of circulation is preferred over the other for a particular enantiomeric intermediate in a particular orientation. The conrotatory ring closure of a substituted butadiene in collinear electric and magnetic fields has been used to illustrate this (25). Consequently, the electric field is not required if the molecules are already oriented. Thus a particularly favourable situation for a breakdown of microscopic reversibility could be reaction, transport or phase transition processes *far from equilibrium* involving chiral molecules aligned in a crystal, on a surface or at an interface in the presence of a magnetic field (26,29). Gilat (45) has suggested that a magnetic field will disturb the equilibrium of enantiomers aligned at the surface of a bulk racemic solution so that one particular enantiomer will tend to concentrate at the surface; however, contrary to what is implied in reference (3), this idea conflicts with the present analysis which requires aligned enantiomers to remain strictly degenerate at equilibrium in a magnetic field.

Since the processes in prebiotic chemical evolution were likely to have been far from equilibrium, the 'ratchet effect' associated with a breakdown of microscopic reversibility, perhaps in conjunction with a chiral autocatalytic process, would greatly enlarge the range of possible chiral advantage factors. Unfortunately, there is as yet no experimental evidence for a breakdown of microscopic reversibility in chemical processes. A good test would be careful experiments with electric and magnetic fields to see if the parallel and antiparallel arrangements will steer asymmetric reactions under kinetic control towards one or other enantiomeric product. A positive result, no matter how tiny (provided it was routinely reproducible), would prove unequivocally that a breakdown of microscopic reversibility has been induced and would thereby initiate a new era in the study of reaction, transport and phase transition processes involving chiral species, and hence of the origin and role of optical activity in nature (29).

There have been speculations about the possible role of T and CP violation in processes involving chiral molecules which might be significant for biology (46-48), in addition to speculations about the more obvious possible role of P violation (37,38,44). However, CP and hence T violation has only ever been observed in particle $\longleftrightarrow$ antiparticle processes in the neutral K-meson system which is peculiar in that an observable species, K_L, bridges the worlds of matter and antimatter. We might be tempted to deduce from this that CP violation and the associated breakdown of microscopic reversibility

179

could occur in principle in processes involving transformations between a chiral molecule made of matter and its currently inaccessible mirror image made of antimatter (*i.e.* between *CP* enantiomers). However, there are several impediments to this idea (40): for example, such molecule-antimolecule transformations would require a gross violation of the law of baryon number conservation which does not arise in the K-meson system because mesons have baryon number zero. Hence the type of *CP* violation observed in the K-meson system seems unlikely to have any *direct* manifestations in molecular physics, although from our incomplete understanding of *CP* violation at this time we cannot completely rule out such a possibility (47,48).

ACKNOWLEDGMENTS

I thank the Engineering and Physical Sciences Research Council for the award of a Senior Fellowship.

REFERENCES

1. Milton, R. C. deL, Milton, S. C. F., and Kent, S. B. H., *Science* **256**, 1445-1448 (1992).
2. Avetisov, V. A., Goldanskii, V. I., and Kuz'min, V. V., *Physics Today*, July 1991, 33-41.
3. Keszthelyi, L., *Quart. Rev. Biophys.* **28**, 473-507 (1995).
4. Lord Kelvin, *Baltimore Lectures*, London: C. J. Clay and Sons, 1904, p. 619.
5. Pasteur, L., *Rev. Scient.* [3] VII, 2-6 (1884).
6. Barron, L. D., *Chem. Soc. Rev.* **15**, 189-223 (1986).
7. Barron, L. D., *Chem. Phys. Lett.* **123**, 423-427 (1986).
8. Barron, L. D., *Molecular Light Scattering and Optical Activity*, Cambridge: Cambridge University Press, 1982.
9. Barron, L. D., *Mol. Phys.* **43**, 1395-1406 (1981).
10. Gottfried, K., and Weisskopf, V. F., *Concepts of Particle Physics*, Vol. 1, Oxford: Clarendon Press, 1984.
11. Curie, P., *J. Phys. (Paris)* (3) **3**, 393-415 (1894).
12. Wagnière, G., and Meier, A., *Chem. Phys. Lett.* **93**, 78-81 (1982).
13. Barron, L. D., and Vrbancich, J., *Mol. Phys.* **51**, 715-730 (1984).
14. Dougherty, R. C., *J. Am. Chem. Soc.* **102**, 380-381 (1980).
15. Wilczek, F., *Fractional Statistics and Anyon Superconductivity*, Singapore: World Scientific, 1990.
16. Halperin, B. I., March-Russell, J., and Wilczek, F., *Phys. Rev.* B **40**, 8726-8744 (1989).

17. Morrison, J. D., and Mosher, H. S., *Asymmetric Organic Reactions*, Washington D. C.: American Chemical Society, 1976.
18. Mason, S. F., *Molecular Optical Activity and the Chiral Discriminations*, Cambridge: Cambridge University Press, 1982.
19. Mason, S. F., *Int. Rev. Phys. Chem.* **3**, 217-241 (1983).
20. Bonner, W. A., *Origins of Life and Evolution of the Biosphere* **20**, 1-13 (1990).
21. Mason, S. F., *Chem. Soc. Rev.* **17**, 347-359 (1988).
22. Bonner, W. A., *Topics in Stereochemistry* **18**, 1-96 (1988).
23. Wagnière, G., and Meier, A., *Experientia* **39**, 1090-1091 (1983).
24. Mead, C. A., Moscowitz, A., Wynberg, H., and Meuwese, F., *Tetrahedron Lett.* pp. 1063-1064 (1977).
25. Barron, L. D., *Chem. Phys. Lett.* **135**, 1-8 (1987).
26. Barron, L. D., Fundamental Symmetry Aspects of Molecular Chirality, in P. G. Mezey (Ed.), *New Developments in Molecular Chirality*, Dordrecht: Kluwer Academic Publishers, 1991, pp. 1-55.
27. Sakurai, J. J., *Modern Quantum Mechanics*, Menlo Park, California: Benjamin/Cummings, 1985.
28. Tolman, R. C., *The Principles of Statistical Mechanics*, Oxford: Oxford University Press, 1938.
29. Barron, L. D., *Science* **266**, 1491-1492 (1994).
30. Onsager, L., *Phys. Rev.* **37**, 405-426 (1931).
31. Svirko, Y., and Zheludev, N., *Faraday Discuss.* **99**, 359-368 (1994).
32. Kolb, E. W., and Wolfram, S., *Nucl. Phys.* B**172**, 224-284 (1980).
33. Aharoney, A., Microscopic Irreversibility, Unitarity and the H-theorem, in B. Gal-Or (Ed.), *Modern Developments in Thermodynamics*, New York: Wiley, 1973, pp. 95-114.
34. Weinberg, S., *The Quantum Theory of Fields*, Vol. 1, Cambridge: Cambridge University Press, 1995.
35. Lee, T. D., *Particle Physics and Introduction to Field Theory*, Chur: Harwood Academic Publishers, 1984.
36. Hegstrom, R. A., Rein, D. W., and Sandars, P. G. H., *J. Chem. Phys.* **73**, 2329-2341 (1980).
37. Mason, S. F., and Tranter, G. E., *Proc. Roy. Soc.* A**397**, 45-65 (1985).
38. MacDermott, A. J., and Tranter, G. E., *Croat. Chem. Acta* **62** (2A), 165-187 (1989).
39. Jungwirth, L., Skála, L., and Zahradnik, R., *Chem. Phys. Lett.* **161**, 502-506 (1989).
40. Barron, L. D., *Chem. Phys. Lett.* **221**, 311-316 (1994).
41. Hegstrom, R. A., Chamberlain, J. P., Seto, K., and Watson, R. G., *Am. J. Phys.* **56**, 1086-1092 (1988).
42. Woolley, R.G., *Israel J. Chem.* **19**, 30-46 (1980).

43. Amann, A., Theories of Molecular Chirality, in W. Gans *et. al.* (Eds.), *Large-Scale Molecular Systems*, New York: Plenum Press, 1991, pp. 23-32.
44. Salam, A., *J. Mol. Evol.* **33**, 105-113 (1991).
45. Gilat, G., *Mol. Eng.* **1**, 161-178 (1991).
46. Garay, A. S., Broken Symmetries in Physics and their Relevance in Chemistry and Biology, in D. C. Walker (Ed.), *Origins of Optical Activity in Nature*, Amsterdam: Elsevier, 1979, pp. 245-257.
47. Quack, M., *J. Mol. Struct.* **292**, 171-196 (1993).
48. Quack, M., *Chem. Phys. Lett.* **231**, 421-428 (1994).

IV. INTERSTELLAR MEDIUM AND CHIRAL SYMMETRY BREAKING

Chirality in Interstellar Dust and in Comets: Life from Dead Stars

J. Mayo Greenberg

Huygens Laboratory, University of Leiden
P.O.Box 9504, 2300 RA Leiden, The Netherlands

Abstract. Interstellar dust grains have mantles of prebiotic organic molecules. A large fraction of the clouds of interstellar dust grains pass close enough to neutron stars for the circularly polarized ultraviolet radiation to produce a 10% or higher enantiomeric excess in the organic grain mantles. The time between such close passages is about ten times larger than the average lifetime of the molecular clouds so that the most prestellar and protostellar clouds contain predominatly left or right handed prebiotic molecules. Comets as agglomerated interstellar dust preserve the initial enantiomeric excess. Even if only 0.1% of the comet material survives as small comet dust particles which preserve their prebiotic molecules, there could be $\sim 10^{25}$ chances for life to orginate from one of these if it lands in water.

INTRODUCTION

The space between the stars is filled with a mixture of gas atoms and molecules in which are suspended small solid particles called interstellar dust grains or just interstellar dust. The distribution of the interstellar matter is extremely inhomogeneous with the concentrations called interstellar clouds. These clouds come in a wide variety of sizes and densities and, when significantly dense, may collapse to form new stars. Along with some of the stars there are expected to occur planetary systems containing not only planets like those in our own solar system but also comets. The chemical and physical properties of comet nuclei are determined by the original interstellar cloud material - both gas and dust - as it has evolved in the final stages of the collapsing cloud.

Even were we to assume that the comet nucleus composition is <u>exactly</u> that of the presolar interstellar medium, the question remains what <u>is</u> the presolar interstellar medium? There exist neither observations of an interstellar cloud at the stage of collapse just <u>before</u> a star is born nor, as yet, any good theories for the chemistry of the gas and dust in such a very dense cloud. Even if the precometary medium composition were known, the comet nucleus composition is not, at least until there has been a successful comet nucleus sample return mission.

The problem, then, of establishing a precise correspondence between presolar chemistry and comet composition involves, in the former, a forward extrapolation from observed conditions before star formation or a backward extrapolation of the medium after star formation, and for the latter, a time reversed extrapolation of the composition in the comet coma.

The correspondence between molecular abundance ratios in comet nuclei and in the interstellar medium out of which they are formed is thus fraught with potential pitfalls. Both the chemistry and the morphology of the interstellar dust are needed to provide the connections to comet nuclei. In spite of all these qualifications, there appears to be a very close resemblance between comets and agglomerated interstellar dust, as we know it in molecular clouds.

The chemical evolution of interstellar dust results from the complex of interactions occurring in the gas, on the dust, and in the dust. This set of processes leads to the formation of complex prebiotic organic molecules as a major fraction of the dust.

During the traversal of a cloud of dust and gas through our galaxy there are occasions when it passes in the neighborhood of a neutron star (like the source of the crab nebula) with which is associated circularly polarized ultraviolet light. Since some of the organic molecules in the dust have mirror image symmetry they will be differently affected by the circularly polarized light suggesting a mechanism for producing an enantiomeric excess of one handedness over the other (left or right) in the dust.

In this paper we shall outline the stages of chemical evolution in space leading first to complex prebiotic organics in interstellar dust, then to a degree of chirality in these organics and finally to a degree of chirality in the chemical components of comet nuclei and comet dust. Finally we will suggest a mechanism for delivery of small aggregates of chiral interstellar dust to a (our) planetary surface as the prebiotic building blocks for the origin of life on the earth.

GENERAL OBSERVATIONAL DEDUCTIONS

From the observations of the way the light from stars is blocked from the infrared to the ultraviolet (see Figure 1) one can deduce that there are at least three populations of dust grains. The "large" ones which are responsible for the approximately linear increase in extinction through the visible are typically of the order of 0.1 μm in size (mean semi-diameter) and are non-spherical and aligned in space as evidenced by the linear polarization of starlight. The other populations consist of small carbonaceous particles and large molecules (probably similar to polycyclic aromatic hydrocarbons - PAH's) whose sizes are less than, or much less than, 0.01 μm (2,3). Most of the mass of the solid particles - approxmately 90% - is contained in the tenth micron particles.

From infrared absorption spectra of the dust it has been deduced that the large particles which occupy the less dense regions of interstellar space consist of cores of an amorphous silicate material (predominantly containing Si, Mg, Fe, O) mantled by an organic material consisting predominantly of carbon but containing significant amounts of nitrogen and oxygen along with the hydrogen (4,5,6,7). In

186

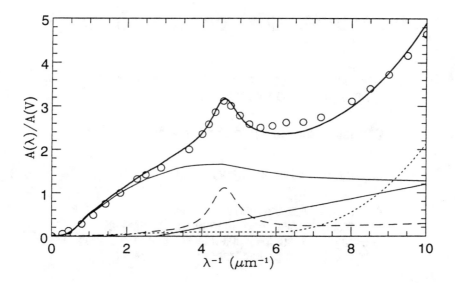

Figure 1. Decomposition of the average interstellar extinction curve into 4 components. Open circles are observations. Thin solid curve is calculated for a population of silicate core - organic refractory particles with mean size ~ 0.1 μm. Linear part may be due to small surface perturbations on core-mantle particles. Dashed curve is due to a population of small ($\leq$ 0.01 μm) carbonaceous particles (once thought to be graphite but this is no longer accepted). Dotted curve is calculated as due to a population of large molecules - PAH's. See ref. 1.

the denser regions of space, the so-called molecular clouds where one observes many molecules in the gas phase, the dust contains outer mantles of frozen volatile molecules such as H_2O, CO, CH_3OH, H_2O, CH_4, NH_3, etc. (8,9).

The spectroscopic identification of the small particles which are responsible for the far ultraviolet extinction is based essentially on a series of infrared emission features which are characteristic of large aromatic molecules classified as polycyclic aromatic hydrocarbons, PAH's (10).

The particles which provide the extinction hump at 216 nm are almost certainly carbonaceous but no specific identification has yet been made.

The cosmic available elements for making solid particles are shown in Table 1. In space some of these elements are in the gas and some in the solids.

Interstellar dust deduced from:

1. Blocking and polarization of starlight: amount + wavelength dependence

2. Infrared spectra (solid state molecules)

3. Particle scattering theory: grain modelling

4. Laboratory experiments on ultraviolet photoprocessing and thermal processing of simple and complex interstellar analog materials (ices). Gas plus dust theoretical modelling

≥ 3 populations of dust grains

a) Tenth micron ($a \approx 0.1\,\mu m$) core mantle particles

Major mass fraction	core	=	silicates
	mantle	=	organic refractory (O.R.) + ices

b) Pure carbonaceous $a \lesssim 0.01\,\mu m$

c) PAHs

etc.

Figure 2. The general observational and theoretical bases for determining the grain populations.

Table 1. Relative cosmic (solar system) abundances of the most common elements

Element	Abundances relative to H
H	1
He	0.079
C	4.90(-4)
N	0.98(-4)
O	8.13(-4)
Mg	0.380(-4)
Si	0.355(-4)
S	0.162(-4)
Fe	0.467(-4)

In Fig. 2 is shown a summary of the grain populations and how they are determined.

BRIEF HISTORY OF DUST GRAINS

There is abundant optical evidence for the variability in the sizes of the large grains which appear to be larger in dense molecular clouds by accretion from the gas. There are also clear indications of the variability in the relative numbers of the different populations of the grains; for example, reduction of the small particles by accretion on the large grains. A detailed description of how these variations occur is beyond the scope of this paper and, in fact, is not yet completely known. But certain main features may be pointed out and, in particular, those which bear most directly on the possible connections between interstellar dust evolution will be emphasized. While most of the chemical and physical evolution takes place in interstellar space there has to be a starting point for the growth of the dust grains because there is no plausible mechanism for creating solid particles directly out of the gas even in the densest molecular clouds - the collision time scales are impossibly large for nucleation to occur. The starting point is the production of silicate cores. From there on one can produce a sequence of possible evolutionary steps leading to all the populations of dust grains. In the following we outline a cyclic picture involving sources and sinks of the dust grains as related to cloud evolution and star formation.

Figure 3 will indicate the sequence of steps and the cyclic evolution of the "large" grains.

Dust in Space
A brief history of a dust grain

Start

a) Small silicate particles condense in the atmospheres of cool stars and are ejected into space (9.7 μm excess infrared emission)

Molecular clouds

b) Silicate particles cool to T=10-15K and act as condensation nuclei for accretion of gas atoms + molecules as a mantle of frost

c) Complex chemical interactions between gas and solid leads to an H_2O and CO dominated grain mantle.

d) Ultraviolet radiation of mantle breaks simple molecules leading to complex molecules (laboratory analog results compared with infrared observations) Photoprocessing $\Big\}$ Lab. analog

Diffuse clouds

e) Star formation blows away clouds. Volatile icy mantles evaporate or are destroyed leaving complex organic refractory mantle only. Pieces of O.R. break off. $\Big\}$ ERA analog

Back to b,c,d,+e and repeat many times until grain consumed by star formation or becomes part of a comet.

Total mean lifetime 5×10^9 yr, = Turnover time for the interstellar medium into and out of stars

Figure 3. Brief history of the cyclic evolution of the "large" interstellar grains.

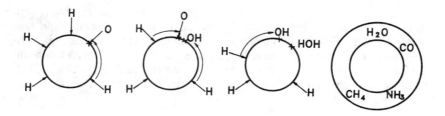

Figure 4. Schematic of the way H_2O mantles are formed by accretion and reactions of oxygen with hydrogen on grain surfaces. Note that a minimum grain size is required for mantle growth because for very small (a $\leq$ 0.01 μm) the reaction energy of O with H will cause the desorption of the molecules (12,13,14,15).

Formation of Nucleation Cores.

From the infrared emission excess of evolved stars at $\sim$ 10 μm and $\sim$ 20 μm which correspond to the Si-O stretch and bend absorptions in silicates we deduce that small silicate particles are nucleated in and are blown out of those stellar atmospheres (11). After emerging in the surrounding interstellar space they immediately cool down to a temperature T $\approx$ 10 K. This temperature is reached as a balance between the absorption of visible and ultraviolet radiation of distant stars and emission in the far infrared. Black bodies as particles placed in the mean interstellar radiation field would be even colder - almost as cold as the 2.7 K residual cosmic black body radiation field - but small particles are a bit warmer because their small size makes it impossible for them to emit efficiently at wavelengths much larger than their linear dimensions.

Accretion of Gas Phase Species.

Ultimately, because of dynamical processes in the interstellar medium, the cold silicate particles find themselves in a cloud of increasing density where the rate of accretion of atoms and molecules becomes much more rapid. Initially the abundant condensable atomic species (O, C, N) attach to the grain surface and as hydrogen atoms impinge, their retention time is long enough for them to traverse the grain surface so many times that they will inevitably make contact and combine with an O, C, or N atom. This process leads rather quickly (in terms of mean cloud lifetimes) to the saturated molecules H_2O, CH_4 and NH_3. See Fig. 4 for a schematic representation of this accretion process. Meanwhile, atomic molecular and ion reactions are taking place in the gas phase leading first to a relatively high abundance of CO, with other molecules generally orders of magnitude less abundant. What is remarkable, is that given the observed ratios of dust to gas, the surface area of the dust grain is such that, in times short compared with cloud life times all the available molecules (and atoms other than H) should have hit and stuck leaving none to be observed in the gas. This is evident for the depletion rate of any molecular species

$$dn(m)/dt = n_d \, \pi a^2 \, v_m \tag{1}$$

from which an e-folding depletion time scale is given by

$$\tau_{dep} = 1/n_d \, \pi \, a_d^2 \, v_m \tag{2}$$

where $n(m)$ is the number density of molecules, n_d is the number density of dust grains, πa^2 is the mean area of a dust grain and v_m is the mean molecular kinetic velocity. For characteristic gas temperatures of about 50 K, $v_m \approx 3 \times 10^4 \mathrm{cm \, s^{-1}}$. Using the mean number density of dust grains with respect to hydrogen, $n_d \approx 10^{-12} \, n_H$, and a mean dust size of $a_d \approx 0.1 \, \mu m$, one derives $\tau \approx 3 \times 10^9/n_H$ yrs. Molecular cloud densities are typically $n_H = 10^4$, and cloud lifetimes are typically $\approx 10^7$ years so that the depletion of all condensable gas phase species occurs well within the cloud lifetime.

Photoprocessing of Mantles.

As already noted there is sufficient ultraviolet radiation in space to heat the dust grains to temperatures $T \approx 10K$. Given the mean ultraviolet radiation field for photons with energy $h\nu > 6eV$ outside molecular clouds of $\Phi_{uv} \approx 10^8 \, h\nu \, \mathrm{cm^{-2} \, s^{-1}}$ (16) we can immediately see that these photons should modify the accreted grain mantles in relatively short times. Given the fact that 6eV is more than adequate to break most molecular bonds, this time can be readily estimated from the assumption that each ultraviolet photon which hits a grain is absorbed; i.e. there is an absorption efficiency of one which is generally true of particle sizes $\approx 0.1 \, \mu m$ for photon wave lengths $\lambda \leq 0.2 \, \mu m$ which corresponds to $h\nu > 6 \, eV$. The number of molecular bonds of size d_m in a spherical grain of radius a is $n_B \sim (4/3 \, \pi a^3)/d_m^3$ and the rate of UV photon absorption is $dn_{h\nu}/dt = \pi a^2 \, \Phi_{h\nu}$. With the given parameters, and using $d_m \approx 3 \times 10^{-8}$ cm we derive a time scale for photoprocessing in the diffuse medium of

$$\tau_{h\nu}^{D.C.} \approx 4/3 \, a_d/d_m^3 \Phi_{h\nu} \approx 160 \, \mathrm{yrs} \tag{3}$$

This is extremely short compared with any cloud time scale but we should more realistically consider the interiors of molecular clouds within which the ultraviolet is substantially reduced by absorption of the dust itself. The UV attenuation (optical depth) at the center of a spherical cloud of radius R=1 parsec (3 x 10^{18} cm ~ 3 light years) and density $n_H \approx 10^4$ cm^{-3} is, in fact, $\tau_{UV} \approx 2 n_d \pi a^2 R \approx 20$ so that very little ultraviolet penetrates to the center of such a cloud. But, there are always other sources of ultraviolet within the clouds themselves. A universal source arises from the penetration of cosmic rays. As a result of such penetration, principally by the protons, the hydrogen molecules are dissociated sufficiently to produce a hydrogen ionization rate within the cloud of $\zeta \approx 5$ x 10^{-17} s^{-1}. Corresponding to this there is an ultraviolet flux (of Lyα photons) of $\Phi_{h\nu} \approx 10^4$ cm^2 s^{-1} which is a base level even in opaque molecular clouds. Given the base level we deduce that the photoprocessing time scale in even dense cloud centers is

$$\tau_{h\nu}{}^{M.C} = 10^4 \, \tau_{n\nu}{}^{D.C.} \approx 1.6 \text{ x } 10^6 \text{ yr} \tag{4}$$

which is still short compared with molecular cloud lifetime.

A general consequence of the photoprocessing is the gradual accumulation of more and more complex molecules in the grain mantles. The earliest stages of this processing are exhibited by the observation of formaldehyde and methanol in dust mantles. Another observational consequence of this photoprocessing is the generation of CO_2 from the accreted CO and of the cyanate ion, OCN$^-$ (see ref. 17 and references therein). But what shows up in the diffuse clouds is the presence of very complex aliphatic and aromatic hydrocarbons to be discussed later.

Grain explosions.

As a consequence of ultraviolet photoprocessing "any" dust mantle molecules initially accreted will have their bonds broken and, because of the very low temperatures, a radical concentration will be built up. The rate of production of radicals depends on the efficiency of radical production; i.e., the fractional number of radicals per ultraviolet photon absorbed. Initially this fraction is close to unity so that the time to achieve, for example, a 1% concentration of radicals in a grain mantle is about one hundredth of the photoprocessing time. We define this as τ_r which in a molecular cloud is then

$$\tau_r{}^{M.C.} \approx 10^{-2} \, \tau_{h\nu}{}^{M.C.} \approx 1.6 \text{ x } 10^4 \text{ yr.} \tag{5}$$

At grain temperatures of 10 K the maximum radical concentration, given by the fractional number of next nearest neighbours is ~ 10%. In practice this concentration is never achieved. However laboratory studies of irradiated low temperature ices have indicated the presence of certainly ~ 1% (5). The importance of this is that the energy stored, if the radicals are allowed to diffuse may recombine by, say, a sudden impulsive heating. The chain reaction releases energy sufficient to blow off the volatile mantle molecules. Laboratory studies show that a sudden temperature rise from 10 K to about 25 K is sufficient to release enough radicals from their potential wells to generate a chain reaction and an explosion (18). In molecular clouds, explosions may be triggered either by turbulent induced collisions between grains at relative velocities of $v_{d-d} \geq 50$ ms^{-1} or by cosmic ray nuclei penetrating the grains. The mean time scale for explosions with the cosmic ray induced process is $\tau_{C.R.} \sim 5 \times 10^5$ yr and for the dust-dust collision is $\tau_{d-d} \approx (5 \times 10^9)/n_H$ yr. Thus the time between explosions in dense clouds ($n_H > 10^4$) decreases in the same way as the accretion time scale (both $\sim n^{-1}$) and there is a state of kinetic equilibrium maintaining the molecules in the gas and preventing them from totally depleting onto the grains. It is important to note that radical diffusion at 10 K is entirely negligible (14).

Organic Refractory Grain Mantle

Referring to the scheme in Fig. 3, the process of accretion, photoprocessing and desorption occurs many times in the molecular cloud phase.

At some point the molecular cloud becomes the site for formation of stars and that part not consumed by star formation is dissipated to return to the diffuse cloud phase. When this occurs all the volatile molecules in the grain mantles are desorbed but what is left is a relatively refractory (by interstellar temperature standards) molecular fraction which is dominated by carbon but which contains oxygen and nitrogen along with hydrogen.

As seen by the comparison of the infrared spectra of the protostars in Orion and of the galactic center in Fig. 5, the former shows the strong H$_2$O ice absorption (OH stretch at 3.1 μm) while the latter has no such feature but _does_ have an absorption at 3.4 μm. This 3.4 μm feature, when studied in detail exhibits the characteristic features of CH stretches in CH$_3$ and CH$_2$ groups in aliphatic hydrocarbons.

An extensive study has been, and is being, made to provide a chemical basis for the now substantially improved observations of the organic refractory grain mantles. In the laboratory, when irradiated ices are warmed up, evaporating away the volatiles, there remains a yellow residue which, by analogy is what has been created in the molecular cloud mantles. A chemical analyusis (GCMS) of these residues exhibits the range of molecules shown in Table 2. We note that many of these are prebiotic and that some have mirror image symmetry. These, of course, are created in equal amounts of left and right handed versions in the laboratory simulation. Further, irradiation of such organics has been provided by sending samples up on the EURECA spacecraft to be exposed for four months to the full effects of the sun's ultraviolet. This is an analog of what happens to the first generation organics in the diffuse cloud phase. A comparison of the infrared spectra of these irradiated residues with that observed in interstellar space is remarkably close (see Fig. 6) and, in fact, is significantly better than any other laboratory created material purported to represent the interstellar mantles.

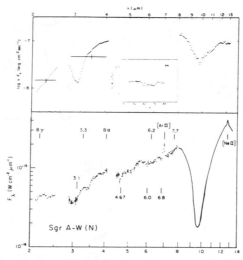

Figure 5. Infrared spectrum of Orion, and the galactic center: the BN source in Orion (upper); the galactic center (lower). Note the absence of an H_2O ice band but the presence of a 3.4 μm feature in the galactic center.

Table 2. GC-MS and HPLC analysis of the products resulting from the simulation of interstellar grain photolysis (taken from ref. 31)

Compound	Amount
Glycolic acid ($HOCH_2CO_2H$)	30.0
3-hydroxy-propionic acid ($HOCH_2CH_2CO_2H$)	1.0
Formamidine ($HCNH(NH_2)$)	0.1
2-hydroxy-acetamide ($HOCH_2CONH_2$)	18
Hexamethylene tetramine ($(CH_2)_6NH_4$)	1.25
Urea (NH_2CONH_2)	2.5
Biuret ($NH_2CONHCONH_2$)	0.07
Oxamic acid (NH_2COCO_2)	0.1
Ethanolamine ($HOCH_2CH_2NH_2$)	0.5
Glycerol ($HOCH_2CH(OH)CH_2OH$)	6.25
Glycine ($NH_2CH_2CO_2H$)	0.4
Oxamide ($NH_2COCONH_2$)	0.4
Glyceric acid ($HOCH_2CH(OH)CO_2H$)	7.5
Glyceramide ($HOCH_2CH(OH)CONH_2$)	18.7
Ethylene glycol* ($HOCH_2CH_2OH$)	6.2

These products were obtained form the photolysis of $H_2O:CO:NH_3$ = 5:5:1. The amount of the average is the average of ten analyses.
* It was not possible to determine whether the ethylene glycol observed using [13]CO contains [13]C.

195

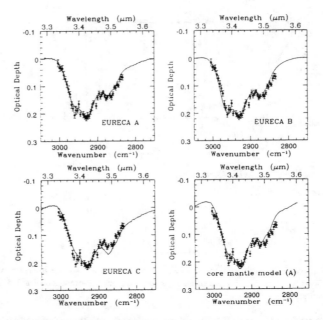

Figure 6. Comparison of the 3.4 μm spectra of EURECA A, B, and C with the Galactic center spectrum of IRS 6E (dots). The lower right is a calculated absorption by a prolate (3:1) spheroidal silicate core - organic refractory mantle particle where the optical constants at 3.4 μm are derived from the EURECA date (20).

The justification for considering both types of organics as representative of interstellar dust organics is that in every molecular cloud phase, the <u>outer</u> layer of organics is first generation while the inner layer is what has remained after about 5×10^7 years in the diffuse cloud phase.

Figure 7 gives a representative model of an average dust grain in a diffuse cloud and an extrapolation to what the average large grain looks like at the last stage of cloud contraction when all condensable molecules and small grains have accreted on the outer mantle.

Table 3 lists the observed molecular species already detected in the grain mantles. The most abundant ones (in the 5-20% range) next to H_2O are CO, CH_3OH, CO_2, H_2CO. The rest are in the < 5% level.

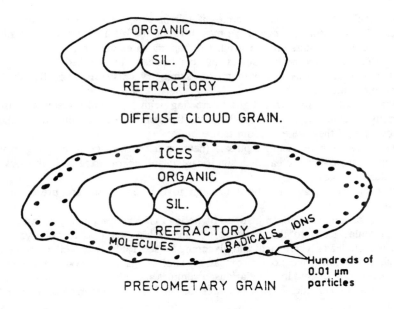

Figure 7. Average diffuse cloud grains and average precometary dust grains on which all volatile ices have accreted and all small particles/large molecules have accreted on/in the ices. The actual relative number of very small particles can range into the thousands.

MIRROR IMAGE MOLECULES IN INTERSTELLAR DUST AND IN COMETS

The evidence that comets are aggregates of interstellar dust (21,22) and that both interstellar dust and comets are largely made of complex organic matter (16,23,24,25) has given a substantial boost to the prescient conjecture of Oro back in 1961 (26) that comets are responsible for life on earth. However, in the *modern* context of the interstellar dust model of comets not just *small* molecules like HCN and H_2CO are present in the comets, but the basic organic building block may already be abundant in the comet nucleus as well as small precursors. Laboratory and observational results show that the organic matter in comets should contain many prebiotic molecules. The ultraviolet photoprocessing of simple ices both in space as well as in the laboratory leads normally to racemic mixtures. A very important suggestion of a possible prebiotic process leading to non-racemic mixtures has been made by Bonner and his colleagues (27,28). They suggest that circularly polarized light from neutron stars on passing clouds of interstellar dust as modelled by Greenberg (28,29) may selectively eliminate one or the other

197

handed molecules in the dust mantles if mirror image molecules are originally present. If this theory is correct we may attribute the birth of life to the death of stars since neutron stars are the residue of supernova explosions.

Bonner's hypothesis requires four conditions to be satisfied:
(1) that prebiotic molecules with mirror symmetry exist on interstellar dust as mantle molecules;
(2) that circularly polarized light interacting with very cold systems (dust is generally at temperatures $T_d \approx 10$ K) has the same effects on mirror symmetry molecules as if they were at room temperature;
(3) that there are enough single pulsars (neutron stars) in space with strong enough UV radiation to have created a significant enantiomeric excess in the dust in individual clouds;
(4) that the dust may have been brought to the earth's surface preserving its chirality.

Criterion (1) has been justified by noting that photoprocessing of interstellar grain mantles leads to the presence of identifiable prebiotic mirror symmetry molecules such as glyceric acid and glyceramide (see Table 2) as well as serine and alanine (30,31,32). We note also that glycerol is made in good abundance by photoprocessing of cold interstellar dust analogues and this may have substituted for ribose in a simpler nucleic acid polymer (33,34,35) as a precursor of a pre-RNA world.

ASYMMETRIC ABSORPTION OF CIRCULARLY POLARIZED LIGHT AT LOW TEMPERATURES - LABORATORY EFFICIENCY.

Criterion 2 has been tested by a laboratory experiment conducted in the same way as interstellar dust photoprocessing and has been fully described elsewhere (30). It consists of a cryogenic cooler which can produce a 10 K temperature at an aluminum block (cold finger) inside a vacuum chamber. Ultraviolet radiation is admitted through a port in the wall of the vacuum chamber. Chemical analysis of the residues shows the presence of such prebiotically relevant molecules as glyceric acid and glyceramide and such amino acids as glycine, alanine and serine (36).

The interstellar dust consists of silicate cores with organic mantles consisting in part of prebiotic molecules and, in denser regions of space, an additional outer mantle of ices accreted from the gas (2). See Table 3.

The question is, can an excess of left- or right handed molecules be produced effectively by circularly polarized light at these low temperatures? We have investigated this by examining the enantiomeric excess produced in a racemic mixture of an amino acid whose vacuum absorbance is located in the vicinity of the Hg resonance line. We deposited a thin layer of tryptophan on the aluminum block, then attached the block within the cryocooler and lowered the temperature to 10 K.

Table 3. Molecules directly observed in interstellar grains and/or strongly inferred from laboratory spectra and theories of grain mantle evolution

Molecule	Comment*	
H_2O	O	M_2
CO	O	M_2
CO_2	O	M_2
CH_3OH	O	M_2
H_2CO	O	M_2
H_2S	O	M_2
NH_3	O	M_2
$(H_2CO)_n$	I	M_2
OCN^-	O	M2
NH_4^+	O	M_2
OCS	O	M_2
CH_4	O	M_2
O_2	I	M_2
S_2	I	M_2
SO_2	I	M_2
complex organic	O	M_1
'silicate'	O	C
'carbonaceous'	(O,I)	B

*O = observed, M_1 = inner mantle, M_2 = outer mantle, B = small bare, I = inferred, C = core

The layer was then irradiated for the order of 100 hours by circularly polarized light at 252.4 nm from a high pressure mercury lamp with a flux at the surface of about 10^{12} photons cm^{-2} s^{-1}. The irradiated samples and an unirradiated sample were then analyzed in an HPLC system, consisting of a Kratos-ABI spectroflow 400 pump, a Rheodyne six-port injector valve with a 100-μl injection loop and a Kratos 773 UV detector operated at 280 nm. Enantiomeric separation was carried out on a 150 x 4.0 mm I.D. stainless Crownpak CR+ (Daicel Industries, Tokyo, Japan) analytical column. The tryptophan enantiomers were eluted from the column with a mobile phase consisting of methanol/perchloric acid (pH 1.4) 25:75, v/v which was degassed ultrasonically before use. The flow-rate was maintained at 1.0 ml/min. Before injection, the sample was dissolved in 50 ml mobile phase and filtered through a 0.2 μm membrane filter.

The relative amounts of D and L tryptophan in each of two irradiated samples were derived using as many as 11 runs per sample through the analytical

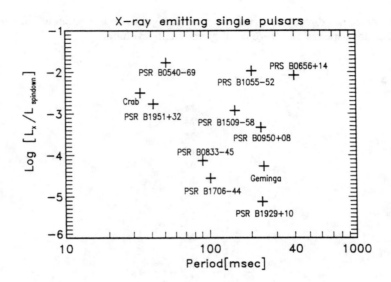

Figure 8. Distribution of pulsars by period.

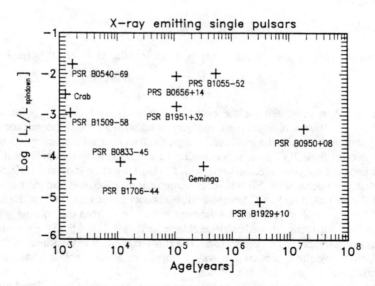

Figure 9. Distribution of pulsars by age.

column to derive the statistics. We found the (relative) increase in the D/L excess (measured by the relative area in the detected column) to be 1.020/.992 = 1.02 or 1.033/.992 = 1.033 for two independent experiments of about 100 and 50 hours each. We have estimated the thickness of the sample to be 5-10 microns which is at least twenty times thicker than the UV penetration depth so that the actual enantiomeric excess in an interstellar dust mantle could be at least 10 times higher; i.e. as high as 50%.These results indicate that the selective destruction (by absorption) of the left relative to the right handed tryptophan is as effective in low temperature solids as in liquids at room temperature. Thus, given an adequate source of circular polarized ultraviolet radiation, we can expect the interstellar grains to store a significant chirality in the organic mantles.

CIRCULARLY POLARIZED ULTRAVIOLET IRRADIATION BY NEUTRON STARS: SPACE EFFICIENCY

Since a neutron star emits left and right circularly polarized radiation equally at opposite poles, if a cloud of dust were to pass by the star symmetrically (equal amounts "above and below") the net chirality in the cloud would be zero; i.e., equal amounts of left handed and right handed excesses.Such symmetrical passage is unlikely so that most clouds of dust would have a net enantiomeric excess.

The question to be answered is whether random passages of clouds past neutron stars will produce a reasonable expectation of a presolar cloud having a significant enantiomeric excess. The answer to the question resolves itself into two parts: (1) What is the total number of active neutron stars in the galaxy? (2) What is the circularly polarized ultraviolet flux from these sources?

It is only recently that we can begin to answer these questions on the basis of observations (37,38). The crab pulsar is the youngest and is *known* to be about 1040 years old. The oldest pulsar detected is approximately 20 million years old (see Figures 8,9). If we assume that type II supernova explosions occur about every 60 years to produce these pulsars we estimate about $2 \times 10^7 / 60 = 3 \times 10^5$ pulsars currently active in the Milky Way with periods extending from 30msec for the crab to about 0.40 sec for PSR B0656 + 14 (see ref. 38 and references therein). The relationship between the age and the period is based on the spin down rate assuming a constant magnetic field. The total luminosity L is proportional to ω^4. We note that, according to the newest information, the ratio of the X-ray luminosity to the spin down power appears to be roughly constant and comparable to that of the crab pulsar. We shall take the crab as our standard pulsar. If we assume that the X-ray luminosity is an extension of the visible-ultraviolet luminosity and if we assume that the spectral distribution follows $\nu F(\nu)$ = constant as is observed for the crab pulsar at a distance of 2kpc then $\nu F(\nu) = 10^{15} H_z \times 10^{-29}$ watts/m$^2 \sim 10^{-11}$erg cm^{-2}s^{-1} which, at 1pc, corresponds to about 4×10^5 erg cm^{-2}s^{-1} or 2.5×10^6 UV photon cm^{-2}s^{-1} in the decade 3-30 ev; i.e. 10ev photons.

In Fig. 10 we picture an interstellar grain (within a cloud) travelling

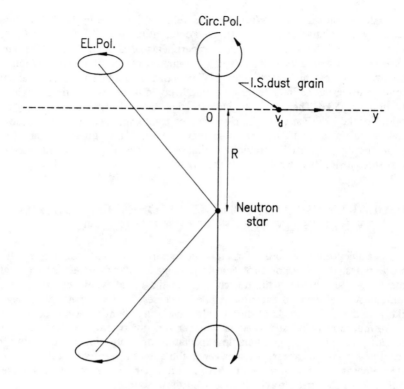

Figure 10. Schematic of an interstellar dust grain passing by a neutron star.

perpendicular to the spin axis of a neutron star at a distance of closest approach R. Along the axis the synchrotron radiation is circularly polarized while at an angle, the light is elliptically polarized. For a dust grain passing above the neutron star the right handed mirror image dust molecules would be selectively destroyed leaving an excess of left handed molecules, and vice versa below. Given an ultraviolet flux of Φ_0 at 1 pc, the net circularly polarized fluence along the path indicated is

$$F = \int_{-\infty}^{\infty} R \, \Phi_0 \, dy/[v_d(R^2 + y^2)^{3/2}]$$

$$= 2 \, \Phi_0(R/pc)^{-2}(R/cm)(v_d/cm \ s^{-1})^{-1} \tag{6}$$

The distance from the crab at which the total fluence, F, acting on a typical grain (relative speed 10 km s^{-1}) corresponds to the laboratory flux acting for about

50 hrs is given roughly by R $\approx$ 40pc.

The neutron stars with fluxes at 1 pc equal to that of the crab nebula at 40 pc are all those with $(\omega/\omega_C)^4 \geq (40)^2$ which leads to all those with $T < 6.3T_C = 210$ ms. Thus all pulsars with ages ≤ 1.7 million years - about 3×10^4 - qualify as candidates to produce a very high enantiomeric excess in a cloud passing within 1 pc. With a mean random cloud speed of 10 km s^{-1} the time between cloud-pulsar collisions (for this number of "effective" pulsars) is $\tau_{C-P} = 3 \times 10^8$ years.

Should a cloud "collide" with a neutron star it will generally only survive an additional few tens of millions of years (well before the next possible encounter) and thus the dust which it contains will either have dissipated, or partaken in the collapse leading to star formation (and possible comet formation), before the next encounter. Thus the probability of finding an individual cloud which has survived a neutron star encounter is $\approx 3 \times 10^7/3 \times 10^8 = 0.1$. In the discussion of neutron star irradiation we have not included the effects of an outer mantle of ices on the dust or of the ultraviolet attenuation in the cloud and/or towards the pulsar. These will introduce some modifications in our predictions but probably not important ones, on the average. The irradiation of the ices by circularly polarized ultraviolet could provide an additional source of enantiomeric excess by photoproduction rather than photodestruction. Furthermore, if we were to relax the enantiomeric excess requirement by a factor of 5, the fraction of clouds with 10% enantiomeric excess increases by a factor of 10; i.e., all presolar clouds would satisfy this requirement. A more precise evaluation of the probability of a presolar cloud having an excess of left (or right) handed molecules can be made but in any case it appears highly probable that a significant fraction of solar systems started off with a significant enantiomeric excess.

PRESERVATION OF CHIRALITY IN COMETS

There exists almost incontrovertible evidence for the preservation of complex molecular interstellar dust ingredients in comets (39). The evidence that the H$_2$O of interstellar dust was not heated during the formation is also well documented and such volatile molecules as CH$_4$ and even CO appear to have been preserved in comets in the kinds of relative abundances deduced from observations of interstellar dust. The model of comets as aggregated interstellar dust leads also to the fact that comets are very fluffy bodies whose overall density is about 0.3 g cm^{-3}. This is consistent with a density of 0.28 g cm^{-3} deduced from non-gravitational motions of comets (40).

The thermal evolution of such a comet has been studied by Haruyama et al (41) and it is generally to be expected that the interior of the nucleus remains at temperatures < 80K during its stay in the Oort cloud and, upon returning as a new comet, is still colder than its initial formation temperature.

DEPOSITION OF CHIRAL INTERSTELLAR DUST ON THE EARTH

There are a number of possible ways to deliver interstellar dust to the earth. One of these is the direct deposition of the dust on the earth while the solar system passes through an interstellar cloud (25,42). The evidence for this possibility is provided by the general one that the solar system had a good chance of passing through such a cloud when the earth was capable of bearing early life forms and secondly by the specific one that recent observations of interstellar dust in the solar system even during our passage through a very tenuous (low density) interstellar medium (43) shows that such dust must certainly have survived while in a passing dense cloud.

However there is no question that comets have provided a major fraction of the organics to the earth's surface and that half or more of the oceans were brought by comets in the first 1000 million years of the earth's existence (44,45,46). Note that the complex organics alone in the interstellar dust model (17,21) ccount for about half the mass of the water (ice) so that one comet of about 1 km radius would deliver about 1% of the earth's total current biomass in the form of these organics. However it must be recognized that a large body of density 1g cm^{-3} impacting an early earth atmosphere even 20 times as dense as the current one would not be cushioned enough to prevent its hitting the earth with a velocity close to its initial one (46).

The heat from such a collision would destroy the initial molecules by pyrolizing them and would almost certainly wipe out any chirality. But comets are not as dense as 1 g cm^{-3} and furthermore they have a morphological structure (in the framework of the interstellar dust model) which gives several reasons to expect far less destructive processes to have occurred. First of all, the low mean density itself of ~ 0.3 g cm^{-3} makes it possible for a (dense) atmosphere to have substantially cushioned the comet nucleus fall. Secondly, the low density loose aggregation leads to the creation within the nucleus of expansion shocks rather than compression shocks (47). This causes the nucleus to fragment and ablate into smaller parts which are better cushioned by the atmosphere. Thirdly, the morphological structure and the physical properties of the core-mantle materials in the individual dust components within the fragments provide a further basis for preserving the non-volatile organics. The heat within the comet fragment is concentrated in the ice mantles of the submicron dust grains because amorphous ice has an extremely low sound speed. The dust ice mantles act like a heat shield for the organics and furthermore, by evaporation, lead not only to cooling of the structure but the gas expansion produces further fragmentation. The final fragments with densities as low as or lower than 0.1 g cm^{-3} like comet dust after all volatiles are removed (22) may float gently to the earth.

Comet Nucleus Impacts on an Early Earth and the Deposition of Prebiotic Seeds

Comet impacts on solar system bodies are well documented and reinforced by the Shoemaker-Levy event of recent years. In the early solar system the frequency of such impacts (by asteroidal debris as well) was such that during th first 5-7 hundred million years, the devastating effects of energy release would unquestionably have prevented the survival of any living creature even if formed. However, in the tail-off period about 3.8 billion years ago the collision frequency (by large nuclei) fell off to about 1 in 6000 years. This still does not leave a very large window for the emergence of life, but the fact that the evidence exists for extensive life already being present at that time (48,49) suggests that given favorable conditions the passage from prebitoic molecules to self replicating organisms must have occurred in a very short time (or times, if a recurrent process was involved).

One favorable condition is that on the early earth there was a dense atmosphere of CO_2 up to as much as 20 atmospheres. This provides a cushioning effect on impacting objects. When taken in combination with the very low density interstellar dust aggregate structure of comet nuclei, this suggests that during an impact a substantial fraction of small comet dust aggregates could have been ablated off the surface. Being of low density, they would then be able to float gently to the surface of the earth and then land in a body of water already present or perhaps brought in by the comet itself. This fact combined with the model of Melosh (50) which predicts the survival of even a small fraction of impactors as unshocked intact fragments, suggests that an enormous quantity of such particles were available for providing a head start on the origin of life. We do not require more than a small fraction of the comet dust particles to survive with their organics and chirality intact. For example, even if only 0.1% of the mass of a comet of 3 Km radius survives intact, as particles like the one shown in Fig. 11 (but with the outer mantles of ice evaporated away) the number landing on the planet surface would be $\sim 10^{25}$!

We now envisage that these comet dust fragments may fulfill the requirements suggested by Krueger and Kissel (51) for chemical thermodynamics to start molecular self-organization. Each comet dust particle consisting of a porous aggregate of submicron silicate cores with organic refractory mantles sitting within a water bath containing nutrients satisfies the conditions for non-equilibrium thermodynamics as in Nicolis and Prigogine (52):

1. Heterogeneous catalytic reaction at large specific area surface (see Fig. 10 and picture the porosity increased from 0.8 to 0.96 after ice sublimation).
2. Localized binding sites of templates and catalysts - the silicate cores.
3. Finite size compartments (holes of $\sim \mu$m) satisfying conditions for stationary state of reaction-diffusion.
4. High concentration of the substrate and products inside the compartments.
5. Lower concentration of substrate and products outside.

Comet Nucleus Impacts on an Early Earth and the Deposition of Prebiotic Seeds

Comet impacts on solar system bodies are well documented and reinforced by the Shoemaker-Levy event of recent years. In the early solar system the frequency of such impacts (by asteroidal debris as well) was such that during th first 5-7 hundred million years, the devastating effects of energy release would unquestionably have prevented the survival of any living creature even if formed. However, in the tail-off period about 3.8 billion years ago the collision frequency (by large nuclei) fell off to about 1 in 6000 years. This still does not leave a very large window for the emergence of life, but the fact that the evidence exists for extensive life already being present at that time (48,49) suggests that given favorable conditions the passage from prebitoic molecules to self replicating organisms must have occurred in a very short time (or times, if a recurrent process was involved).

One favorable condition is that on the early earth there was a dense atmosphere of CO_2 up to as much as 20 atmospheres. This provides a cushioning effect on impacting objects. When taken in combination with the very low density interstellar dust aggregate structure of comet nuclei, this suggests that during an impact a substantial fraction of small comet dust aggregates could have been ablated off the surface. Being of low density, they would then be able to float gently to the surface of the earth and then land in a body of water already present or perhaps brought in by the comet itself. This fact combined with the model of Melosh (50) which predicts the survival of even a small fraction of impactors as unshocked intact fragments, suggests that an enormous quantity of such particles were available for providing a head start on the origin of life. We do not require more than a small fraction of the comet dust particles to survive with their organics and chirality intact. For example, even if only 0.1% of the mass of a comet of 3 Km radius survives intact, as particles like the one shown in Fig. 11 (but with the outer mantles of ice evaporated away) the number landing on the planet surface would be ~ 10^{25}!

We now envisage that these comet dust fragments may fulfill the requirements suggested by Krueger and Kissel (51) for chemical thermodynamics to start molecular self-organization. Each comet dust particle consisting of a porous aggregate of submicron silicate cores with organic refractory mantles sitting within a water bath containing nutrients satisfies the conditions for non-equilibrium thermodynamics as in Nicolis and Prigogine (52):

1. Heterogeneous catalytic reaction at large specific area surface (see Fig. 10 and picture the porosity increased from 0.8 to 0.96 after ice sublimation).

2. Localized binding sites of templates and catalysts - the silicate cores.

3. Finite size compartments (holes of ~ μm) satisfying conditions for stationary state of reaction-diffusion.

4. High concentration of the substrate and products inside the compartments.

5. Lower concentration of substrate and products outside.

Figure 11. A piece of a fluffy comet: Model of an aggregate of 100 average interstellar dust particles. Each particle consists of a silicate core, an organic refractory inner mantle, and an outer mantle of predominantly water ice in which are embedded the numerous very small ($\leq$ 0.01 μm) particles responsible for the ultraviolet 216 nm absorption and the far ultraviolet extinction. Each particle as represented corresponds, in reality, to a size distribution of thicknesses starting from zero. The packing factor of the particles is about 0.2 (80% empty space) and leads to a mean comet mass density of 0.28 g cm^{-3}, and an aggregate diameter of 3 μm. With the ices removed, the dust mass density is $\approx$ 0.1 g cm^{-3} and the porosity is $\approx$ 0.96.

6. Semipermeable boundary - permeable for water and small molecules but closed for large ones.

7. High reaction affinities for the reactants and (auto-)catalysts to form simple RNA's in the compartment.

8. Near equilibrium conditions at moderate temperatures (250-300 K) in the environment to prevent RNA from melting, and substrate species from decay.

The discussion by Krueger and Kissel (44) points out particularly to the chemical composition of the organics predicted by Greenberg (31) - nitrogen as a heteroatom and as reactive precursors of purines and pyrimidines - possibly cytosine. The oxygen is present in aldehydes and carboxylic acids, carbon oxides and, of course, water. Most of the hydrocarbons are unsaturated and the aldehydes may polymerize to form sugars given the right conditions.

The complex chemistry is beyond the scope of this report. But what we want to emphasize is that, because each of the vast multitude of comet dust particles provides its own basis for chemical evolution and because each of them provides an individual molecular and chiral basis for a head start it becomes conceivable that the probability for developing self-organized replicating systems is multiplied enormously - factors of as high as $\sim 10^{25}$ are implied.

We suggest that because of this the time scale for the formation of living organisms may have been much smaller than the 6000 year window between comet impacts and that we may be dealing with time scales encompassed within a human lifetime. If for a particular comet dust grain only years were required we would be able to say that it was from this one seed that all life emerged. As speculative as this appears, it is a way we can imagine that what is a highly improbable event for one of the ensemble of comet dust particles becomes a possible event when available to 10^{25} of them! Just as fish lay millions of eggs so that just a few may survive to adulthood, so we imagine that only a few of the 10^{25} comet dust seeds need become self replicating systems and only one of these in favorable nutrient conditions would have been needed to quickly populate its environs.

ACKNOWLEDGEMENTS

We are grateful for the support by NASA grant NGC 33-018-148 and by a grant from the Netherlands Organization for Space Research (SRON) for research on the organic refractories in space. I thank Aigen Li for his help and Ms. R.E. de Kanter for her patience with me while producing this manuscript.

REFERENCES

1. Greenberg, J.M., Li, A., Evolution of cold, warm and hot dust populations in diffuse and molecular clouds, to be published in: New Extragalactic Perspectives in the New South Africa (eds. D. Block, J.M. Greenberg), Kluwer, Dordrecht, 1996.
2. Greenberg, J.M., Interstellar dust: an overview of physical and chemical evolution, in: Evolution of interstellar dust and related topics (eds. Bonetti, A., Greenberg, J.M., Aiello, S.), Amsterdam, North Holland, 1989, pp. 7-51.
3. Greenberg, J.M., Scient. Amer. **250**, 124-135 (1984)
4. Greenberg, J.M., Chemical and physical properties of interstellar dust, in: Molecules in the Galactic Environment (eds. M.A. Gordon and L.E. Snyder), Wiley, 1973, pp. 94-124.
5. Jenniskens, P., Baratta, G.A., Kouchi, A., de Groot, M.S., Greenberg, J.M., Strazzulla, G., Astron. Astrophys. **273**, 583-600 (1993)
6. Schutte, W.A., The evolution of Interstellar Organic Grain Mantles, PhD Thesis, Leiden, 1988
7. Pendleton, Y.J., Sandford, S.A., Allamandola, L.J., Tielens, A.G.G.M., Sellgren, K., Astrophys. J. **437**, 683-696 (1994)
8. Whittet, D.C.B., in: Dust and Chemistry in Astronomy (Millar, T.J., & Williams, D.A., eds.), IOP Publ., Bristol, 1993, pp. 9.
9. Schutte, W.A., in: The Cosmic Dust Connection, Greenberg, J.M.(ed.), Kluwer, Dordrecht, 1996, in press
10. Léger, A., d'Hendecourt, L.B., Boccaro, N. (eds.). Polycyclic Aromatic Hydrocarbons and Astrophysics. Reidel, 1987.

11. Woolf, N.J., Circumstellar Dust, in: The Dusty Universe (G.B. Field, A.G.W. Cameron, eds.), Watson publ., NY, 1975, pp. 59-69

12. Greenberg, J.M. Interstellar grains, in: Stars and Stellar Systems vol. VII Chapter 6 in Nebulae and Interstellar Matter (eds. B.M. Middlehurst and L.H. Aller), University of Chicago Press, 1968, pp. 221-364.

13. Greenberg, J.M. Interstellar dust: a sink and source of molecules, in: Les Spectres des Molecules Simples au Laboratoire et en Astrophysique, Universite de Liege, 1980, pp. 555-560.

14. Greenberg, J.M., Hong, S.S. The chemical composition and distribution of interstellar grains, in: IAU symp. 60: Galactic radio astronomy, (eds. F.J. Kerr and S.C. Simonson III), Reidel, 1974, pp. 153-177

15. Aannestad, P.A., Kenyon, S.J., Astrophys. J. **230**, 771-781 (1979)

16. Greenberg, J.M., Interstellar dust, in: Cosmic dust (J.A.M. McDonnell, ed.), Wiley, 1978, pp. 187-294.

17. Greenberg, J.M., The interstellar dust model of comets: post Halley, in: Dust in the Universe, (eds. M.E. Bailey, D.A. Williams), Cambridge Univ. Press, 1988, pp. 121-143.

18. Schutte, W.A., Greenberg, J.M. Explosive desorption of icy grain mantles in dense clouds. Astron. Astrophys. **244**, 190-204 (1991).

19. Greenberg, J.M., Mendoza-Gómez, C.X., de Groot, M.S., Breukers, R. Laboratory dust studies and gas-grain chemistry. In: Dust and chemistry in astronomy (eds. T.J. Millar, D.A. Williams) (Proc. conf. Manchester, Jan. 1992), IOP publ. Ltd., 1993, pp. 265-288.

20. Greenberg, J.M., et al., Astrophys. J. **455** (1995) L177-L180.

21. Greenberg, J.M., What are comets made of - a model based on interstellar dust, in: Comets (ed. L. Wilkening), Univ. of Arizona press, 1982, pp. 131-163.

22. Greenberg, J.M. & Hage, J.I., Astrophys. J. **361**, 260-274 (1990)

23. Kissel, J., Krueger, F.R., Nature **326**, 755-760 (1987)

24. Greenberg, J.M., Dust in dense clouds. One stage in a cycle, in: Submillimetre wave astronomy (ed. J.E. Beckman and J.P. Phillips), Cambridge University Press, 1982, pp. 261-306

25. Greenberg, J.M. Chemical evolution of interstellar dust - a source of prebiotic material? In: Comets and the origin of life (ed. C. Ponnamperuma), Reidel, Dordrecht, 1980, pp. 111-127.

26. Oro, J., Nature **190**, 389-390 (1961).

27. Bonner, W.A., Rubinstein, F. in: Prebiological Self Organization of Matter, eds. C. Ponnamperuma, F.R. Eirich, 1990, pp. 35.

28. Bonner, W.A. Origins of Life and Evolution of the Biosphere **11**, 59 (1991).

29. Greenberg, J.M. The chemical composition of comets and possible contribution to planet composition and evolution. In: The galaxy and the solar system (eds. R. Smoluchowski, J.N. Bahcall and M.S. Matthews), Un. of Arizona Press, 1986, pp. 103-115.

30. Agarwal, V.K., Schutte, W., Greenberg, J.M., Ferris, J.P., Briggs, R., S. Connor, C.P.E.M. van de Bult, Baas, F. Origins of Life **16**, 21-40 (1985).

31. Briggs, R., Ertem, G., Ferris, J.P., Greenberg, J.M., McCain, P.J., Mendoza-Gomez, C.X., Schutte, W., Origins of Life & Evolution of the Biosphere **22**, 287-307 (1992)

32. Cronin, J.R., Chang, S., Organic matter in meteorites: molecular and isotopic analysis of the Murchison meteorite, in: The Chemistry of Life's Origins, Greenberg, J.M., Mendoza-Gómez, C.X. & Pirronello, V.(eds.), Kluwer, Dordrecht, 1993, pp 259-299

33. Orgel, L., Quant. Bio. **52**, 9-16 (1987).

34. Marcus, J.N., Olsen, M.A., in Comets in the Post Halley Era. (Newburgn, R.L. et al., eds.), Kluwer, Dordrecht, 1990, pp. 439

35. Weber, A.L., Origins of Life **19**, 317 (1989).

36. Mendoza Gómez, C.X. Complex irradiation products in the interstellar medium. PhD Thesis Leiden, Jan. 1992.

37. Greenberg, J.M., et al., Interstellar dust, chirality, comets and the origins of life: life from dead stars? in: Chemical Evolution: the structure and model of the first cell. (Ponnamperuma, C., Chela-Flores, J., eds.). J. Biol. Phys. **20**, 61-70 (1994).

38. Hermsen, W. et al, Astrophys. J. Suppl. **92**, 559 (1994).

39. Mumma, M.J., Stern, S.A., Weissman, P.R., Comets and the origin of the solar system: Reading the Rosetta stone, in: Planets and Protostars III (E.H. Levy, J.I. Lunine, M.S. Matthews, eds), Univ. of Arizona Press, Tucson, 1993, pp. 1177-1252.

40. Rickman, H., Adv. Space Res. **9**, nr. 3, 59-71 (1989)

41. Haruyama, J., Yamamoto, T., Mizutani, H., Greenberg, J.M., J. Geophys. Res. **E8** 15.079-15.090 (1993).

42. Greenberg, J.M. The largest molecules in space: interstellar dust. In: Cosmochemistry and the origin of life (ed. C. Ponnamperuma), Reidel, 1983, pp. 71-112.

43. Grün, E. et al., Nature **362**, 428-430 (1993).

44. Chyba, C.F., Nature **343**, 129-133 (1990).

45. Chyba, C.F., Sagan, C., Nature **355**, 125-132 (1992).

46. Chyba, C.F., Thomas, P.J., Brookshaw, L., Sagan, C., Science **249**, 366-373 (1990).

47. Zeldovich, Ya.B., Raizer, Yu.P. Physics of shock waves and high-temperature hydrodynamic phenomena, Academic Press, New York, 1967.

48. Schidlowski, M., Nature **333**, 313 (1988)

49. Moorbath, S., J. Biol. Phys. **20**, 85 (1994)

50. Melosh, H.J., Geology **13**, 144 (1985)

51. Krueger, F.R., Kissel, J., Origins of Life etc. **19**, 87 (1989)

52. Nicolis, G., Prigogine, I., Self-organization in nonequilibrium systems, Wiley, 1977

COLD PREBIOTIC EVOLUTION, TUNNELING, CHIRALITY AND EXOBIOLOGY

Vitalii I. Goldanskii

N.N.Semenov Institute of Chemical Physics Russian Academy of Sciences Ul.Kosygina 4, 117334 Moscow, Russia

ABSTRACT. The extra-terrestrial scenario of the origin of life suggested by Svante Arrhenius (1) as the `panspermia` hypothesis was revived by the discovery of a low-temperature quantum limit of a chemical reaction rate caused by the molecular tunneling (2). Entropy factors play no role near absolute zero, and slow molecular tunneling can lead to the exothermic formation of quite complex molecules. Interstellar grains or particles of cometary tails could serve as possible cold seeds of life, with acetic acid, urea and products of their polycondensation as quasi-equilibrium intermediates. Very cold solid environment hinders racemization and stabilizes optical activity under conditions typical for outer space. Neither `advantage factors' can secure the evolutionary formation of chiral purity of initial prebiotic monomeric medium - even being temporary achieved it cannot be maintained at subsequent stages of prebiotic evolution because of counteraction of `enantioselective pressure'. Only bifurcational mechanism of the formation of prebiotic homochiral - monomeric and afterwards polymeric - medium and its subsequent transformation in `homochiral chemical automata' (`biological big bang' - passage from `stochastic' to `algorithmic' chemistry) is possible and can be realized. Extra-terrestrial (cold, solid phase) scenarios of the origin of life seem to be more promising from that point of view than terrestrial (warm) scenarios. Within a scheme of five main stages of prebiological evolution some problems important for further investigation are briefly discussed.

INTRODUCTION

Life can be defined as the form of existence of complex polymeric systems (proteins, polynucleotides etc) able to self-replicate under the conditions of permanent exchange of energy and substance with the surrounding medium (and in accordance with the universal genetic code). Age of life at the earth is restricted at one end by the age of our planet - ca. 4.5 Bln. years - and even more rigidly by the age of the earth's solid crust, ca. 4 by the age of cell-like fossils (3, 4) and the genetic code (5) as ca. 3.8 Bln. years. Thus, the duration of chemical (prebiological) evolution at the earth could not exceed 200 million years and was possibly even much shorter. Two main classes of scenarios of the origin of life on the earth have been suggested.

The first of these, based on the pioneer ideas of Oparin (6) and Haldane (7) and classical experiments of Urey (8), Miller (9) and their followers is the so-called terrestrial or warm scenario. This 'standard model' of the origin of life deals with chemical conversion in primitive atmosphere and/or 'primordial soup' initiated by various physical agents, e.g. light, ionizing radiation, electric discharges, shock waves etc and resulted in the formation of basic building blocks of biopolymers - amino acids, carbohydrates, purines, pyrimidines etc. The hypothesis of the extra-terrestrial origin of life dates back to the 'panspermia' hypothesis of Arrhenius (1). Its modification by Crick (10) as a 'directed panspermia' variant is based on idea of transportation of the genetic material from a certain source populated by a supposed supercivilization.

Discovery of the quantum low-temperature limit of achemical reaction rate and its explanation as the manifestation of molecular quantum mechanical tunneling (2, 11) lead us to the idea of a 'cold prehistory of life' which also belongs to the class of extra-terrestrial scenarios.

However, neither the terrestrial nor extraterrestrial scenarios of the origin of life, in the early stages of development, included any attempts to combine the explanation for the existence of two main properties of living species which are unique from the standpoint of physics - namely, the functional property of ability for self-replication and structural property of their homochirality; that is, the chiral purity of amino acids in all proteins (L-enantiomers) and of sugars - ribose and deoxyribose - in RNA and DNA (D-enantiomers).

The absence of such attempts was particularly surprising in viewof the firm conviction already expressed in 1860 by the discoverer of dissymmetry, Pasteur - 'Homochirality is the demarcation line between living and non-living matter'.

Now, it seems obvious that just the coexistence of these two (and only these two) above properties may serve as Ariadne's thread in the labyrinth of hypotheses of the origin of life, and predetermines the path of prebiological evolution (12-14 and references therein). Detailed combined analysis of two properties of living matter lead us to the following three conclusions:
1)   Chiral purity, typical for the stage of prebiotic evolution was a necessary condition for the subsequent development of self-replication.
2)   Chiral purity of the bio-organic world was achieved neither by the continuous (gradual, evolutionary) accumulation of fluctuations nor as a result of the systematic production of enantiomeric excess by some global external 'advantage factor' (AF) but by a bifurcational type breaking of the mirror symmetry in nature.

3) The sign of chiral purity in the earth's bio-organic world (L-amino acids and D-sugars) is random rather than predetermined by some global AF (e.g. by the non-conservation of parity in weak interactions - particularly weak neutral currents, (WNC).

This paper will not give a detailed description of the arguments which lead us to the above conclusions, nor do we treat here the fundamentals of stereochemistry of optical antipodes (enantiomers), nor are we going to expound the origin and physical sense (and particularly any mathematical formula) of such widely known phenomenon of quantum mechanics as tunneling of particles through potential barriers. The article is devoted mainly to the connections between the peculiarities of chemical reactions at low temperatures (in particular, in space conditions) and the origin of life on earth. It touches also briefly some topics which seem to be among the most interesting subjects for further investigations.

TUNNELING PHENOMENA IN CHEMICAL CONVERSIONS

One of the most important consequences of the wave properties of matter is tunneling - the ability of particles to penetrate the potential barriers whose height exceeds the particles' kinetic energy: classically, the sub-barrier region is forbidden for such particles.

Tunneling starts to become important when the so-called de Broglie wavelength of a particle becomes comparable with the barrier width in such a way that the probability of tunneling decreases with an increase of the width, the height of the potential barrier, and the
mass of the tunneling particle.

The tunneling concept had its first successful applications in nuclear physics (alpha-decay, spontaneous fission, thermonuclear reactions) and later greatly contributed to solid-state physics, electronics and even to cosmology.

Since the late 1920s and early 1930s numerous articles devoted to the role of tunneling in kinetics of chemical reactions have appeared (15, 16). However, for several decades the experimental search for chemical tunneling was restricted to the reactions in solutions, i.e. at comparatively high temperatures. Whereas, the role of tunneling is particularly significant for cryochemical reactions - below the so-called `crossover temperature' $T_c$ (17, 18) which corresponds to qual contributions of classical `over-barrier' transitions described by the Arrhenius law and quantum tunneling, `under-barrier' transitions. For $T<T_c$ tunneling starts to dominate the over-barrier transitions, and the rate of exothermic reactions in many typical cases gradually reaches its low-temperature plateau. Thus, one could expect the existence of significant temperature-independent chemical reactivity of various substances even in the vicinity of absolute zero.

The low-temperature limit of the rate of chemical reactions in the full sense (which implies the rearrangement of atoms and/or molecules, changes of nature, lengths and angles of the valence bonds) was predicted (15-18) and then discovered in 1973 (2) in the case of radiation-induced polymerization of formaldehyde and explained as molecular tunneling (2, 11).

At present, scores of examples of low-temperature kinetical plateaus in processes of various classes are known: intramolecular and intermolecular transfer of hydrogen, transfer of heavy particles or groups in monomolecular (e.g. isomerization) and bimolecular reactions (e.g. numerous chain reactions), rotational tunneling and quantum diffusion. The theory of tunneling in chemical conversion is also well developed. However, these problems lay outside the framework of this article, and it would be better to address the readers to various reviews (11, 19-23) and monograph (24).

COLD PREHISTORY OF LIFE: INTERSTELLAR POLYMERIZATION

The last paragraph of the article (2) on the discovery of a quantum low-temperature limit of a chemical reaction rate reads: `Near absolute zero, entropy factors play no role, and all equilibria are displaced to the exothermic side, even for the formation of highly ordered systems. Therefore, it would be of interest to establish the role of slow chemical reactions at low and ultralow temperatures in chemical and biological evolution (cold prehistory of life?)'.

Indeed, although polymerization or polycondensation processes with the formation of such complex products as polypeptides and polynucleotides should be thermodynamically profitable when chemical equilibria are determined exclusively by changes of enthalpy, i.e. by the thermal effects of chemical reactions, the very problem of approaching chemical equilibrium near absolute zero seemed to have no sense in the framework of classical chemical kinetics. In contrast, tunneling phenomena open up broad possibilities for exothermic cryochemical reactions; in particular, chain reactions triggered by light or ionizing radiation.

The abovementioned hypothesis of cold prehistory of life, i.e. synthesizing rather complex molecules under the combination of space cold and various radiations of cosmic origin postulated in 1973 (2) found the support of Wickramasinghe and Hoyle (25-28) whose interpretations were widely disputed. Starting with claims that formaldehyde undergoes polymerization in interstellar space with the formation of polyoxymethylene and even of polysaccharides, these authors soon came to the hypothesis of `living interstellar clouds' (29) and even of the extra-terrestrial origin of some viruses (30) invading the Earth, e.g. the virus of the influenza pandemia in 1919. These claims met various reasonable objections. Convincing evidence of the existence of polyoxymethylene in space should be dated much later (31-33) and referred to the comparison of data obtained for the coma of comet Halley by heavy-ion analyzers aboard the Giotto spacecraft, laboratory mass spectra and infrared absorption spectra of formaldehyde polymerized at the surface of silicate grains under the irradiation by protons at 20K. Molecular tunneling has been regarded as the most likely mechanism (31-36). Thus, the main stage of the formation of prebiotic polymers (cold seeds of life) is hypothetically represented by the grains of dust of dense interstellar clouds and more explicitly, by the dirty ice mantles surrounding the cores of these grains. The temperature of these grains is estimated as $T_g \div 10 - 20$ **K**. Polymerization can be triggered by ultra-violet radiation - in the outer region of dark clouds - and by long-range cosmic protons - in their depths. The main components of dirty ice mantles, listed in order of their decreasing volatility are: ethane, ammonia, formaldehyde, hydrocyanide, hydrogen isocyanide, water and, finally, polyformaldehyde. The rate of condensation does not depend on volatility while the rate of sublimation (both spontaneous and radiation - induced) decreases rapidly with diminishing volatility. Therefore, the above sequence corresponds also to the increasing steady-state enrichment of dirty ice by various components compared with the gas phase. While the relative abundance of formaldehyde in the gas phase is not larger than several tenths of a per cent of sum of other listed compounds, its abundance in the solid phase - if in the polymer form - could be much higher. Moreover, one should take into account that the cosmic abundance of various elements provides for water and formaldehyde only, the possibility to form the whole mass of dust grains while the maximum total mass of interstellar ethane is half, that of HCN and HNC a quarter, and ammonia one sixth.

Estimates of the rate of gas $\leftrightarrow$ solid processes in both directions (condensation and sublimation) lead to the conclusion that the lifetime of clouds is sufficient to ensure `shuffling of the deck of cards' - a kind of numerous repetitions of condensation and evaporation of the molecules physically absorbed (adsorption heat **D** $\div 0.05$-0.1 eV) which leads to the enrichment of the solid phase by less volatile molecules. The evaporation of the mantle would be strongly hindered by its conversion to a crystal with larger binding energy. There will be no evaporation at all of molecules incorporated into polymer chain. Therefore, one can expect to find a particularly strong increase of formaldehyde abundance caused by its polymerization in the outer layers of interstellar grains. Criteria of the possibility (or impossibility) of the effective accumulation of polymers in the interstellar dust grains include the interplay of three characteristic times - the time of the addition of one new link to the growing polymer chain - τ_0 ; the average time of the absorption of an ultra-violet quantum by each molecule of the outer region of the dark cloud ($\tau_{uv} \div 100$ years) because such absorption can lead not only to the initiation of the polymer chains but also to their rupture, and the

215

life-time of the clouds determined by their collisions or gravitational collapse ($\tau_{cloud} \div 10^5 - 10^7$ years).

Extrapolation to $T_g = 20K$ of our kinetic data on the polymerization of solid formaldehyde at higher temperature (up to 140K) by an Arrhenius plot gives the duration of the addition of several hundred links to the growing polymer chain which exceeds, by many orders of magnitude, not only the τ_{uv}, but also the τ_{cloud} values. The situation is the complete opposite for the experimental data on the rate of polymerization of formaldehyde at the low-temperature kinetic plateau caused by molecular tunneling - τ_0 is shorter than τ_{uv} by about ten orders of magnitude. Thus, tunneling could significantly increase the number of possible low-temperature reactions in dense clouds. One case would be, for example, the possibility of tunneling polymerization at the surface of dust grains with the formation of a very thin (several molecular layers) polymer film around the inner region of dirty ice. Such a film could protect the surface of this inner region from both condensation and sublimation.

For chemical and prebiotic evolution, the reactions of polycondensation in dirty ice mantles with the participation of CH_2O, HCN, HNC, NH_3 and H_2O are of interest. Such reactions could lead to the formation of amino acids, polypeptides, sugars and nucleotide bases (purines and pyrimidines), they are exothermic, but not chain-type.

There are no reasons why there should be a `pure' molecular tunneling mechanism of such reactions - the rate of tunneling falls steeply to a vanishingly small limit with increasing barrier widths and tunneling masses. However, each single step of chemical conversion which represents an elementary gas phase process (such as the reaction $H_2C{=}O + NH_3 \rightarrow H_2C{=}NH + H_2O$), proceeds in the solid as a sequence of many individual or collective conformational rearrangements of molecules, complexes or indeed whole regions of molecular crystals. The collision of a dust grain with a cosmic proton or ultra-violet quantum, or the release of recombination energy at the grain surface can induce the transfer of the `driving' particle, such as the electron, which determines the number of conformational rearrangements. As long as quantum effects open the possibilities of various low temperature chemical conversions it seemed to be of interest to calculate the equilibrium composition of cold interstellar dust grains. In fact, there was no need to take into account the entropy in such calculations, they were based exclusively on enthalpies and the cosmic abundance of H, C, N and O atoms. The maximum release of heat was found to correspond to the formation of acetic acid, urea and certainly also the products of their exothermic polycondensation (37). As gravitational instability develops in the dark dust-gas cloud, a differentiation of matter occurs, and protostar forms. Planetesimals accrete from the dust-gas disc forming around the star, and enlarge to planets together with the formation of meteorites and comets. Consequently, the organic compounds which had formed in the dust-gas cloud can reach the planet by two processes: first, during the accretion of the planet; second, after the planet had formed, through the adsorption of these compounds on the surface of the planet from the surrounding space. The organic compounds which reached the planet in this fashion might then have served as the raw materials for the formation of the `primordial soup'.

The formation of even the most complex biopolymers in interstellar dust grains does not guarantee their onservation during the later formation of new stars and planetary systems, when the dense clouds collapse. One cannot exclude, for example, the strong heating at the surface of planet at certain stages in their evolution. Nevertheless, one should also keep in mind the variety of possible chemical reactants and reactions in `warm' systems depending on their history. For example, polymers created and stabilized by endcapping at low temperatures could survive the consequent warming, but they cannot be formed directly at higher temperatures. Meanwhile the existence of such polymers opens additional higher temperatures. Meanwhile the existence of such polymers opens additional possibilities of chemical conversions not only in cold, but also in warm systems, such as the integration of molecules in shock-wave induced solid-phase reactions (38, 39) and in the vicinity of phase transitions, which proceed during the alternative heating and cooling of reactants (40). Favourable conditions for the latter type would be provided, for example, by multiple transportation of stable organic compounds between circumstellar discs and interstellar clouds (41) and in comets with extended orbits which alternately suffer short heating when approaching the Sun, and prolonged deep cooling, away from the Sun.

COLD PREHISTORY OF LIFE:
FACTORS FAVOURING AN ENANTIOMERIC EXCESS

The introduction emphasized the inseparable ties between the appearance of chiral purity of the predecessors of the bio-organic world and the subsequent origin of life. Although there is no evidence (and no definite conjecture) of the achievement of chiral purity under extra-terrestrial conditions it should be noted that some of the above-mentioned factors typical for such conditions (in particular, the combination of deep space cold and various radiations) work in favour of the achievement of some enantiomeric excess during the cold prehistory of life.

For the better understanding of the differences between the problems of the breaking of mirror symmetry in warm and cold scenarios it would be worthwhile considering the picture of two equal potential wells separated by a barrier (Fig. 1) - one of them is the mirror reflection of the other, so we can

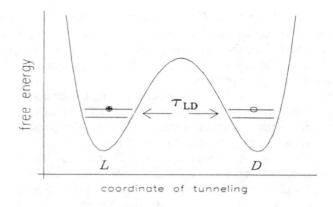

FIGURE 1. Tunneling between **L** and **D** states of chiral molecule: τ_{LD} is the average lifetime of the chiral state.

call one of them the **L**-well and the other the **D**-well and treat them as analogous of enantiomers with **L** and **D** chirality. Neither of these two wells (or two molecules with opposite chirality) represents the stationary state of the system, because tunneling leads to spontaneous **L↔D** conversion, i.e. the system is delocalized between two equal potential wells (the so-called Hund's paradox). In a warm scenario, chemical processes go quite rapidly on the evolutionary time scale and the sign of the chirality is conserved (i.e. **LD**-delocalization can be neglected) during the chemical transformations of the enantiomer molecules. In a `cold' scenario the rates of chemical transformations are exceedingly low, and molecules may repeatedly undergo **L↔D** conversions - either via tunneling or as radiation-induced processes, for which the time scale are comparable with the rates of the chemical processes themself. Accordingly, the **LD**-delocalization in a `cold' scenario may lead to a situation in which the concept of the certain sign of the chirality of an isomer molecule `gets lost' over the time scales of the physico-chemical processes. The first problem associated with the breaking of mirror symmetry in a `cold prehistory of life' is thus the problem of the stabilization of chirality of the molecules of two mirror isomers-enantiomers.

It has been shown (42-45) that if chiral molecules interact with the opticallyinactive medium consisting of a strongly cooled gas of low density, stabilization of the chirality of isomer molecules would be possible over times much longer than the tunneling oscillation time τ_{LD} (Fig.2). Consequently, although racemization does occur, the time scale for the processes increases sharply compared with the racemization time for an isolated particle.

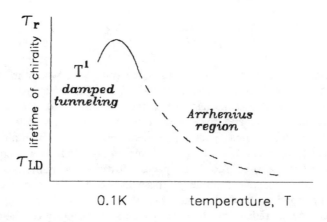

FIGURE 2. Stabilization of chirality in low-temperature gases (racemization time $\tau_r > \tau_{LD}$).

The very fact that the chirality of molecule is stabilized at low temperatures by the interaction with the medium is, understandably, an extremely attractive aspect of a cold scenario but one should also keep in mind that the problem of the deracemization of the medium as a whole arises here. Specifically, each of the isomer molecules is initially in a state with a definite chirality, i.e. in either the **L**-state or the **D**-state. The ensemble of such molecules, however, is probably in racemic state. Therefore, it is quite important to analyze the problem of the stabilization of the optical activity of the ensemble of molecules incorporated in solid low-temperature matrices taking into account the contributions of various types of relaxation processes.

Such analysis shows (46) that at very low temperatures the time τ_S of **L**↔**D** transitions in solid strongly exceeds the time τ_{LD} of 'free' tunneling oscillations, and moreover, τ_S rises in the vicinity of absolute zero with the increase of temperature, and the most effective suppression of racemizing processes in molecular ensembles should be observed at $T_c \div 10$ - $20K$, i.e. just under the conditions typical for dirty-ice mantles or interstellar dust grains (Fig.3).

One of the most widely discussed problems of the origin of life is the search for various local or global advantage factors (**AF**) (12, 14, 47) which can work in favour of preferential accumulation of one of two enantiomers. By definition, the advantage factors (**AF**) characterize the relative difference in the rate constants k^L and k^D for mirror conjugated reactions of two enantiomers:

219

$$g = \left| \frac{k^L - k^D}{k^L + k^D} \right| \quad , \quad 0 \le g \le 1 \ .$$

Several authors have made suggestions concerning AFs which are specific just for the extra-terrestrial (cold) scenario of the origin of life. Gladyshev et al. (48-50) invoked the role of such factors as magnetic, electric and gravitational field of various cosmic objects and the orientation of reacting molecules absorbed at the surface of dust grains. They also mentioned that chiral molecules might persist unaltered because of low temperature and low frequency of intermolecular (or grain) collisions in space. However, no numerical estimations were given and, moreover, none of these hypothetical extra-terrestrial AFs, belong to the range of 'true' physical AFs as defined by Barron (51-53) (i.e. such physical fields, radiations etc which possess the property of 'helicity', which can exist in two enantiomeric forms and transform into its antipode under the effect of spatial inversion, but do not change upon time inversion in combination with any spatial rotation).

Demands for such properties of 'true' AF are satisfied for circularly polarized light as well, of course, as for circularly polarized ultra-violet synchrotron radiation emitted by the neutron star remnants of supernovae. According to the hypothesis of Bonner et al. (14, 54, 55) this radiation has produced chirally enriched organic mantles of interstellar grains in the asymmetric photolysis of 'dirty ice'. As mentioned above there is still no evidence of chiral purity of any space object. However, at least some of the 74 amino acids identified

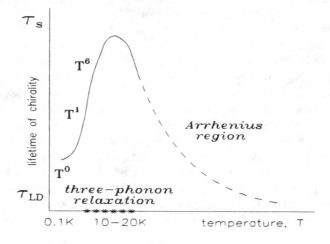

FIGURE 3. Stabilization of chirality in low-temperature solids ($\tau_S > \tau_{LD}$).

220

in the widely described Murchison meteorite (which fell in Australia in 1969) and just those which belong to amino acids typical for proteins of the earth's biosphere (e.g. alanine and glutamic acid) were found to be not racemic but slightly enriched by L-enantiomers (56).

Suspicion that such enrichment was caused by contamination of terrestrial origin seem to be excluded by the recent observation of ^{13}C abundance in these amino acids (57), i.e. the L-enantiomeric excess was accompanied by a ^{13}C enrichment typical of extra-terrestrial organic materials (up to 3%). This result was reasonably interpreted as an argument in favour of the presence of deracemized compounds in the early Solar System (57, 58). Moreover, if the partially deracemized state represents an intermediate stage in the transition to a chirally pure state, the abovementioned results can provide an indication of the timescale required for the process of spontaneous breaking of mirror symmetry (12): according to our estimations (59, 60) it could be as short as 10^6 - 10^7 years, whereas the whole prebiotic stage of the earth's history lasted several hundreds of millions of years.

COLD PREHISTORY OF LIFE: THE BIOLOGICAL BIG BANG

Our detailed analysis of the connections between chirality, the origin of life and evolution (12, 13, 36, 61, 62) is, to a large extent, based on the use of two main parameters - chiral polarization $\eta = (x_L - x_D)/(x_L + x_D)$ (x_L and x_D are the concentrations of L and D enantiomers, and thus η is the dimensionless normalized enantiomeric excess) and enantioselectivity γ of the incorporation of optical antipodes into the growing polymeric chains:

$$\gamma = \omega_{LL} - \omega_{LD} = \omega_{DD} - \omega_{DL}$$

(the first of the two indexes below w designates the end link of the chain; the second the added link).

The treatment of enantioselectivity - as a parameter additional to chiral polarization - is quite important for the analysis of the possibility of a specific (enzyme -like) activity not only for homochiral, but also for regular heterochiral polymers.

It was concluded that heterochiral polymers cannot have any specific activity, either because of strong structural limitations or because of strong kinetic limitation (for a certain 'unique' sequence of chiral fragments). It is even more important that the takeover of organic medium by homochiral chains which possess biochemical functions involves overcoming two successive critical points (13, 62). The first is the formation of chirally pure organic medium ($\eta_{pur} \cong 1$) for providing of a selection of any homochiral informational chains in abiogenic condition, while the second point connects to the appearance of chemical functions with very high specifity, in particular, absolute enantioselective polymers assemblage ($\gamma_{abs} \cong 1$). Overcoming both critical points in some local "seat of life" is necessary for the origin of self-replicating macromolecular patterns. The exponentially fast propagation of such patterns looking like an explosive chain reaction (so-called "Biological big bang") might be a reason for the formation of the early biosphere.

Taking into account all the above circumstances one can represent the main stages of prebiological evolution by the scheme (Fig. 4).

Two features are fundamentally important: first, a strong mirror symmetry breaking in the organic medium preceded the polymeric takeover and predetermined the formation of omochiral polymers. Second, the chiral purity of the medium had to be maintained not only at the stages of primary homochiral polymers but also subsequently, during the formation of structures and functions possessing the biochemical level of complexity. Only after the appearance of structures having enantiospecific activity the requirement of a chirally pure medium can be dropped.

Just the need to reach the second critical point is the main cause which makes the maintaining of chiral purity of a medium impossible, due to any AFs, however strong, because while the AF enriches the monomeric environment by some enantiomer, the enantioselective polymeric takeover has the opposite effect (the so-called stereoselective or enantioselective pressure (13)) that is, the impoverishment of a monomeric system by that enantiomer, inhibition and finally complete blocking of the effect of AF.

Thus, processes in which symmetry breaking depends exclusively on the action of an AF and occur by gradual accumulation of asymmetry, could not be crucial for early steps of prebiotic evolution. One needs a fundamentally different type of processes that can effect a strong mirror symmetry breaking even without an AF and can withstand the enantioselective pressure throughout the stages of formation of enantiospecific activity.

Processes of the `bifurcation` type, well known from the theory of non-equilibrium phase transition, possess the required properties. In this case the connection between chiral polarization η of monomeric medium and enantioselectivity γ of the polymeric takeover is described by universal equation of bifurcation type (62):

$$-\eta^3 + (1-\rho)\eta - \gamma(1-\eta^2) = 0$$

where $\rho \geq 0$ depends on kinetics of transformations of L- and D-monomers in some organic area.

222

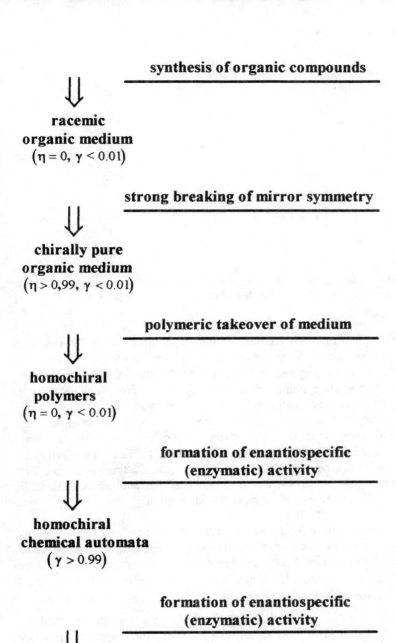

FIGURE 4. Five main stages of prebiological evolution.

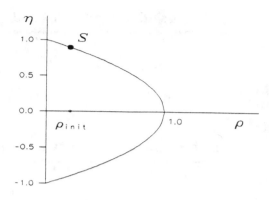

Figure 5. Bifurcation diagram before polymeric takeover. The point **S** illustrates the set ρ_{init} of starting conditions for the beginning of prebiotic transition.

When $\gamma=0$, the first two terms in the equation describe the symmetry breaking of monomeric medium before its polymeric takeover. Truly, the racemic state of chemical system becomes unstable at the bifurcation point $\rho_c=1$ and for $\rho_c<1$, there are two stable states with broken symmetry. Such behavior of a chemical system can be illustrated by bifurcation diagram (Fig.5).

Note, that the value of controlling parameter ρ_{init} should be much less than unity for strong mirror symmetry breaking connected to overcoming of the first critical point of prebiotic evolution (point S on Fig.5). Since the parameter ρ describes the degree of deviation of chemical transformation from thermodynamic equilibrium, by this is meant that the starting conditions for primary polymeric takeover of organic medium should be strongly non-equilibrium.

The last term in our bifurcational equation corresponds to enantioselective pressure which looks as some compensating "chiral field" (Fig.6). As the result, the enantioselective pressure shifts the position of the bifurcation point ρ_c towards ρ_{init}. The shift depends on the selectivity of a polymeric takeover and when γ trends to absolute enantioselectivity $\gamma_{abs} \cong 1$, the bifurcation point $\rho_c(\gamma)$ shifts to 0.

Therefore for any $\rho_{init} \ll 1$ such a value of γ_{cr} exists when the bifurcation point ρ_{init} reaches ρ_{init}. This is most critical situation for prebiotic transition, because the chirally pure state, which had been formed at the beginning of the prebiotic transition, becomes here unstable and the monomeric environment is racemized. Could an enantiospecific function be formed before this critical moment? This is the major question now.

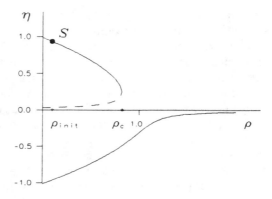

Figure 6. Bifurcation diagram under evolution of specific catalytic functions.

The analysis shows (13, 61) that if the condition for overcoming the first critical point is satisfied, the chirally pure state loses its stability just at the moment when enantioselectivity of polymeric takeover reaches the value γ_{abs}.

It is not less important that biological big bang is accompanied by the racemization of the monomeric environment while the broken mirror symmetry for the living matter is preserved in the form of homochirality of macromolecules possessing an enantiospecific activity. Just this specifity of Biosphere is observed on the earth.

Thus we come to the conclusion that the concept of prebiotic evolution based on the bifurcation with symmetry breaking far from the thermodynamic equilibrium satisfies both important features of the prebiotic transition.

The appearance of strongly non-equilibrium conditions resulting in the bifurcation with symmetry breaking in some "warm" or "cold" organic area seem any principle restrictions.

COLD PREHISTORY OF LIFE: PUZZLES AND CONSIDERATIONS FOR THE FUTURE

Here we come to the first of the still unsolved key problems of any scenario of the origin of life - what was the mechanism of bifurcation with mirror symmetry breaking in the racemic organic medium? This needs a deracemizing process based on the co-operative (non-linear) interactions of enantiomers which can lead to the self-organization of chirality in the system; mirror-symmetry breaking occurs spontaneously as soon as critical conditions are reached. Although many schemes of interactions have been suggested (i.e. the

225

pioneering scheme of Frank (63) and the generalized treatment of this problem by Morozov (64)) which fulfil such requirements, neither of these schemes have suggested any concrete, specific chemical reactions for a liquid phase. The situation looks more promising in the solid phase since cooperative nteractions in solids are more pronounced than in liquids. Let us refer here to the most detailed example of spontaneous generation of optical activity in 1,1'-binaphthyl (65,66). This chiral hydrocarbon exists in two crystalline forms, one of them an eutectic mixture of individual optically pure crystals, the other a racemic compound. The generation of optical activity was found to be a strongly non-equilibrium process. Recently it has been proposed that a phenomenological description of these results might be caused by the non-linear kinetics of the inversion of mirror isomers (67), and the experimental data were reproduced by computer simulation. Kondepudi (68) have observed and investigated a peculiar example of a total spontaneous resolution of L-and D-crystals of sodium chlorate when their solutions were stirred during crystallization.

Thus, the strong contribution of co-operative interactions to various chemical and/or physico-chemical conversions in solids (particularly, at low temperatures - contrary to warm liquids) speaks in favour of the extra-terrestrial (cold) scenario of the origin of life. One should note here that the crucial role of the driving force of causality in co-operative stochastic systems can be played by correlated fluctuations treated in detail in recent publications (69-72). However, in view of the above scheme of prebiological evolution there remains an important intriguing problem. It is not sufficient just to find the mechanisms of spontaneous breaking of mirror symmetry in any scenario of the origin of life, it is also necessary to explain the way of formation of enantiospecific enzymatic activity of homochiral polymers. Indeed, even if the homochiral polymers of extra-terrestrial origin ('cold seeds of life' (34)) arrived at the racemic earth before acquiring enzyme-type activity (biological **AF**), they would have been racemized in the same way as amino acids from proteins after the death of their possessors (73-75). Therefore, any comprehensive hypotheses of `panspermia' should include some ideas about the mechanism of extra-terrestrial formation of structures having enantiospecific activity and their subsequent delivery to the earth (whose prebiotic chemical medium could, at that stage, be racemic). Two more points need mentioning. What were the chronologically first prebiotic compounds which did acquire chiral purity and, afterwards, enzymatic activity? Were they amino acids or carbohydrates (sugars)? Priority of chiral purity of amino acids would mean the priority of proteins as enzymes while the priority of chiral purity of sugars (e.g.ribose) would serve as a serious argument in favour of a RNA-world, with the ribozyme as an initial enzyme.

From my point of view the priority of chiral purity of carbohydrates seems the more probable. Indeed, among the 74 amino acids found in the Murchison meteorite (i.e. of natural origin) only 20 are found on the earth as components of proteins and thus chirally pure (in fact 19 of them because glycine is optically inactive). On the contrary, practically all natural carbohydrates possess chiral purity and, moreover, they have several asymmetric carbon atoms per molecule, which could widen the possibilities of co-operative interactions in the solid phase.

Another point is connected with the problem of the sign of chirality of the earth's biosphere. As was briefly mentioned in the introduction, we assert that the choice of this sign was random rather than predetermined by some global AF (e.g. by weak neutral currents). However, one can put a question, why - in spite of the accidental choice of the sign of chirality in the course of prebiotic spontaneous breaking of mirror symmetry - this sign turns to be the same all over the earth.

This question has been discussed in detail (12) and the answer is very simple, it is determined by the interplay of two characteristic times ,- the time required for the takeover of the entire biosphere by the very first domain of life and the expectation time τ_{ex} for the formation of the embryo of the new phase with the opposite symmetry.

The expectation time is a criterion which expresses the possibility that life arises under some certain conditions or others, on some certain celestial objects or others. Specifically, if we know the parameters of the medium which are characteristic of the given celestial object (e.g. a planet, a dust-gas cloud, etc) we can estimate the expectation time for the beginning of an irreversible deracemization of the medium. If we find that τ_{ex} exceeds the age of celestial object, we should acknowledge that the appearance of life is impossible in this case, since a eracemization of the medium - a necessary prerequisite for the appearance of living structures - does not occur.

The concept of an expectation time for the beginning of a breaking of mirror symmetry makes it possible to move on the solution of yet another problem which is being widely debated; was the appearance of life on the earth a consequence of single event or the result of competition among several prebiospheres which arose independently? We can approach the resolution of this question along the path of a deracemization of the prebiosphere. The appearance and coexistence of natural habitats within which the chirality of the organic matter has different signs are equivalent to the appearance of a set of competing prebiospheres, i.e. to a multiplicity of nucleation events. If, on the other hand, the deracemization process generated by a critical fluctuation spanned the entire planet, the 'act of organization' was unique.

Let us consider a gedanken experiment: we assume that the racemic primeval soup' has reached a critical state required for a transition to chiral order and occupies two habitats which communicate with each other (e.g. northern and southern hemispheres). We also assume that an advantage factor is operating in each habitat and that nature and measure of the advantage factors are identical in two regions, but the signs of the advantage factors are opposite. The action of the advantage factor leads to an excess of one of the antipodes

- but different ones in the two habitats. If the mixing of matter is sufficiently intense, however, the medium as a whole will remain racemic. Since such a medium is in a state which is unstable with respect to fluctuations of the chiral polarization, the very first critical fluctuation which appears after a time τ_{ex}, marks the beginning of a deracemization process.

Since the appearance of a critical fluctuation is equally probable in each of the two habitats, we should recognize that the sign of the chirality of the prebiosphere will be determined with equal probabilities by the sign of the advantage factor of each of the habitats. In other words, it will be random for the prebiosphere as a whole. It is not difficult to see that the result of this gedanken experiment does not depend on the number of habitats which we consider or on the number of the local advantage factor. Since the `colonization' of a medium by a critical fluctuation occurs in a time of 10^2 - 10^4 years (mixing due to flows, etc) and since this time is substantially shorter than the expectation time for the critical fluctuation, which is the next to follow the first, 1-10 million years (12, 59), we can confidently say that the deracemization of the prebiosphere was a result of a single event, rather than a consequence of a set of local deracemization events (this is true regardless of whether there is an advantage factor).

ACKNOWLEDGMENT

The author is greatly indebted to his colleague and friend Dr. Vladik A. Avetisov for valuable remarks and advices.

REFERENCES

1. Arrhenius, S., *World in the Making,* New York: Harper and Row, 1908.
2. Goldanskii V.I., Frank-Kamenetskii, M.D. and Barkalov, I.M., *Science* 182, 1344 (1973).
3. Schopf, J.W. (Ed), *Earth's Biosphere. Its Origin and Evolution,* Princeton, NY: Princeton University Press,.1983.
4. Schidlowskii, M., *Nature* 333, 313 (1988).
5. Eigen, M., Lindemann, B.F., Tietze, M, Winkler-Oswatitsch, R., Dress, A. and Von Haeseler, A., *Science* 244, 673 (1989).
6. Oparin, A.I., *The Origins of Life on the Earth,* Edinburgh: Oliver and Boyd, 1957.
7. Haldane, J.B.S., *Ration Ann.* 148, 3 (1929).
8. Urey, H.C., *Proc. Nat. Acad. Sci.* 38, 351 (1952).
9. Miller, S.L. and Orgel, L.E. , *The Origins of Life on the Earth,* NY Prentice-Hall: Englewood Cliffs, 1974.
10. Crick, F., *Life Itself: Its Origin and Nature,* NY: Simon and Schuster (1981).
11. Goldanskii, V.I., *Nature* 279, 109 (1979).
12. Goldanskii, V.I., Kuz'min, V.V., *Sov. Phys. Uspekhi* 32, 1 (1989).
13. Avetisov, V.A., Goldanskii, V.I., *BioSystems* 25, 141 (1991).

14. Bonner, W.A., *Orig. Life Evol. Biosphere* **21**, 59 (1991).
15. Bell, R.P., *The Proton in Chemistry*, (2nd Edn) Ithaca, NY: Cornell University Press, 1973.
16. Bell, R.P., *The Tunnel Effect in Chemistry*, London: Chapman and Hall,. 1980.
17. Goldanskii, V.I., *Doklady AN SSSR* **124**, 1261 (1959).
18. Goldanskii, V.I., *Doklady AN SSSR* **127**, 1037 (1959).
19. Goldanskii, V.I., *Chem. Scripta* **13**, 1 (1978).
20. Goldanskii V.I., Trakhtenberg, L.I. and Fleurov, V.N., *Sov. Sci. Rev. B (Chem.)* **9**, 59 (1987).
21. Goldanskii, V.I., Benderskii, V.A. and Trakhtenberg, L.I., *Adv. Chem. Phys.***75**, 349 (1989).
22. Benderskii, V.A. and Goldanskii, V.I., *Int. Rev. Phys. Chem.* **11**, 1 (1992).
23. Benderskii, V.A., Goldanskii, V.I. and Makarov, D.E., *Physics Reports* **233**, 195 (1993).
24. Goldanskii, V.I., Trakhtenberg, L.I. and Fleurov, V.N., *Tunneling Phenomena in Chemical Physica*, Gordon and Breach., 1989.
25. Wickramasinghe, N.C., *Nature* **254**, 452 (1974).
26. Wickramasinghe, N.C., *Mon. Not. R. Astr. Soc.* **170**, 111 (1975).
27. Hoyle, F. and Wickramasinghe, N.C., *Nature* **258**, 610 (1977).
28. Hoyle, F. and Wickramasinghe, N.C., *Nature* **306**, 420 (1983).
29. Hoyle, F. and Wickramasinghe, N.C., *Lifecloud*, London: Dent and Sons, 1978.
30. Hoyle, F. and Wickramasinghe, N.C., *Diseases from Space*, NY:Harper and Row,.1979.
31. Mitchell, D.L., Lin, R.P., Anderson, K.A., Carlson, C.W., Curtis, D.W., Korth, A., Reme, H., Sauvaud, J.A., D'Uston, C. and Mendis, D.A., *Science* **237**, 626 (1987).
32. Huebner, W.F., *Science* **237**, 628 (1987).
33. Moore, M.H. and Tanabe, A.T., *Astrophys. J.* **365L**, 39 (1990).
34. Goldanskii, V.I., *Nature* **258**, 612 (1977).
35. Goldanskii, V.I., *Nature* **269**, 583 (1977).
36. Goldanskii V.I., Kuz'min, V.V. *J. Brit. Interplanet Soc.* **43**, 31 (1990).
37. Goldanskii, V.I., Gurvich, L.V., Muzylev, V.V, and Strelnitskii, V.S., *Scientific Informations of the Astronomical Council of the Academy of Sciences of the USSR* **47**, 3 (1981).
38. Barkalov, I.M., Adadurov, G.A., Dremin, A.N., Goldanskii, V.I., Ignatovich, T.N., Mikhailov, A.N, Talrose, V. L. and Yampolskii, P.A., *J. Polym. Soc. C*, **16**, 2597.
39. Goldanskii, V.I., Ignatovich, T.N., Kosygin, M.Yu.and Yampolskii, P.A., *Doklady Biochemistry* (Proc. Acad. Sci. USSR) **297**, 218 (1972).
40. Kargin, V.A., Kabanov, V.A. and Papisov, I.M., *J. Polym. Sci.*, **4**, 767 (1964).
41. Sagan, C. and Khare, B.N., *Nature* **277**, 102 (1979).
42. Simonius, M., *Phys. Rev. Lett.* **40**, 980 (1978).
43. Harris, R.A. and Stodolsky, L., *Phys. Lett. B* **78**, 313 (1978).
44. Harris, R.A. and Stodolsky, L., *Phys. Lett. B* **116**, 464 (1982)..
45. Harris, R.A. and Stodolsky, L., *J. Chem. Phys.* **78**, 7330 (1983).
46. Berlin, Yu.A., Gladkov, S.O., Goldanskii, V.I., and Kuz'min, V.V., *Sov. Physics - Doklady* **306**, 844 (1989).
47. Goldanskii, V.I., Kuz'min, V.V., *Nature* **356**, 114 (1991).
48. Gladyshev, G.P., *The Moon and the Planets* **19**, 89 (1978).
49. Gladyshev, G.P. and Khasanov, M.M., *J. Theor. Bio.* **90**, 191 (1981).
50. Gladyshev, G.P. and Kitaeva, D.Kh., *Zh. Vses. Khim. Obshch. im. Mendeleeva* **5**, 625 (1990).
51. Barron, L.D., *J. Am. Chem. Soc.* **108**, 5539 (1986).
52. Barron, L.D., *Chem. Phys. Lett.* **123**, 423 (1986).
53. Barron, L.D., *Chem. Phys. Lett.* **135**, 1 (1987).
54. Rubenstein, E., Bonner, W.A., Noyes, H.Y. and Brown, G.S.,: *Nature* **306**, 118 (1983).

229

55. Bonner, W.A. and Rubenstein, E., *BioSystems* **20**, 99 (1987).
56. Engel, M.H. and Nagy, B., *Nature* **296**, 837 (1982).
57. Engel, M.H., Macko, S.A. and Silfer, J.A., *Nature* **348**, 47, (1990).
58. Chyba, C.F., *Nature* **348**, 113 (1990).
59. Morozov, L.L., Kuz'min, V.V. and Goldanskii, V.I., *Doklady-Biophys. (Proc. Acad. Sci. USSR)* **274**, 55 (1984).
60. Morozov, L.L., Kuz'min, V.V. and Goldanskii, V.I., *Doklady-Biophys. (Proc. Acad. Sci. USSR)* **275**, 71 (1984).
61. Avetisov, V.A., Goldanskii, V.I., *Phys. Lett. A* **172**, 407 (1993).
62. Avetisov, V.A., Goldanskii, V.I. , Kuz'min, V.V. *Physics Today* **44**, 33 (1991).
63. Frank, C.F., 1953: Biochem. Biophys. Acta., 11, 459.
64. Morozov, L.L., *Origin of Life* **9**, 187 (1979).
65. Pincock, R.E. and Wilson, K.R., *J. Am. Chem. Soc.* **93**, 1291 (1971).
66. Pincock, R. E. and Wilson, K.R., *J. Am. Chem. Soc.* **97**, 1474 (1975).
67., *Chem. Phys. Lett.* **184**, 526 (1991).
68. Kondepudi, D.K., Kaufman, R.J. and Singh, N.. *Science* **250**, 975 (1990).
69. Berlin, Yu.A., Drobnitsky, D.O., Goldanskii, V.I., and Kuz'min, V.V., *Phys. Rev. A*, **45**, 3547 (1992).
70. Berlin, Yu.A., Drobnitsky, D.O., Goldanskii, V.I., and Kuz'min, V.V., *BioSystems* **26**, 165 (1992).
71. Berlin, Yu.A., Drobnitsky, D.O., Goldanskii, V.I., and Kuz'min, V.V., *Chem. Phys. Lett.* **189**, 316 (1992).
72. Goldanskii, V.I. and Mikhailov, A.S., *Phys. Lett. A.* **176**, 6 (1993).
73. Bada, J.L., *Adv. Chem. Ser.* **106**, 309 (1971)
74. Bada, J.L. and Schroeder, R., *Naturwissenschaften* **62**, 71 (1975).
75. Bada, J.L., in *Chemistry and Biochemistry of the Amino Acids*, London: Chapman and Hall,. 1985.

On the Possibility that Supernovae Influence Homochiral Organic Molecule Formation

David B. Cline

Univ. of California Los Angeles, Dept. of Physics and Astronomy

405 Hilgard Avenue, Los Angeles, CA 90095-1547 USA

Abstract. The origin of the homochirality of the organic molecules in living material remains a mystery. We propose here that the supernova(s) that initiated the solar system also produced the mechanism for the chiral symmetry breaking in the organic material in the interstellar medium. The major process would involve either $\bar{\nu}_e$ from a supernova explosion or the $A\ell^{26}$ produced in the debris in conjunction with the weak neutral current process. Both give rise to chiral positrons and interactions with the organic material in the prebiotic medium. We describe some possible tests of this concept. We note that in this picture, all life forms in the Universe based on hydrocarbons should have the same chiral structure.

I INTRODUCTION

There is increasing evidence that the formation of the solar system was initiated by one or more nearby supernovas. This evidence is obtained largely from the isotopes found in pre-solar grains in old meteorites and from dynamical models of the formation of stars and planetary systems. In this paper, we address the question of the origin of a homochiral prebiotic organic medium in connection with supernova emissions. We find that most of the emission ($\bar{\nu}_e, A\ell^{26}$) would provide a mechanism for selectively inducing a homochiral structure in the organic materials in the interstellar medium (ISM) and elsewhere.

For more than a century, there has been evidence for the chiral nature of life forms on earth. Pasteur was among the first to point this out (1848-1880), and the universal nature of chiral symmetry breaking in DNA and RNA is now very well established for all life forms. With the discovery of parity violation in charged current reactions in 1956 and of the weak neutral current

(WNC) in 1973, two universal symmetry-breaking processes (WNC and β-decay) were uncovered that could have effected the handedness of DNA and RNA. The main problem is the extremely small symmetry-breaking effects ($\Delta E/k_B T \sim 10^{-17}$). There are, however, plausible non-linear mechanisms that could have amplified this small symmetry-breaking phase transition up to the full symmetry-breaking level observed in life forms. However, there is a long-standing controversy as to whether these non-linear effects are actually large enough to have determined the selection of the handedness of life [1–11].

Recently there has been increasing interest in the chiral nature or handedness of biomolecules. In fact there are some who claim that the complex biomolecular structure of life must have arisen from a "chiral pure" medium [12,13]. This concept, combined with the likelihood that the period on Earth for life to have originated seems to be sometime between 3.8 and 3.5 billion years ago, leaves a small window of 300 million years or less for life to have emerged from the prebiotic medium. Indeed, some speculate the time could be less than 10 million years.

In this paper, we discuss the possible effects of a chiral impulse that could have come from the intense pulse of neutrinos or radioactive debris interaction from a very close supernova explosion or a sudden release of a large amount of radioactive material near Earth to aid in the formation of a chirally pure "prebiotic" medium for the origin of life [14].

II CHIRAL SYMMETRY BREAKING AND BIFURCATION

In contrast to the gradual build up of symmetry breaking, we propose here another mechanism resulting from the effects of a nearby supernova or an impulse of β-emitters ($A\ell^{26}$). Initially this might seem to be an unimportant effect that has been ignored in the past literature on this subject but, with increasing interest currently, more attention should be paid to this issue. Here we present some results based on our previous work [8,14]. We believe it is most likely that this interaction occurred in the ISM where organic-loaded dust particles helped form the enormous cloud that subsequently formed the solar system.

To simulate this intense pulse interaction combined with some reasonable condition, we added a time-dependent quasi δ-pulse, which gives very large, but not infinite, amplitude to the small bias g. We are assuming that both the WNC and the impulse of chiral symmetry-breaking effects occurred. Then the first-order stochastic equation [1,8]

$$\frac{d\alpha}{dt} = -A\alpha^3 + B(\lambda - \lambda_c)\alpha + \epsilon^{1/2} f(t) + Cg \quad , \tag{1}$$

becomes

$$\frac{d\alpha}{dt} = -A\alpha^3 + B(\lambda - \lambda_c)\alpha + \epsilon^{1/2} f(t) + Cg + d\delta_{\lambda\lambda'} \quad , \tag{2}$$

where α is the amplitude of the symmetry-breaking solution, λ the control parameter, λ_c the symmetry-breaking transition (critical) point, $\epsilon^{1/2}$ is the rms value of fluctuation (noise), $f(t)$ is the normalized fluctuation (noise), g is the interaction or bias symmetry-breaking selector, d is the amplitude of a quasi δ-pulse, and $\delta_{\lambda\lambda'}$ is the δ-function which gives $\delta_{\lambda\lambda'} = 1$ when $\lambda = \lambda'$ and $\delta_{\lambda\lambda'} = 0$ when $\lambda \neq \lambda'$. Therefore,

$$\lambda = \lambda_c + \gamma t \quad , \tag{3}$$

where λ_c is the initial value of λ, γ the evolution rate, and t the evolution time.

To solve the first-order stochastic equation numerically, we designated the initial symmetry-breaking amplitude, α, as zero at time $t = 0$, but the $f(t)$ is a random number generated by computer within [-1, 1]. To obtain the trace of α at a different time for each trial, we just sampled it at different times. (Our studies confirm the results of Kondepudi and Nelson [1].) One simulation assumes that the impulse takes place near and after the critical point ($\lambda_c = 1.0$) at $\lambda = 1.5$, and $\delta = d/g = 100$ is shown in Fig. 1. In this case, the impulse plays a dominant role to push the process toward a favored state! We have studied over 40,000 trials and find a trend that the early chiral impulse produces a significant bias towards the preferred state. Furthermore, we found an 80% probability of the favored state being selected [8,15]. This illustrates the effect of a small symmetry-breaking process on a slowly evolving system. In essence, the small bias is magnified by the process of signal averaging, where the noise effect essentially cancels out. This process is sometimes called the Kondepudi effect [1]. Over a very long period preceding the formation of the solar system, there could have been many supernovas generating both $\bar{\nu}_e$ and $A\ell^{26}$ "impulses" into the organic medium, as shown schematically in Fig. 2. The β^+ from the $A\ell^{26}$ could cause an asymmetry in the interaction with the organic molecules.

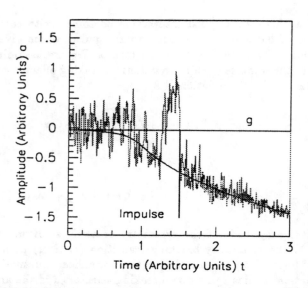

FIGURE 1. Single trial with an impulse at $\lambda = 1.5$ where $\delta/g = 100$. The impulse strongly affects the phase transition.

III SOURCE OF THE CHIRAL IMPULSE

There are three effects from a Type-II supernova in the primordial cloud that may have formed the solar system [16]:

1. $\bar{\nu}_e$ emission (with $\bar{\nu}_e + p \rightarrow e^+ + n$ interaction in the hydrocarbon),

2. ν_e, $\bar{\nu}_e$, ν_x, $\bar{\nu}_x$ ($x = \mu, \tau$) neutrino emission and subsequent coherent interaction with the nuclei of the organic materials,

3. Intense β emittances like $A\ell^{26}$ formed by supernovas that emit polarized e^+ particles that interact with the organic materials.

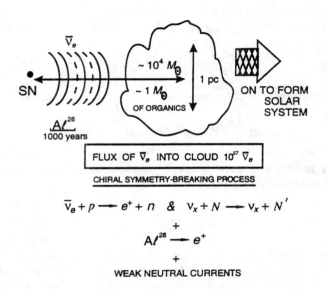

$$\bar{\nu}_e + p \longrightarrow e^+ + n \quad \& \quad \nu_x + N \longrightarrow \nu_x + N'$$

$$+$$

$$A\ell^{26} \longrightarrow e^+$$

$$+$$

WEAK NEUTRAL CURRENTS

FIGURE 2. Schematic effect of a Type-II supernova on the organic molecules in a dense molecular cloud (possibly one that went on to participate in the formation of the solar system). This could have produced an intense chiral impulse.

Of course, all this time the effect of the WNC can be driving the system towards a homochiral state, as depicted in Eqs. (1) and (2).

Let us consider the rate of these effects:

1. For $\bar{\nu}_e$ absorption and a supernova 1 parsec away (or inside a 1-pc dense cloud), the number of interactions will be $\sim 10^4$/kg of material for 1.0 $M_\odot$ of organic material (which would be 6×10^{34} interactions in the organic matter that is *active*); therefore, the positrons from the $\bar{\nu}_e$ interactions would lose energy at a rate of 10^{-19} MeV/cm and thus travel nearly a parsec. An asymmetry of 10^{-17} could be amplified in this situation.

2. For the coherent $\nu_x + N \rightarrow \nu_x + N$, and for the carbon in the hydrocarbons, we would have $\sim 10^2$ more or 6×10^{36} interactions in the organic material.

3. For the $A\ell^{26}$ over the half life, there would be $\sim 10^{50}$ decays producing $\sim 10^{50}$ positrons that lose energy at the rate of 10^{-19} MeV/cm; for MeV positrons, the range of the positrons would be of order of a parsec (ignoring possible magnetic-field effects).

Consider the example where 0.001 $M_\odot$ of $A\ell^{26}$ is produced and assume, for simplicity, that the energy of the e^+ is 1 MeV and is contained in the gas cloud. Assume the cloud has a density of 10^4 atoms/cm^3 and that 10^{-3} of the atoms are organic, the stopping power for e^+ will be

$$\frac{dE}{dx} \sim \text{MeV/g/cm}^3 \ , \tag{4}$$

and for a density of $\rho = 10^4$ atoms/cm $^3 \sim 10^{-19}$ g/cm^3 we find,

$$dx \simeq \frac{dE}{\rho}(\text{MeV}) \sim 10^{19} \text{ cm } (\sim 3 \text{ ps}) \tag{5}$$

(we neglect radiation ρ processes and magnetic fields), and for an average energy exchange of 10 eV, we have

$$\frac{10_e^5 \text{ collisions}}{A\ell^{26} \text{ decay}} \ . \tag{6}$$

For 0.001 M$_\odot$ of Aℓ^{26} and 10^{-3} organic fraction, we obtain a total of $\sim 10^{55}$ collisions of polarized positrons with organic materials in the cloud assuming all of the e^+ stop in the cloud. (We assume that only one of the collisions can result in spin exchanges.) There will also be the same order of polarized photons from the $e^+e^- \rightarrow \gamma\gamma$ annihilation. It is estimated in Refs. [14,17] that the asymmetry due to the weak interaction would be of order 10^{-11} to 10^{-6}, depending on the positron energy $\{$it scales like $\alpha^2[\alpha/(\nu/c)]^2\}$. Thus it takes $N \sim 10^{22}$ interactions for the asymmetric to become statistically important. In this example there are far more interactions.

It is clear that the largest effect in all of these processes will be the result of the interaction of the polarized e^+. It is plausible that these e^+ give rise to a reasonable fraction of chiral interactions and, thus, classify as a chiral impulse, as shown schematically in Fig. 1.

IV CONCLUSIONS

The combined effect of the WNC in the organic material in a primordial gas cloud and a nearby supernova explosion, which gives rise to the different interactions described here, could give appreciable chiral symmetry breaking.

Experimental measurements of the effects of chiral (polarized) electrons and $\bar{\nu}_e$ (from a reaction, for example) could give input to a model calculation of the onset of homochirality in organic materials.

REFERENCES

1. Kondepudi, D. and Nelson, G.W., *Nature* **314**, 438-441 (1985).
2. Zel'dovich, Y.B. and Mikhailov, A.S., Sov. Phys. Usp. **30**, 977-992 (1987).

3. Avetisov, V.A., Goldanskii, V.I., and Kuz'min, V.V., *Physics Today* (July), 33-41 (1991).
4. Salam, A., *J. Mol. Evol.* **33**, 105 (1991).
5. Avetisov, V.A., Goldanskii, V.I., and Kuz'min, V.V., *Physics Today* (April), 199 (1992).
6. Hegstrom, R.A., Rein, D.W., and Sandars, P.G.H., *J. Chem. Phys.* **73**, 2329-2341 (1980).
7. Mason, S.F. and Tranter, G.E., *Proc. R. Soc. London* **A397**, 45 (1985).
8. Liu, Y., Wang, H., and Cline, D., in *Discovery of Weak Neutral Currents: The Weak Interaction Before and After*, Mann, A.K. and Cline, D.B. (eds.), New York: AIP Press, Conf. Proc. 300 (1994) 499-505.
9. Ulbricht, T., *Quart. Rev.* **13**, 48 (1959).
10. Vester, F. *et al.*, *Naturweissenschaften* **40**, 68 (1958).
11. Yamagata, Y., *J. Theor. Biol.* **11**, 495 (1966).
12. Bonner, W., *Origins of Life* **21**, 59-111 (1991).
13. Goldanskii, V.I. and Kuz'min, V.V., *Sov. Phys. Usp.* **32**, 1-29 (1989).
14. Cline, D., Liu, Y., and Wang, H., *Origins of Life* **25**, 201 (1995).
15. Park. J. *et al.*, in *Photoelectronic Detectors, Cameras,and Systems*, *SPIE* **2551**, 40-52 (1995).
16. Greenberg, J.M. and Mendoza-Gomez, C.X., in *The Chemistry of Life's Origin*, New York: Kluwer Acad. Pub. (1993) 1-32.
17. Meiring, W.R., *Nature* **329**, 712-714 (1987).

V. FUTURE TESTS

The Weak Force and SETH: the Search for Extra-Terrestrial Homochirality

Alexandra J. MacDermott

Department of Chemistry, University of Cambridge
Lensfield Road, Cambridge CB2 1EW, UK

Abstract. We propose that a search for extra-terrestrial life can be approached as a Search for Extra-Terrestrial Homochirality — SETH. Homochirality is probably a pre-condition for life, so a chiral influence may be required to get life started. We explain how the weak force mediated by the Z^o boson gives rise to a small parity-violating energy difference (PVED) between enantiomers, and discuss how the resulting small excess of the more stable enantiomer may be amplified to homochirality. Titan and comets are good places to test for emerging pre-biotic homochirality, while on Mars there may be traces of homochirality as a relic of extinct life. Our calculations of the PVED show that the natural L-amino acids are indeed more stable than their enantiomers, as are several key D-sugars and right-hand helical DNA. Thiosubstituted DNA analogues show particularly large PVEDs. L-quartz is also more stable than D-quartz, and we believe that further crystal counts should be carried out to establish whether reported excesses of L quartz are real. Finding extra-terrestrial molecules of the same hand as on Earth would lend support to the universal chiral influence of the weak force. We describe a novel miniaturized space polarimeter, called the SETH Cigar, which we hope to use to detect optical rotation on other planets. Moving parts are avoided by replacing the normal rotating polarizer by multiple fixed polarizers at different angles as in the eye of the bee. Even if we do not find the same hand as on Earth, finding extra-terrestrial optical rotation would be of enormous importance as it would still be the homochiral signature of life.

Homochirality is one of the most characteristic signatures of life. If we *do* find homochirality elsewhere in the universe, it will almost certainly indicate either life or life-in-the-making (i.e. advanced pre-biotic chemistry). So a search for extra-terrestrial life could be approached as a Search for Extra-Terrestrial Homochirality — **SETH**. Such an approach is independent of terracentric preconceptions as to what sort of molecules extra-terrestrial life might be based on, and is thus more general than searches for amino acids, water, oxygen, etc..

Only the very smallest and simplest molecules are achiral. As soon as molecules reach an even moderate level of complexity the possibility arises of two mirror image chiral forms. So chirality is bound to appear at a very early stage in the history of life, and from Fisher's lock and key hypothesis (1) there can be no doubt that if there is chirality there must be homochirality for an efficient biochemistry. So homochirality is certainly at least a *consequence* of life, and furthermore is now believed to be probably also a *pre-condition* for life. This is because polymerization to give the necessary long-chain stereoregular polymers (e.g. all-L polypeptides) will not go in racemic solution — addition of the wrong hand tends to terminate the polymerization (2). An almost homochiral monomer solution is apparently needed for efficient polymerization, and stereochemical purity is certainly needed for template reactions, so a homochiral pre-biotic chemistry may be a pre-condition for life. The homochirality requirement could perhaps be circumvented with "prochiral" life, e.g. nucleic acid analogues based on glycerol instead of ribose (3,4); but although the monomer units are achiral, the hand of the helix still has to be chosen, so a symmetry-breaker is needed sooner or later.

But once the symmetry is broken, initial selection of one hand in ancestral biomolecules fixes the handedness of the rest of biochemistry through diastereomeric connections, such as that between the L-amino acids and D-sugars (5). But D-amino acid/L-sugar "mirror life" should be just as efficient as L-amino acid/D-sugar terrestrial life, so the question arises as to *why* terrestrial life is based on L-amino acids and D-sugars and not D-amino acids and L-sugars. Was it a frozen accident or the result of a chiral influence? Many classical "chiral" influences have now been shown to be not really chiral at all (e.g. magnetic fields, the Coriolis force, etc.) (6), and the only *universal* truly chiral influence is the weak force: this affects all chiral molecules at all times, and no specific conditions or reaction mechanisms are required.

The weak force is one of the four forces of nature — electromagnetic, strong, weak and gravitational — and it is the only one of the four that is chiral: it can feel the difference between left and right. Fermions exist in two states of opposite helicity, corresponding to spin and momentum vectors parallel (right-handed) or anti-parallel (left-handed). The two helicity states participate equally in the electromagnetic, strong and gravitational interactions, but they do not participate equally in the weak interaction, which is therefore said to violate parity. Left-handed electrons participate in the weak interaction preferentially compared with right-handed electrons; similarly right-handed positrons are preferred over left-handed positrons in the weak interaction to the same degree. So although there are normally equal numbers of left and right-handed electrons, the weak force can feel the left-hand ones better, and so electrons can be viewed as left-handed as far as the weak force is concerned; similarly positrons can be viewed as right-handed.

The handedness of elementary particles means that L and D molecules are really diastereoisomers, not enantiomers: the true enantiomer of an L-amino acid is the D-amino acid made of anti-matter (7). Left and right-handed molecules should therefore differ very slightly in many properties, e.g. NMR chemical shifts (8), and most importantly in energy. This parity-violating energy difference between enantiomers — the "PVED" or ΔE_{pv} — arises from weak neutral current interactions, mediated by the Z^o boson, between electrons and neutrons. These interactions impart a parity-violating energy shift (PVES), E_{pv}, to the energy of a chiral molecule, and an equal and opposite shift, $-E_{pv}$, to that of its enantiomer, giving a parity-violating energy difference (PVED) of $\Delta E_{pv} = 2E_{pv}$. The PVED produces a very slight excess of the more stable enantiomer, which can be amplified to produce the observed homochirality.

To calculate the PVED (9-12) we start from the parity-violating Hamiltonian density for the weak neutral current interaction between electrons and nuclei. This is obtained from the Feynman diagram

by associating a current J^μ with each fermion involved, a coupling constant g_N with each vertex, and a propagator (here $(1/M_Z^2)$) with each virtual boson exchanged, giving

$$\mathcal{H}_{pv} = J_\mu(e)g_N(1/M_Z^2)g_N J^\mu(n)$$

The weak coupling constant is not small ($g_N \approx e$), but the mass M_Z of the Z^o boson is large, which is why the weak force is weak.

The neutron current $J^\mu(n)$ is broken down into currents of the constituent quarks, and the current for each fermion — electron or quark — is broken down into sums of left and right-handed fermion currents. The handedness of the weak force enters through the unequalness of the coefficients of left and right-handed fermion currents, e.g. for electrons

$$c_L = -(1/2) + sin^2\Theta_W$$

$$c_R = sin^2\Theta_W$$

Using the theoretical value of 30° for the Weinberg angle Θ_W the Hamiltonian becomes

$$\mathcal{H}_{pv} = -(G_F/2\sqrt{2})J_\mu(e)J^\mu(n)$$

where $J_\mu(e) = \bar{e}\gamma_\mu\gamma_5 e$, $J^\mu(n) = \bar{n}\gamma^\mu n$ and G_F is the weak coupling constant. On reduction to non-relativistic quantum mechanics by explicitly multiplying out the γ-matrices, the Hamiltonian density $\mathcal{H}_{pv}$ becomes the parity-violating Hamiltonian

$$\hat{H}_{pv} = -(\Gamma/2) \sum_a \sum_i N_a \{ \mathbf{s}_i \cdot \mathbf{p}_i, \delta^3(\mathbf{r}_i - \mathbf{r}_a) \}_+$$

The sums over i and a are over all electrons and nuclei respectively in the molecule, and N_a is the neutron number of nucleus a. This beautifully elegant expression summarizes the physical origin of the PVES. The term in $\mathbf{s} \cdot \mathbf{p}$ represents the projection of the spin onto the direction of momentum, thus touching directly on the left-handedness of the electron. $\hat{H}_{pv}$ is of opposite sign for enantiomers: $\mathbf{p}$ changes sign under parity, being a polar vector, while $\mathbf{s}$ remains the same, being an angular momentum and therefore an axial vector. The delta function expresses the contact nature of the weak force. The smallness of the PVES arises from the smallness of the constant Γ, which contains the very small weak coupling constant G_F.

One evaluates the PVES E_{pv} by taking matrix elements of $\hat{H}_{pv}$ over the ground state wave functions. In the absence of spin-orbit coupling one can assume separability of the wave function into spin and orbital parts, which results in E_{pv} being identically zero because $\mathbf{p}$ is imaginary. If the ground state wave function is corrected for the effect of spin-orbit coupling, the PVES no longer vanishes, and we obtain

$$\Delta E_{pv} = 2 \sum_j^o \sum_k^u P_{jk} (\varepsilon_j - \varepsilon_k)^{-1}$$

where

$$P_{jk} = Re \langle \psi_j | \hat{V}_{pv} | \psi_k \rangle \cdot \langle \psi_k | \hat{V}_{so} | \psi_j \rangle$$

is the "parity-violating strength". The sums are over all occupied (o) MOs j and all unoccupied (u) MOs k, ε_j and ε_k are the energies of the MOs ψ_j and ψ_k respectively, $\hat{V}_{pv}$ is a one-electron version of the parity-violating Hamiltonian, and $\hat{V}_{so}$ is a spin-orbit coupling operator.

The parity-violating strength P_{jk} is closely analogous to the rotational strength

$$R_{OA} = Im \langle \psi_O | \hat{\mu} | \psi_A \rangle \cdot \langle \psi_A | \hat{\mathbf{m}} | \psi_O \rangle$$

in optical activity. $\hat{V}_{pv}$ and μ are both parity-odd (being polar vectors), while $\hat{V}_{so}$ and $\mathbf{m}$ are both parity-even (being axial vectors), with the result that both P_{jk} and R_{OA} are oppositely signed for enantiomers.

The PVED can be calculated by *ab initio* methods (11-13), using the LCAO approximation

$$| \psi_j \rangle = \sum_c \sum_\gamma C_{c\gamma}^j | \phi_{c\gamma} \rangle$$

244

to express the molecular orbitals $|\psi_j\rangle$ as linear combinations of atomic orbitals $|\phi_{c\gamma}\rangle$ of type γ on nucleus c. To evaluate the PVED, we need the MO cofficients $C^j_{c\gamma}$ and the MO energies ε_j (the matrix elements of $\hat{V}_{pv}$ and $\hat{V}_{so}$ over the atomic orbitals are pre-evaluated for the required elements using a 6-31G basis set). The computation is therefore divided into two stages. First we obtain the MO coefficients and energies from GAUSSIAN92 on the University of London Convex. Second, we use our own program PVED84 on the University of Oxford Convex to combine the MO coefficients and matrix elements to give ΔE_{pv}.

The calculated PVEDs are of the order of 10^{-20} to 10^{-17} hartree, about 10^{-17} to $10^{-14} kT$ at room temperature, for typical biomolecules, giving, from the Boltzmann distribution, an enantiomeric excess $|L - D|/L + D = \Delta E_{pv}/kT$ of 10^{-17} to 10^{-14}. These small excesses need amplifying to produce the observed homochirality, and the possible mechanisms fall into three classes: (a) the Yamagata cumulative mechanism, applicable to crystallizations or polymerizations of optically *labile* or achiral monomers; (b) the Kondepudi catastrophic mechanism, applicable to polymerizations of optically *non-labile* molecules; and (c) the Salam phase transition.

An example of the Yamagata cumulative mechanism (14) is the crystallization of quartz, which consists of helical crystals made of achiral silica units. During crystal growth an achiral unit A may add on to a growing crystal of either hand:

$$L_{n-1} + A \rightleftharpoons L_n$$

$$D_{n-1} + A \rightleftharpoons D_n$$

But owing to the PVED the corresponding free energy changes are unequal,

$$\Delta G_L \neq \Delta G_D$$

which means that A will add preferentially to one hand of crystal rather than the other, resulting in a fractional excess at each of N stages of crystallization which is small but multiplicative, leading cumulatively to an excess of one hand. This is equivalent to the PVED of the crystal being N times the PVED of the individual units within the crystal,

$$\Delta E_{pv}(N \ unit \ crystal) = N\Delta E_{pv}(one \ unit)$$

resulting in a greatly amplified enantiomeric excess in the macroscopic crystal. Under conditions of kinetic as opposed to thermodynamic control, a similar amplification effect can occur due to enantiomeric differences in activation energy for addition of A to L_{n-1} or D_{n-1}:

$$\Delta G_L^\ddagger \neq \Delta G_D^\ddagger$$

245

The Kondepudi catastrophic mechanism is based on a kinetic scheme involving autocatalysis and enantiomeric antagonism, i.e. the presence of one enantiomer encourages production of itself but inhibits production of its mirror image. Many biopolymerization reactions have precisely these characteristics. Using the methods of non-equilibrium statistical thermodynamics, Kondepudi showed (15) that for $\Delta E_{pv} > 10^{-17} kT$ amplification to homochirality will eventually occur, but for smaller ΔE_{pv} the amplification effect will be overcome by thermal fluctuations. The amplification time for a PVED of $10^{-17} kT$ is just 10^4 years, a very short time on an evolutionary timescale. However, large volumes are needed: the figure of 10^4 years is based on a volume of $4 \times 10^9 dm^3$, corresponding to a small lake 1km by 1km and 4m deep. But the amplification time and lake size can be reduced dramatically for larger PVEDs, e.g. a PVED of $10^{-14} kT$ (the largest value calculated so far — see later) can be amplified to homochirality in just 1 year in $4 \times 10^5 dm^3$, corresponding more closely to the proverbial warm little pond.

The Salam mechanism (16) is somewhat speculative, and postulates that below a certain critical temperature quantum mechanical tunnelling to the more stable enantiomer should occur in a cooperative phase transition effect. This mechanism could supposedly work at the low temperatures pertaining in space (where the Kondepudi mechanism would go far too slowly). Figureau (17) tried to verify the Salam mechanism by cooling racemic cystine crystals in the hope that optical rotation would appear at low temperatures — but it did not. However, Figureau may have been examining the wrong system: he used cystine because this contains heavy sulphur atoms which give an enhanced PVED (see later), but in fact tunnelling between the D and L forms is at its most efficient if they are close in energy, and would almost certainly be suppressed (18) for PVEDs as large (19) as that of cystine.

Some amplification has been achieved by Kondepudi-type processes in the laboratory (20), starting from initial enantiomeric excesses of the order of a few percent, but it is difficult to test amplification from the 10^{-17} level on Earth because of the long times and large volumes required, and also the danger of biochiral contamination on a planet with life. This is where SETH could be particularly valuable. A world with no life, undergoing purely chemical evolution, would provide an excellent laboratory in which to test for pre-biotic emergence of chiral bias. Where should we look for homochirality? One of the best places to look for pre-biotic homochirality is Saturn's giant moon Titan. It has a rocky core covered with a thick layer of water ice, and may have a hydrocarbon "ocean" on top of this. The Voyager 2 fly-by showed that the atmosphere is 80-90 % N_2, with the rest CH_4, Ar, and small amounts of C_2H_6, C_3H_8, C_2H_2, HCN, and cyanoacetylenes, many of which will have condensed to form the ocean. Brack and Spach (21) have pointed out that the compounds detected by Voyager could readily react to give a range of chiral photoproducts such as 2,3-dimethylpentane, 2-amino-butane and many more. The dirty orange "photochemical smog" seen from Voyager on its fly-by suggests the presence of complex molecules, e.g. polymers of HCN and C_2H_2. Clearly any chiral molecules could also polymerize, and could reach homochirality through Kondepudi amplification. Polymers have also been found in comets. The various Halley spacecraft detected HCN, CH_3CN, NH_3, H_2CO, and $(H_2CO)_n$ in the coma. The comet's dark coating is though to be $(H_2CO)_n$ and other polymers, e.g. the $(HCN)_n$ polymers suggested by Matthews (22). HCN readily forms various polymers, ranging in colour from orange to black, and Matthews (23) believes that they may be responsible for the orange haze seen on Titan and several other planets, as well as the black crust on Halley. Many of these polymers are chiral, some containing repeating chiral $(HCN)_3$ units. HCN polymers are readily hydrolysed to give amino acids, and can also act simultaneously as condensing agents in the production of nucleic acids in life, thus explaining the association of proteins and nucleic acids in life, and avoiding the problem of which came first — they came together! $(HCN)_n$ polymers are thus potentially of immense importance in the origin of life, and since they are also chiral we believe comets could be a very promising place to look for the evolution of homochirality through Kondepudi-type polymerization of HCN.

On Mars we may find homochirality as a signature of extinct life. Earth and Mars had similar conditions around 3.8 billion years ago, so life may have started on both but gone extinct on Mars (probably around 3.5 billion years ago) as it cooled more quickly. Homochirality may therefore be preserved beneath the surface layers as a relic of this past life. Bada and McDonald (24) have studied the racemization half-lives of amino acids, and shown that although these are only about 10^6 years in wet conditions, they are as high as 10^{13} to 10^{27} years in dry or frozen conditions at Martian temperatures (150-220K). The homochiral signature of life could therefore be preserved even if the extinction occurred 3-4 billion years ago. It would, however, be erased by prolonged exposure to liquid water after extinction, but could be preserved in the cold dry conditions of the Martian polar regions.

We now turn to our calculations of the PVED (25). The natural L-amino acids L-alanine, L-valine, L-serine and L-aspartate were all found to be more stable than their "unnatural" D-enantiomers, in their solution conformations and also the α-helix and β-sheet conformations, by $10^{-17}kT$ (11, 26). Turning to the natural D-sugars, it was found that D-glyceraldehyde, the parent of the higher sugars, is indeed PVED-stabilized, by about $10^{-17}kT$ (27). β-D-deoxyribose, in the C2-endo conformation found in DNA, is similarly PVED-stabilized, but β-D-ribose in the C3-endo form found in RNA is *less* stable than its enantiomer (28). This latter result appeared discouraging until we realized that the precursor of β-D-nucleotides is not β-D-ribose but α-D-ribose (via an S_N2 Walden inversion at C1), which we have now shown *is* PVED-stabilized.

Thus the most important biomolecules — amino acids and sugars — have PVEDs of order $10^{-17}kT$, which is only just ampifiable by the Kondepudi mechanism. If the PVED were a little larger, Kondepudi amplification would be very much easier. The PVED is proportional to the sixth power of the atomic number (largely due to the importance of spin-orbit coupling), so attention was turned next to molecules incorporating second-row heavy atoms such as P, Si, S, in the hope of finding larger PVEDs (29). We examined fragments of the sugar-phosphate backbone of DNA without the bases (which are anyway planar and in themselves achiral, and so should contribute little to the PVED), and the right-hand B double helix showed a PVED-stabilization of about $10^{-17}kT$ per sugar-phosphate unit. For the more primitive glycerol-based polymer (in which the monomer units are achiral and the chirality comes only from the helical conformation, as described earlier) the right-hand helix was again found to be PVED-stabilized, in both A and B conformations, by $10^{-17}kT$ per sugar-phosphate unit. It had been hoped that larger PVEDs would be obtained with the heavy phosphorus atom. The fact that the phosphate PVED was still only $10^{-17}kT$ was traced (29) to the phosphorus atom being very electropositive in phosphates due to the electron-withdrawing effect of the four oxygens, resulting in too little electron density on the phosphorus to feel the potentially larger parity-violating effect.

Clearly more electronegative heavy elements such as sulphur would be better candidates for a large PVED, so we turned next (30) to thiosubstituted DNA analogues. Whereas the normal DNA backbone with $-O-PO_2-O-$ linkages has a PVED of $10^{-17}kT$ per unit, thiosubstitution of the side oxygens to give $-O-PS_2-O-$ gave a PVED of $10^{-16}kT$ per unit, and thiosubstitution in the helix itself to give $-S-S-CH_2-$ links gave the enormous PVED of $10^{-14}kT$ per unit. This is our most exciting result yet: whereas PVEDs of $10^{-17}kT$ take 10^4 years to amplify in a large lake, PVEDs of $10^{-14}kT$ take only 1 year in a small pond. This means that homochirality could be attained much more quickly — and so life could get started much more easily — with a thiosubstituted primitive replicator. This ties in with the view that life may have originated in the highly sulphurous environment of deep-sea thermal vents.

Turning now from polymers to crystals, L-quartz consists of right-hand 3-fold helices of silica tetrahedra. As a chiral mineral made of achiral units it can undergo Yamagata amplification, and indeed a 1.4% excess of L-quartz has been reported in a large collection of crystals from all over the world (31). Our calculations (30) show that L-quartz is PVED-stabilized by $10^{-17}kT$ per SiO_2 unit. This was again disappointingly small for the heavy Si atom: but silicates have the same problem as phosphates in that the electron-withdrawing effect of the oxygens leaves the Si atoms very electropositive. However the PVED does not need to be larger than $10^{-17}kT$ to account for the alleged 1% excess of L-quartz: according to the Yamagata mechanism,

$$\Delta E_{pv}(crystal) = N\Delta E_{pv}(1 \ unit)$$

so $\Delta E_{pv}(crystal) = 10^{-2}kT$ (corresponding to a 1% enantiomeric excess) can be obtained from $\Delta E_{pv}(unit) = 10^{-17}kT$ if $N = 10^{15}$, which corresponds to a realistic small crystal of side 0.1 mm (32).

These results for quartz thus predict almost exactly the reported 1% excess of L-quartz. But are the reported enantiomeric excesses real? Although the original reports (31, 33) appear to show an excess of L in several separate samples from different locations all over the world, a later report (34) disputes the earlier statistics and shows a (statistically insignificant) excess of D quartz. More fundamentally, it is possible that any quartz enantiomeric excesses, even if real, may reflect not the PVED but *chiral nucleation*: if one small crystal happens to be L, it acts as a seed and causes other crystals nearby to be L also. Furthermore, the seed may break up if the solution is stirred, producing lots more L seeds which then become spread around the solution, with the result that the whole solution crystallizes in the L form. This has been observed in solution in the famous "Kondepudi effect" (33) (not to be confused with the Kondepudi chiral amplification mechanism discussed earlier). Kondepudi obtained equal numbers of L and D sodium chlorate crystals from an unstirred solution, but if the solution was stirred he always got a large excess of either L or D (but the direction of stirring did not affect the hand of the crystals). The explanation is that stirring causes the seed which happens to form first to break up and spread around the solution. The crucial question as far as natural chiral crystals are concerned is therefore whether this chiral nucleation effect occurs under geological conditions: if no, an enantiomeric excess could reflect the PVED, but if yes, any enantiomeric excess will merely reflect the hand of the seeding crystal. However, due to the PVED there will be slightly more L seeds than D, so one should find chirally nucleated samples with an excess of L quartz slightly more often than samples with an excess of D. So we are now counting not the excess of L crystals in any one pot, but the excess of pots with an excess of L. The size of this excess of L pots depends on the size of the nucleating crystal: from the thermodynamic version of the Yamagata mechanism we have

$$\Delta E(seed \ crystal) = N\Delta E(1 \ unit)$$

which will give an enantiomeric excess of the more stable form in the seed crystals (and a corresponding excess can also be obtained under kinetic con-

249

trol due to the corresponding enantiomeric difference in crystallization rates). We know that for quartz $\Delta E(1\ unit)$ is $10^{-17}kT$, so the crucial question is how large is N for the nucleating crystal. Kondepudi suggests (36) that the side of the nucleating crystal is of the order of 0.5 mm. This would correspond to $N \approx 10^{17}$. If this is true, it would seem that there could indeed be a detectable enantiomeric excess in the seed crystals, and therefore a detectable excess in the number of pots with a chirally nucleated enantiomeric excess of the PVED-determined form. Kondepudi's estimate of the size of the nucleating crystal may of course be too large, but even if N were only 10^{15} it would still be feasible to detect the resulting excess in samples with an excess of the PVED-determined form.

We therefore believe that much larger samples of natural chiral crystals — especially those containing heavy atoms, bearing in mind the Z^6 dependence of the PVED — should be examined to see if there is any enantiomeric excess. It is worth looking because of the great efficiency with which any chiral bias in minerals can be transferred to biology through chiral surface catalysis. For example, L-quartz absorbs L-amino acids preferentially from a racemic mixture, with a 1% selectivity (37); if this were combined with a 1% excess of L-quartz it would produce an overall electroweak enantioselectivity of 10^{-4}, which is dramatically larger than the PVED of individual molecules and would be correspondingly more easily amplified. Even a considerably smaller excess of L-quartz would still produce a substantial enhancement of the electroweak effect as compared with individual molecules.

The weak force appears to predict the correct hand whatever came first — nucleic acids or proteins — in the origin of life. For nucleic acids first, D-glyceraldehyde and α-D-ribose are more stable, so that the D-sugars and D-ribonucleotides would be selected, and the right-hand helical backbone of DNA is also PVED-stabilized. For proteins first, the L-amino acids are more stable in most cases. To confirm the enantioselective role of the weak force we need SETH: finding molecules of the same hand in many different extra-terrestrial systems would lend support to the weak force as a universal symmetry breaker. Even finding the "wrong" hand in extra-terrestrial systems would be enormously significant, because it would at least be the homochiral signature of life! So it is of tremendous importance to search for extra-terrestrial homochirality whether or not one believes in its electroweak origin.

Optical activity is the natural choice for a SETH instrument: it offers an entirely general signature of homochirality, and does not rely on finding specific target molecules, or target molecules with specific properties such as fluorescence. Optical activity works for *all* chiral molecules, in contrast to methods such as GC, which requires different columns for different types of molecule. Conventional optical activity instrumentation such as circular dichroism and optical rotatory dispersion can identify molecules as well as detect their optical activity. But these techniques are much too bulky for space: they contain complex variable wavelength sources and require motors to rotate a moving polarizer to determine the sign. But just to detect optical rotation, one can use a much more minimal apparatus — no bigger than a cigar — without moving parts which might jam in space. The principle is illustrated by an apparatus we call the 'A' Cigar (38), which has the sample solution between fixed polarizers, with the second polarizer at 90° to the first. With miniaturized diode emitters and detectors, and a 10 cm pathlength through the sample solution, this set-up is about the size of a cigar. The idea is that with racemic compounds there is zero optical rotation, which with crossed polarizers gives no light at the detector. If any light at all reaches the detector, there must be optical rotation, i.e. an enantiomeric excess — the signature of life.

But this simple set-up cannot give the all-important *sign* of the optical rotation, and is vulnerable to artifacts, especially false positives due to scattering, necessitating careful filtration to remove dust. We have therefore produced a new design (39), which gets around the problem of not being able to have a rotating second polarizer to get the sign by instead having many polarizers at different angles, used with a diode-array detector. Miniaturization with existing technology would permit the use of a 6 to 8 element diode array within a device that is still only the size of a (Churchillian) cigar. Each diode in the array would have in front of it a polarizer at a different angle, e.g. -60, -30, 0, +30, +60 and 90 degrees to the first polarizer. Whereas with the second polarizer at 90° one gets no light coming through for an optically inactive or racemic sample, with the second polarizer at some other angle one gets *some* light coming through, the amount being given by a cosine squared dependence on the angle between the polarizers. For an optically active sample one can obtain the magnitude of the angle of rotation by seeing how much the actual intensity of light coming through differs from that expected for a racemic mixture. To see how to get the *sign*, consider two second polarizers at $+30^o$ and -30° to the first. If the sample is optically inactive or racemic, one should get the same intensity through both $+30^o$ and -30° polarizers. But if the sample has a small positive rotation, there will be an increased intensity through the $+30^o$ polarizer and a decreased intensity through the -30° polarizer (whereas it would be the other way round for a small negative rotation).

The principle of using two or more fixed polarizers (rather than a rotating one) to determine the polarization characteristics of light is used by the honey bee (40), hence we call this new design the Bee Cigar. Bees do a "dance" to indicate to other bees the direction of flowers in relation to the position of the sun. They therefore need to be able to detect the position of the sun (even on a cloudy day), and they do this by examining the polarization of sunlight scattered from the sky. Whereas sunlight itself is unpolarized, the scattered light shows a different polarization all over the sky according to the position of the sun. Bees have polarized detectors inside their head made of oriented photo-absorbing rhodopsin molecules. The minimum number of independent polarized detectors needed is two, and bees have just two, at $40°$ to each other. We could use just this minimum number in the Bee Cigar, with two second polarizers at say $+30°$ and $-30°$ to the first, but artifacts will be reduced if we average over readings from further polarizers at say $+60°$ and $-60°$, etc..

A Bee Cigar at a single wavelength gives the magnitude and sign of the optical rotation, but does not identify the substance. But if two or more alternative sources at different wavelengths were used (with light directed into the polarimeter with fibre optics), finding say a positive optical rotation at one wavelength and a negative optical rotation at another would give strong clues as to the identity of the molecule; further clues from other techniques aboard the spacecraft, such as gas chromatography, would clinch the identification of the molecule.

Our simple compact Cigar mimics a full bulky optical rotatory dispersion spectrometer: firstly, the array of polarizers mimics a rotating polarizer, giving the sign; and secondly, two or three fixed wavelengths provides a rough approximation to a variable wavelength and helps to identify the compound. These two features also greatly reduce artifacts. Firstly, use of the polarized diode array eliminates false positives. With a single final polarizer, as in the 'A' Cigar, false positives could arise due to scattering, etc.. But sophisticated averaging over elements in the Bee Cigar's diode array will eliminate this. Secondly, use of two or more wavelengths eliminates false negatives, which could arise due to fortuitous cancellation of rotations from different compounds. But it is unlikely that exact cancellation would occur at two or three different wavelengths.

We have already built a simple prototype of the 'A' Cigar, and are now developing the Bee Cigar at the Rutherford Appleton Laboratory with a view to future Mars missions. The sensitivity is about $0.1°$ with our current crude mock-up of the 'A' Cigar, which can readily be increased to $0.001°$ in the Bee Cigar with improved sources and detectors using existing technology. This corresponds to about 10 to 100 micrograms of typical chiral molecules being present in the polarimeter. Optical rotation is thus not as sensitive as biosensors or GC, but has the advantage of being entirely general, and not specific to particular target molecules.

The SETH Cigar is small and cheap enough to take routinely on all solar system missions to test for the homochiral signature of life, and perhaps identify sites worthy of further missions with more sophisticated apparatus or facilities for sample return.

ACKNOWLEDGEMENTS

I would like to acknowledge the contribution of my co-investigators on the SETH Cigar Team: L.D.Barron (Glasgow), A.Brack (Orleans), T.Buhse (Bremen), A.F.Drake (London, R.J.Emery (Rutherford Appleton Laboratory), G.Gottarelli (Bologna), J.M.Greenberg (Leiden), R.Haberle (NASA Ames), R.A. Hegstrom (Wake Forest), K. Hobbs (Glaxo Wellcome), D.K. Kondepudi (Wake Forest), C.McKay (NASA Ames), S.Moorbath (Oxford), F.Raulin (Paris 12), M.C.W.Sandford (Rutherford Appleton Laboratory), D.W.Schwartzman (Howard), W.Thiemann (Bremen), G.E.Tranter (Glaxo Wellcome), J.C.Zarnecki (Kent).

REFERENCES

1. Fischer, E., *Chem. Ber.* **27** 2795 (1894).
2. Joyce, G.F., Visser, G.M., van Boeckel, C.A.A., van Boom, J.H., Orgel, L.E. and van Westresen, J., *Nature* **310** 602 (1984).
3. Bada, J.L. and Miller, S.L., *BioSystems* **20** 21 (1987).
4. Schwartz, A.W. and Orgel, L.E., *Science* **228** 585 (1985).
5. Ponnamperuma, C. and MacDermott, A.J., *Chem. in Brit.* **30** 487 (1994).
6. Barron, L.D., *Chem. Soc. Rev.* **15** 189 (1986).
7. Barron, L.D., *Mol. Phys.* **43** 1395 (1981).
8. Barra, A.L., Robert, J.B. and Wiesenfeld, L., *BioSystems* **20** 57 (1987).
9. Hegstrom, R.A., Rein, D.W. and Sandars, P.G.H., *Phys. Lett. A* **71** 499 (1979).
10. Hegstrom, R.A., Rein, D.W. and Sandars, P.G.H., *J. Chem. Phys.* **73** 2329 (1980).
11. Mason, S.F. and Tranter, G.E., *Proc. R. Soc. Lond. A* **397** 45 (1985).
12. MacDermott, A.J. and Tranter, G.E., *Symmetries in Science IV* (ed. Gruber, B. and Yopp, J.H.), Plenum, NY, p.67 (1990).
13. MacDermott, A.J. and Tranter, G.E., *Croatica Chemica Acta* **62** 165 (1989).
14. Yamagata, Y., *J. Theor. Biol.* **11** 495 (1966).
15. Kondepudi, D.K., *BioSystems* **20** 75 (1987).
16. Salam, A., *J. Mol. Evol.* **33** 105 (1991).
17. Figureau, A., Duval, E. and Boukenter, A., *Orig. Life and Evol. Biosphere* **24** 130 (1994).
18. Chela-Flores, J. and MacDermott, A.J., ms submitted to *Chirality* .
19. MacDermott, A.J., in *Chemical Evolution: Self-Organization of the Macromolecules of Life* (Proc. Trieste Conference on Chemical Evo-

lution and Origin of Life, October 1993, eds. Chela-Flores, J., Chadha, M., Negron-Mendoze, A., Oshima, T.) p.237 (1995).

20. Brack, A. and Spach, G., *J. Mol. Evol.* **15** 231 (1980).
21. Brack, A. and Spach, G., *BioSystems* **20** 95 (1987).
22. Matthews, C.N. and Ladicky, R.A., *Adv. Space Res.* **12** 21 (1992).
23. Matthews, C.N., *Orig. Life* **12** 281 (1982).
24. Bada, J.L. and McDonald, G.D., *Icarus* **114** 139 (1995).
25. MacDermott, A.J., *Origins of Life and Evolution of the Biosphere* **25** 191 (1995).
26. Mason, S.F. and Tranter, G.E., *Mol. Phys.* **53** 1091 (1984).
27. Tranter, G.E., *J. Chem. Soc. Chem. Commun.* p.60 (1986).
28. Tranter, G.E., MacDermott, A.J., Overill, R.E. and Speers, P.J., *Proc. R. Soc. Lond. A* **436** 603 (1992).
29. MacDermott, A.J. and Tranter, G.E., *Chem. Phys. Lett.* **163** 1 (1989).
30. MacDermott, A.J., Tranter, G.E. and Trainor, S.J., *Chem. Phys. Lett.* **194** 152 (1992).
31. Palache, C., Erman, G.B. and Frondel, C., *Dana's System of Mineralogy* 7th ed. Vol. III, Wiley, New York, p.16 (1962).
32. Tranter, G.E., *Nature* **318** 172 (1985).
33. Vistelius, A.B., *Sapiski Vsyesoyuz Mineral. Obsh.* **79** 191 (1950).
34. Frondel, C., *Am. Mineral.* **63** 17 (1978).
35. Kondepudi, D.K., Kaufman, R.J. and Singh, N., *Science* **250** 975 (1990).
36. Kondepudi, D.K., Bullock, K.L., Digits, J.A. and Yarborough, P.D., *J. Am. Chem. Soc.* **117** 401 (1995).
37. Kavasmaneck, P.R. and Bonner, W.A., *J. Am. Chem. Soc.* **99** 44 (1977).
38. MacDermott, A.J. and Tranter, G.E., *J. Biol. Phys.* **20** 77 (1994).
39. MacDermott, A.J., Barron, L.D., Brack, A., Buhse, T., Drake, A.F., Emery, R., Gottarelli, G., Greenberg, J.M., Haberle, R., Hegstrom, R.A., Hobbs, K., Kondepudi, D.K., McKay, C., Moorbath, S., Raulin, F., Sandford, M., Schwartzman, D.W., Thiemann, W.H.-P., Tranter, G.E. and Zarnecki, J.C., *Planetary and Space Science,* in press (1996).
40. Wehner, R., *Scientific American* **235** 106 (1976).

Asymmetric scattering of polarized photons and electrons by chiral molecules

M.E.Pospelov [1]

Budker Institute of Nuclear Physics, 630090 Novosibirsk, Russia

Abstract

We review the recent theoretical activity connected with the problem of the scattering of polarized particles, photons or electrons and positrons, by chiral molecules. It is shown that the asymmetry of the total ionization cross-sections $(\sigma_+ - \sigma_-)/(\sigma_+ + \sigma_-)$ for slow, $v \sim \alpha$, longitudinally polarized charged particles appears in the order α^2 from the degree of molecule geometrical asymmetry even in neglect of the exchange interaction and the spin-orbit interaction inside a molecule (Pospelov, 1993). It contradicts to the common believe that the asymmetry in the scattering of polarized positrons is suppressed by a factor α^4 (Zel'dovich and Saakyan, 1980). In the case of the electron scattering the exchange interaction combined with the specifics of electron state inside chiral molecule leads to the further Z^2-enhancement of the effect (Hegstrom, 1982). This specifics, the helicity density of the molecular electron, manifests also in the high-frequency asymptotics of the scattering of circularly polarized light (Khriplovich and Pospelov, 1992). The matrix element determining this asymptotics practically coincides, up to an overall factor, with that of the weak interaction responsible for the P-odd energy difference of right- and left-handed molecules.

1   Introduction

Since the discovery of parity nonconservation in weak interactions the possibility of using this phenomenon as an explanation of the origin of optical activity in living matter has been discussed extensively (see, for ex., [1]). Longitudinally polarized β-rays interacting with a primarily racemic mixture of different isomers produce a small but systematic shift of concentration in favor of definite handedness. The excess of concentration is proportional to the asymmetry of decomposition cross-section of different isomers by β-particles of definite helicity:

$$\eta = \frac{\sigma_L - \sigma_R}{\sigma_L + \sigma_R}. \tag{1}$$

---

[1] E-mail:pospelov@inp.nsk.su

From simple symmetry arguments this ratio is equal to the asymmetry of the cross-section of definite isomer decomposition by particles of opposite helicities:

$$\eta = \frac{\sigma_+ - \sigma_-}{\sigma_+ + \sigma_-}. \tag{2}$$

Various experimental attempts to observe this effect produced negative results. The most recent experiment [2] aiming to detect the asymmetry of orthopositronium rate formation in the process of scattering of slow polarized positrons by optically active media put 10^{-4} as the upper limit for this effect.

It should be mentioned here that there is no doubt in the existence of such correlations in principle. A complete analogue of the effect discussed, the chiral selection induced by circularly polarized UV light, was established many years ago by Kuhn and Braun [3] (see also [4] and Refs. therein). The magnitude of that asymmetry for $\omega \sim \mathrm{Ry}$ could be estimated as

$$\eta \sim \alpha \xi, \tag{3}$$

where the fine structure constant originates from the relative smallness of M1-amplitude; ξ denotes the degree of chiral distortion of a molecule and constitutes numerically $10^{-2} - 10^{-1}$.

2 Scattering of positrons and electrons

Theoretical analysis of this phenomenon was done by different authors [5, 6, 7]. The cross-section of molecule decomposition in these works is assumed to be equal to the total ionization cross-section. There are two main situation to be considered in this problem. First is the case of fast, $v \sim 1$, electrons and positrons when the degree of polarization in β-decay may reach 100%. This is a case which was presumably realized in nature. Another case of interest is the scattering of slow, $v \sim \alpha$, particles which is important in the light of experimental attempts to find the connection between the chirality of molecules and the helicity of incoming particles.

In the earliest work on this problem Zel'dovich and Saakyan [5] gave the estimate of the effect caused by the direct interaction between magnetic moments of incoming particle and molecular electron. For slow, $v \sim \alpha$, particles they found this correlation to arise in the relative order α^4 from the degreee of geometrical asymmetry. Their conclusions were modified by Hegstrom [6], who found much bigger effect caused by a spin-helical structure of molecular electron state for the scattering of slow electrons. The exchange interaction induces the asymmetry in the order $(Z\alpha)^2$, where Z is the characteristic charge of the nuclei in chiral compound. This factor reflects the actual size of spin-orbit interaction inside molecules which is responsible for the nonzero density of helicity of the molecular electron. For the scattering of slow polarized positrons the exchange interaction is absent and the conclusion of Hegstrom for the asymmetry of the total ionization cross-section coincides with that of Zel'dovich and Saakyan.

The suppression of the asymmetry at $v \sim \alpha$ by four powers of fine structure constant in neglect of exchange contradicts, however, to the naive expectation for this effect. The latter would be rather of order α^2 to reflect only the relative smallness of the magnetic moment interaction in comparison with Coulomb one. The absence of the effect in this order, if it is indeed the case, must be a consequence of some symmetry arguments or selection rules. This was an original reason which stimulated our interest to this problem. Our attempt to estimate the asymmetry of ionization cross-section and connect it, if possible, with optical rotatory properties of these molecules leads to the surprisingly bigger result. It turns out that the absence of the effect in order α^2 characterizes the leading order of the Born series only and does not show up for higher orders and, therefore, for the exact solution. It means, in particular, that in comparison with the naive expectation the asymmetry is suppressed not by an additional factor α^2 but rather by some power of α/v. It causes the fast decrease of the effect with the growth of energies of incoming particles but yields the same naive estimate at $\alpha \sim v$.

Let us explain our assertion in more details (For more extended discussion see Ref.[8]). We assume that the velocity of particles is much less than that of light, but big enough to treat this process by mean of Born approximation:

$$\alpha \ll v \ll 1 \tag{4}$$

This condition, in principle, allows us to make an order of magnitude estimation at $v \sim \alpha$ and find the v-dependence of the effect.

The interaction between scattering particle and molecular electron is given by the formula:

$$V = -\frac{\alpha}{|\vec{r}_e - \vec{r}|} + \frac{\alpha}{4m^2|\vec{r}_e - \vec{r}|^3}\{-(\vec{\sigma}[\vec{r}_e - \vec{r}, \vec{p}]) + 2(\vec{\sigma}[\vec{r}_e - \vec{r}, \vec{p}_e])\} + \sum_a \frac{Z_a\alpha}{|\vec{r}_a - \vec{r}|} \tag{5}$$

where $\vec{r}$, $\vec{p}$, $\vec{\sigma}$ are the coordinate, momentum and spin of scattering particle; $\vec{r}_e$, $\vec{p}_e$ are the coordinate and momentum of molecular electron. The first term in (5) is the usual Coulomb interaction, σ-dependent terms are the part of Breit potential. We restrict ourselves now to the lowest possible order in v/c and therefore omit other terms of Breit Hamiltonian. The sum over a represents the interaction of the scattering particle with static charges inside the molecule. We consider a one-electron case for intrinsic molecular dynamics and assume that the total electric charge of molecule is equal to zero, i.e. $1 + \sum_a Z_a = 0$.

The initial spin density matrix for longitudinally polarized particles is given by the expression:

$$\rho_i = \frac{1}{2}(1 + \lambda\vec{\sigma}\vec{n}).$$ (6)

Here $\vec{n}$ is the unit vector of initial momentum and λ is the helicity. The interaction (5) transfers the molecular electron from the initial state i to the final f while changing the momentum of the scattering particle on the value $\vec{q} = \vec{p'} - \vec{p}$. The matrix element corresponding to this transition reads in first Born approximation as follows:

$$M^{(1)} = < f|V(\vec{q})\exp(-i\vec{q}\vec{r}_e)|i> = < f|4\pi\alpha\exp(-i\vec{q}\vec{r}_e)[\frac{1}{q^2} + \lambda\frac{i(\vec{n}[\vec{q}\vec{p}_e])}{2m^2q^2}]|i> \simeq$$

$$< f|4\pi\alpha[\frac{-i(\vec{q}\vec{r})}{q^2} - \lambda\frac{i(\vec{q}[\vec{n}\vec{p}_e])}{2m^2q^2}]|i> .$$ (7)

The square of this amplitude averaged over a random orientation of the scatterer gives the differential cross-section of the process. The total ionization cross-section is mainly determined by the characteristic momentum transfers $m\alpha^2 \ll |\vec{q}| \ll m\alpha$ (see, for ex., [9]). The second line in the formula (7) corresponds to that limit. The λ-independent part of the cross-section to logarithmic accuracy is given by:

$$\sigma_+ + \sigma_- = \frac{8\pi}{3}(\frac{\alpha}{v})^2 \sum_{f(E_f>0)} |<f|\vec{r}|i>|^2 \ln(v/\alpha).$$ (8)

For a reasonable estimation of (8) we take

$$\sigma_+ + \sigma_- \sim \frac{1}{(m\alpha)^2}(\frac{\alpha}{v})^2 \ln(v/\alpha).$$ (9)

As to linear part in λ, it vanishes after the integration over random orientation of molecules. Indeed, let us write down the part of the square of amplitude which should be averaged over directions of $\vec{q_\perp}$ and $\vec{n}$:

$$\lambda(\vec{p}_e[\vec{n}\vec{q}])\exp\{i\vec{q_\perp}(\vec{r}_e - \vec{r'}_e) + iq_\parallel\vec{n}(\vec{r}_e - \vec{r'}_e)\},$$ (10)

where $\vec{r}_e$ and $\vec{r'}_e$ are coordinates taken from different matrix elements. After all angular integrations only two vectors are left in this expression, p_e and $\vec{r}_e - \vec{r'}_e$, which cannot form a superscalar and, therefore, this expression vanishes

258

identically. This vanishing occurs independently on the particular view of the wave functions of the electron inside chiral molecule.

Now let us turn back to the work of Zel'dovich and Saakyan. Their estimation for the asymmetry is $\eta \sim \alpha^4(v/\alpha)\xi$ [5]. In our approach this result could not arise unless we would add to the interaction (5) other terms of Breit Hamiltonian. The corresponding term in the amplitude reads as:

$$M = ... - < f|4\pi\alpha \exp(-i\vec{q}\vec{r}_e)\frac{v(\vec{n}\vec{p}_e)}{mq^2}|i > \qquad (11)$$

After trivial calculation we get the asymmetry identical to that obtained in [5]:

$$\sigma_+ - \sigma_- \sim \frac{8\pi}{3}\lambda(\frac{\alpha}{v})^2\alpha^3 v \sum_{f(E_f>0)} \frac{< i|l|f > < f|\vec{r}|i >}{m\alpha} \qquad (12)$$

At $v \sim \alpha$ the asymmetry is suppressed by four powers of α. We shall demonstrate below that this contribution is not leading at the region of small v. It should be mentioned in the conclusion, that chirality-helicity correlation in the cross-section induced by interaction (5) in the particular case of molecules oriented with respect to direction of β-radiation does not vanish even in the leading Born approximation and can be estimated as $\eta \sim \alpha^2(\alpha/v)\xi$.

The cancellation of the effect for disoriented media is connected with the specificity of the first Born approximation and is not necessary for exact solution of this problem. In other words, the first Born approximation is not sufficient to give the complete picture of scattering by such a complicated object as a chiral molecule is.

Now we proceed to the investigation of the λ-dependent part of the cross-section in the interference between amplitudes in first and second approximations. The general expression for the amplitude in the second Born approximation is given by:

$$M^{(2)} = \sum_n \int \frac{d\vec{k}}{(2\pi)^3} \frac{< f,p'|V|n,k > < n,k|V|i,p >}{E_i + p^2/(2m) - E_n - k^2/(2m) + i0}. \qquad (13)$$

In the intermediate state $|n,\vec{k} >$ the energy of molecular electron is E_n and the energy of the scattering particle is $k^2/(2m)$. The biggest contribution to this amplitude comes from intermediate states with $E_n - E_i \sim Ry$ and momentum transfers $|\vec{k} - \vec{p}| \sim |\vec{p'} - \vec{p}|$. The claim to calculate this amplitude seems to be unrealistic. For our purposes, however, it is sufficient to show the existence of nonvanishing λ-dependence in the interference with (7) and estimate its behaviour with the growth of v. As a result of the calculation we

259

expect the appearance of new structures, which are absent in the first order amplitude:

$$M^{(2)} \sim < f|\lambda A(\vec{l}\vec{n}) + iB(\vec{l}[\vec{n}\vec{q}])|i >, \qquad (14)$$

where A and B are scattering angle dependent quantities, $\vec{l}$ is the orbital momentum operator taken about the point around which the multipole expansion was done. The first term here is the Faraday-type of correlation between orbital momentum of molecular electron and the helicity of incoming particle; the second is the correlation of the scattering plane with the orbital momentum operator of the target electron. Both terms in (14) give a nonzero interference with (7) and λ-dependent cross-section as a result. As we could see earlier the cancellation in the leading order occurs independently on the view of molecular electron wave functions. The challenging character of the second-order Born amplitude lies in the appearance of the "third body" effect in the problem and the result appears to be sensitive to the molecular specifics. We would like to skip here the detailed consideration of this amplitude and simply quote the answer for the asymmetric part of cross-section from the work [8]:

$$\sigma_+ - \sigma_- = \frac{8\pi}{3}\lambda\alpha^2(\frac{\alpha}{v})^2 \sum_{f(E_f>0)} \frac{\mathrm{Im} < i|\vec{l}|f >< f|\vec{r}|i >}{m\alpha} \frac{\pi\omega_{fi}}{v^2 m} \int \frac{d\theta}{\theta}, \qquad (15)$$

where θ is the scattering angle. This formula gives the relation of the effect with the optical rotatory strength. The simple estimate for the λ-dependent cross-section now is:

$$\sigma_+ - \sigma_- \sim \frac{1}{(m\alpha)^2}\lambda\alpha^2(\frac{\alpha}{v})^4 \log(\alpha/v)\xi. \qquad (16)$$

The logarithmic factor here originates from the final integration over θ. Therefore, the asymmetry in the ionization cross-section appears first in the order:

$$\eta \sim \alpha^2(\frac{\alpha}{v})^2\xi. \qquad (17)$$

3   Exchange interaction and the density of helicity

Exchange interaction requires a separate treatment in the case of electron scattering. To take it into account we should consider the dynamics of the spin of the molecular electron. The spin-orbit interaction plays a crucial role [11] and leads to correlations between momentum and spin; so called "density of helicity" [12]. These correlations show up in the interference between direct and exchange diagrams and determine the asymmetry of ionization. According to

[6], in the case of slow electron scattering, the effect constitutes $\eta \sim (Z\alpha)^2\xi$. In the extension of this idea Meiring proposed the method to calculate the ionization asymmetry in the wide range of energies [7]. It has been assumed that the λ-dependent part of the cross-section could be done by simple factorization of the molecule matrix element of the helicity in the initial state and the cross-section of the free lepton-lepton scattering. In the particular case of the slow positron scattering this method gives $\eta \sim \alpha^4\xi$ whch is not correct as we could see from the previous section. Now we would like to point out that this factorizaton is not valid even in the case of electron scattering.

Usually chiral molecules consist of light atoms ($Z \leq 8$). Spin-orbit interaction in this case can be treated as a small perturbation. Its general form is given by [10]:

$$V_{SO}(\vec{r}) = -\frac{1}{4m^2}\left([\vec{\sigma}\vec{p}]\nabla U(\vec{r})\right). \qquad (18)$$

Because of the absence of spherical symmetry in molecular potential the orbital momentum does not exist as a quantum number. The interaction (18) does not remove any degeneracy in the system. The initial density matrix, linear in spin-orbit coupling, now is [6, 10]:

$$\rho = |i><i| + \sum_n{}' \vec{\sigma}\frac{|n><n|[\vec{p}\nabla U(\vec{r})]|i><i| + |i><i|[\vec{p}\nabla U(\vec{r})]|n><n|}{4m^2(E_i - E_n)}; \qquad (19)$$

σ-dependent term here represents the density of helicity. It is responsible for the number of theoretically predicted phenomena.

At the electron scattering with large energy transfer the expansion in $\vec{p}_e\vec{p}/p^2$ can be performed. Indeed, there is a scale separation with the factorization of the free electron-electron scattering cross-section and the molecule matrix element of helicity as a result:

$$\sigma_+ - \sigma_- \sim Sp[\rho\frac{\vec{\sigma}\vec{p}_e}{p}]\,(\sigma_+ - \sigma_-)_{free}. \qquad (20)$$

However. this molecule matrix element vanishes in the first order in spin-orbit interaction. It is clear from the following simple consideration. To our accuracy $\vec{\sigma}\vec{p}_e$ is proportional to the time derivative of $\vec{\sigma}\vec{r}_e$ and should be zero for finite motion. In other words, momentum operator can be rewritten via the commutator with Hamiltonian. After that the sum over intermediate states drops out and the final commutator obtained vanishes:

$$Sp[\rho\vec{\sigma}\vec{p}] = \sum_n 2\frac{<i|[\vec{p}\nabla U(\vec{r})]|n><n|\vec{p}|i> + <i|\vec{p}|n><n|[\vec{p}\nabla U(\vec{r})]|i>}{4m^2(E_i - E_n)}$$

$$= \frac{i}{2m}<i|\epsilon_{klm}\nabla_k U(\vec{r})[p_l, r_m]|i> = 0 + O(\alpha^4). \qquad (21)$$

This vanishing leads to the suppression of the effect at large energy transfers. Therefore, the main contribution comes again from the domain of energy transfer comparable with characteristic atomic frequencies $\sim Ry$ and the specific properties of molecule potential are important. Thus, the asymmetry of the cross-section cannot be obtained universally for all types of chiral media by factorizing the initial "helicity" of molecular electron.

Z^2-enhancement of the effect in the scattering of electron is not the only possible manifestation of the spin-helical structure inside chiral molecule. It is necessary to produce P-odd energy shift between opposite isomers caused by weak interaction [12]. Now we would like to point out that the density of helicity determines also the high-frequency asymptotics of optical activity.

4 Asymptotics of optical activity at $\omega \gg$Ry.

The refraction index $n(\omega)$ is related to the forward-scattering amplitude $f(\omega)$ through the well-known formula:

$$n(\omega) = 1 + \frac{2\pi}{\omega^2}\frac{N}{V}f(\omega), \tag{22}$$

where N/V is the concentration. The usual asymptotics of the refraction index is determined by the Thomson amplitude $f = -\alpha/m$. Optical activity is caused by the term in $f(\omega)$ linear in the degree of circular polarization $\lambda = -i([\vec{e}^*\vec{e}]\vec{n})$. At characteristic atomic frequencies, $\omega \sim Ry$, the forward amplitude responsible for the rotation of the polarization plane f_λ could be estimated as

$$f_\lambda \sim \frac{\alpha}{m}\alpha\xi. \tag{23}$$

The spin-orbit contribution to the effect at these frequencies is very small and, which is more important, cannot be distinguished from the main effect.

It turns out, however, that the growth of ω causes surprisingly fast decrease of the effect in the neglect of the spin-orbit interaction. The asymptotic expansion of $f_\lambda(\omega)$ in the series of inverse powers of ω was developed first in [13]. It can be easily shown that the coefficients in this expansion are related with the optical rotatory strength in the transition weighted with some powers of frequency of that transition and summed over all possible intermediate states. It turns out that few first of these spectral sums vanish [14, 13] as well as corresponding contributions to ω^{-1} and ω^{-3} asymptotics of $f_\lambda(\omega)$. It should be noted here that the real part of optical activity amplitude must be an odd function of ω. The main reason for these cancellation could be presented as follows. By virtue of completeness relation these spectral sums are equivalent to matrix elements over the initial state from some *local* operators. These operators must be of superscalar character, i.e. they contain odd numbers of ϵ_{ijk} contracted with a tensor composed from the momentum operator and derivatives of the potential. The first possible operator of that kind is of very high dimension [13], $\epsilon_{kmn} p_k \nabla_i \nabla_m U \{ p_j, \nabla_i \nabla_j \nabla_n U \}$, which affects on the suppression of optical activity at $\omega \gg$ Ry:

$$ n_+ - n_- \sim \frac{N}{V} \frac{2\pi\alpha}{m\omega^2} Z^2 \alpha \xi \left(\frac{Ry}{\omega} \right)^5 . \tag{24} $$

The singled out relative magnitude of the effect is $Z^2 \alpha \xi (Ry/\omega)^5$. The Z^2 factor here is due to the singular behaviour of this operator. It can be shown that this result change neither by going beyond the one-electron approximation, nor by relativistic corrections to the dispersion law and to electron-electron interaction.

The spin of electron contributes to f_λ at the second order in spin-orbit interaction. At high frequencies its contribution to the optical activity arises in the first possible order, ω^{-1}, and is determined by the matrix element from the operator $\{ \vec{\sigma}\vec{p}, \Delta U \}$. The analogue of (24), the spin-orbit interaction induced asymptotics, now reads as:

$$ n_+ - n_- \sim \frac{N}{V} \frac{2\pi\alpha}{m\omega^2} \alpha^4 Z^5 \alpha \xi \frac{Ry}{\omega} . \tag{25} $$

Therefore. the optical activity asymptotics switches from ω^{-7} to ω^{-3} at $\omega \sim m Z^{-3/2}$. The most interesting is perhaps the fact that the optical activity there is directly related to the intriguing problem of the energy difference of right- and left-handed molecules due to parity nonconservation in weak interactions. Indeed, with good accuracy, especially for high Z, one can substitute $4\pi Z \alpha \delta(\vec{r})$ for $\Delta U(r)$ in the effective operator $\{ \vec{\sigma}\vec{p}, \Delta U \}$. But after it this operator coincides up to an overall factor with that of the P-odd weak interaction. Therefore. the measurement of optical activity in the region $\omega > m\alpha Z^{-3/4}$ may constitute an essential preliminary stage of an experiment aimed at the discovery of the mentioned P-odd energy difference. In the case of the success of such an experiment the knowledge of the OA would allow one to extract from it a reliable quantitative information on the weak interactions. At least, the optical activity measurement would be a reliable test of the accuracy of the theoretical calculations of P-odd effects in chiral molecules.

5 Conclusions

Staying away from the direct answer on the question of whether or not the origin of optical activity is connected with parity nonconservation in weak interactions, we tried to reanalyze physical aspects of the problem of scattering by chiral target. The calculation of the effect in sec.2 was done in the leading order in v/c. The reason being that for slow particles the cross-sections are biggest and therefore it is a more convenient situation for experimentalists. It was shown that the asymmetry appears in relative order $\alpha^2(\alpha/v)^2$ even in neglect of exchange interaction. The effects of exchange related with the density of helicity are important in the case of electron scattering because of Z^2-enhancement associated with the spin-orbit interaction. Numerically this asymmetry could be estimated as $10^{-5} - 10^{-4}$ which looks not to be hopeless for measuring in future experiments.

It is clear that for a real situation in nature the maximum of polarization of β-particles corresponds to $v \sim 1$. In this case relativistic corrections cannot be considered as small ones. At the same time the effects of exchange and the interference between first and second Born amplitudes rapidly decrease with the growth of v. In this situation the result of Zel'dovich and Saakyan, $\eta \sim \alpha^4(v/\alpha)\xi \sim \alpha^3\xi$, looks as a reasonable estimate. So, at $v \sim 1$ the asymmetry may reach 10^{-7} in the most favorable case.

The optical activity in the scattering of the high-frequency photons by chiral molecules was shown to be determined by the spin-orbit interaction. It gives also a reliable quantitative information about the P-odd energy difference of chiral isomers caused by weak neutral currents.

References

[1] S.F.Mason, Nature 311 (1984) 19.

[2] R.A.Hegstrom, A.Rich, J.Van House, Nature 313 (1985) 391.

[3] W.Kuhn, F.Braun, Naturwissenchaften 17 (1929) 227.

[4] W.A.Bonner, J.Mol.Evolut 4 (1974) 23.

[5] Ya.B.Zel'dovich, D.B.Saakyan, Soviet Phys. JETP 51 (1980) 2329.

[6] R.A.Hegstrom, Nature 297 (1982) 643.

[7] W.J.Meiring, Nature 329 (1987) 712.

[8] M.E.Pospelov "On the scattering of polarized particles by chi- ral molecules", Preprint DOE/ER405661-108-INT93-10-0

[9] L.D.Landau, E.M.Lifshitz. Quantum Mechanics (Pergamon Press, 1965)

[10] M.E.Pospelov, ZhETF 104 (1993) N8(2) (Soviet Phys. JETP).

[11] A.Garay, P.Hrasko, J. Mol. Evolut. 6 (1975) 77.

[12] R.A.Hegstrom, D.W.Rein, P.G.H.Sandars, J. Chem. Phys. 73 (1980) 2329.

[13] I.B. Khriplovich, M.E. Pospelov Phys. Lett. 171A (1992) 349.

[14j A. Hansen, Mol.Phys. 33 (1977) 483.

ON THE DETERMINATION OF THE PHYSICAL ORIGIN OF HOMOCHIRAL-ITY IN LIFE

David B. Cline
University of California Los Angeles. Dept. of Physics and Astronomy
405 Hilgard Avenue, Los Angeles, California 90095-1547 USA

ABSTRACT

There are many who believe that the observed molecular homochirality in all living things is actually required for life and must have preceeded it in the prebiotic medium. While the sign of the homochirality could be the result of chance, it is also possible that it is produced by the weak interactions being chirally broken. We consider three stages of this problem: (1) Fundamental interactions that may cause asymmetries; (2) Amplification and experimental tests of the effects through bifurcation processes; and (3) The possible influence of these effects on prebiotic molecules in pre-solar system (dense) molecular clouds. We critically review the experimental evidence for chiral-symmetry-breaking effects and discuss a possible new round of experiments using powerful GaAs polarized RF electron guns.

1. Homochirality of Life

For more than a century, there has been evidence for the chiral nature of life forms on Earth. Pasteur was among the first to point this out (1848-1880), and the universal nature of chiral symmetry breaking in DNA and RNA is now very well established for all life forms. Figure 1 shows how the homochirality is manifest at the molecular level. With the discovery of parity violation in charged current reactions in 1956 and of the weak neutral currents (WNCs) in 1973, two universal symmetry-breaking processes (WNC and β-decay) were uncovered that could have determined the handedness of DNA and RNA. These processes are shown at the fundamental level in Table 1. The main problem is the extremely small symmetry-breaking effects ($\Delta E/k_B T \sim 10^{-17}$). However, there are plausible non-linear mechanisms that could have amplified this small, symmetry-breaking phase transition up to the full symmetry-breaking level observed in life forms. There is, nevertheless, a long-standing controversy as to whether these non-linear effects are actually large enough to have determined the selection of the handedness of life [1-11].

Recently there has been increasing interest in the chiral nature or handedness of biomolecules. In fact there are some who claim that the complex biomoleculer structure of life must have arisen from a "chiral pure" medium [12,13] and may be a precondition for the emergence of self-replicating biomolecular systems. Table 2 lists some of these ideas, which have been put forward largely by W. Bonner and V. Goldanskii [12,13]. We also point out a recent review in Ref. [14]. This concept, combined with the likelihood that the period on

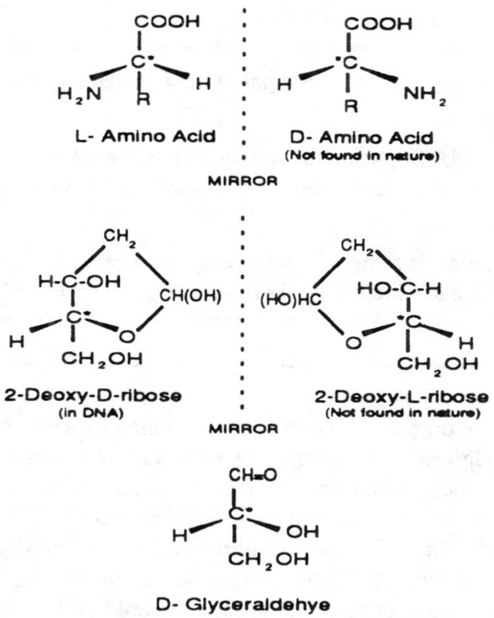

Figure 1. Examples of molecules that are isomers, some of which exist in nature and some that do not.

TABLE 1. History of physical chiral symmetry-breaking observation

Period	Observation	Consequences
1953-1955	τ-θ puzzle in Strange particles	Hints of parity violation
1957	Parity violation prediction/observation in Co^{60} and $\pi^{\pm}/\mu^{\pm}$ decays	V–A theory of weak interactions; ν_e, $\bar{\nu}_e$, $e^{\pm}$ have definite chirality
1964	CP violation ($K_L^0 \to \pi^+\pi^-$ observed)	$K_L^0 \to \pi^+e^- \bar{\nu}_e$ and $K_L^0 \to \pi^-e^+ \nu_e$ have different rates
1973	Observation of WNC	Electroweak unification
1976-1978	Parity violation in WNC	Atomic parity violation
1983	$W^{\pm}/Z^0$ observation at CERN–UA1	Electroweak theory
1987	$\bar{\nu}_e$ detected from SN1987A	--
1986-1994	Precision measurements of electroweak parameters	--
FUTURE: Test of CPT; Origin of CP violation		

TABLE 2. Importance of Homochirality in Living Systems [12,13,15]

1. DNA: Self-replication would not work with heterochiral systems (50% L and 50% D).

2. Errors in DNA Replication: Without a pure chiral structure, the error rate in replication would be unacceptable to long-lived systems (higher animal forms, trees, *etc.*).

3. In a Prebiotic Medium: Homochirality must have been either
   (A) Established in a very short time on earth ($\leq$ 100 Myr), or
   (B) Existed in ISM organic materials near the solar system.

Earth for life to have originated seems to be sometime between 3.8 and 3.5 billion years ago, leaves a small window of 300 million years or less for life to have emerged from the prebiotic medium. Indeed, some speculate that the time could be less than 10 million years.

The viewpoint of this review is to consider three issues: (1) Fundamental interactions that could cause homochirality, (2) Amplification mechanisms, and (3) Application to the pre-solar system interstellar medium (ISM), which may be the natural origin of the prebiotic molecules. In this paper we refer to a recent conference, sponsored by UCLA, on the "Physical Origin of Homochirality in Life," held in Santa Monica, California in 1995 [15], and provide a critical review of the experimental evidence for asymmetric interactions in biological materials.

2. The Molecular Parity Selection of Charged and Weak Neutral Currents

Charged-current effects that violate parity is one candidate for the symmetry-breaking process [9,10]. The WNC, which interferes with the electromagnetic field and gives rise to weak distortions of virtually all electromagnetic effects, could be one of the candidates that causes symmetry breaking in the biosystem [11]. It could make the broken symmetry of the micro world translate to the macro world and, finally lead protein and both DNA and RNA to display, respectively, left-handed (L) and right-handed (D) forms. For organic molecules, however, the difference between the ground state energies of the L and D types is about $\Delta E \sim 10^{-17} k_B T$ of the thermal energy at room temperature [6,7].

There have been various experimental tests of the effects of weak interactions on biological materials, which have all been null to date. (See Refs. [12] and [13] for a review and for an excellent introduction to this subject.) We will return to this issue later when we discuss possible new experiments.

There are other processes that select chiral states, for example absorption of circularly polarized light [2,3]. For some time, there has been a debate as to whether combinations of electric and magnetic fields could cause a chiral symmetry breaking [3,13]; for a recent discussion see Barron [15], which was presented at the meeting by A. MacDermott, Oxford University.

In this review, we consider methods to investigate the origin of the homochirality by studying the internal properties of the biomolecules (illustrated in Fig. 2). These studies will amount to the detection of some sort of macroscopic form of parity violation using different types of probes.

(A) SPATIAL ONLY

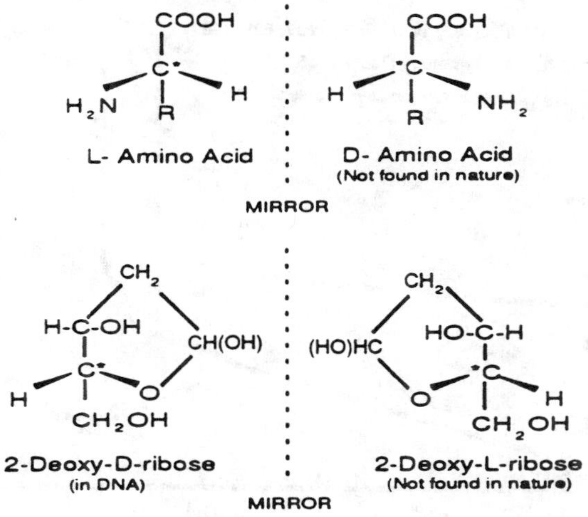

(B) ELECTRON SPIN ALIGNMENT

(SPIN-ORBIT INTERACTION IN BIOLOGICAL MATERIALS)

POSSIBLE TESTS FOR (A) OR (B) OR BOTH
1. CPL Destructive
2. Spin-Orbit Asymmetry of $e^{\pm}$ Beam Interactions

Figure 2. The various components of symmetry breaking in molecules.

These studies will amount to the detection of some sort of macroscopic form of parity violation using different types of probes. Rather than stress a personal opinion on the origin of homochirality, we intend here to point out the past and possible future experiments that may be relevant in answering this question.

The calculations of the effect of weak interactions in the scattering of $e^{\pm}$ in organic molecules are very incomplete. The most recent calculations of Hegstrom [16] provide an excellent set of references to the early work, and the most complete recent calculations have been carried out by W. R. Meiring with the mediated symmetry shown in Fig. 3 [17]. (At the end of this review, we will indicate some criticism of this calculation.) At high energy, the $e^{\pm}$ asymmetries are the same, since the product $q\vec{p} \cdot \vec{\sigma}$ is the same sign for $e^{\pm}$ emitted in β-decay. At low energy, these calculations are more difficult, since the effects of e^{-} exchange diagrams are very important. If these calculations are correct, at 100 eV the e^{-} asymmetry could be very large (10^{-2} - 10^{-4}).

We now turn to the experiments. Over the past 20 years, many experiments have been carried out (see Ref. [12] for a nearly complete list). However, it appears that nearly every positive effect that was observed has turned out to be incorrect. In Table 3, we list some

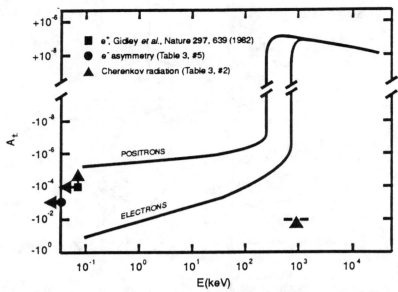

Figure 3. Most recent published calculation of the expected asymmetry for the scattering of $e^\pm$ from chiral molecules [17]. We also include some recent measurements or limits on such an asymmetry.

TABLE 3. Experiments where a chiral effect has been observed (and not refuted)

	Experiment	Beam/ Source	Target	Detection	Comments	Ref.
1	CPL on L/D photo absorption	UV/keV radiation	DC tartaric acid, DL alanine, DL glutamic	Observe destructive difference in one chirality	To be expected from optical activity	Norden [19]
2	e^- target → Č light	P^{32} source and Ca^{137} (no chiral electrons)	R- or S-PBA	Observe different Č light intensity due to chiral electrons	Effect too large to be due to Č radiation from spin effects	Garay et al. [20]
3	Co^{60} → γ + [L,D]	Co^{60} γs	D or L alanine	Observe different amounts of L, D after irradiation	Other experiments did not produce this effect	Akabosh et al. [21]
4	Introduction of L/D chiral particles	Sr^{90}, Y^{90} β-decay	Y^{90} D or L alanine	Detect effects by electron spin resonance technique	The electron spin resonance may be sensitive to spin dependence	Conte et al. [22]
5	Low-energy polarized e^- beam	GaAs polarized source, 5 eV	Camphor L, D	Observe different electron polarization and beam attenuation in L/D	At such low energy, asymmetry may be too large	Campbell et al. [23]

experimental results that are not yet refuted or in direct conflict with previous null effects. We will discuss two of these experiments here. The observation that circularly polarized light destroys L and D isomers selectively (entry 1 in Table 3) is now well established. Figure 4 shows a fairly recent observation, and there is no doubt that this effect is real. (We understand from a private communication from L. Keszthely that there are now experiments that seem to contradict the results of entry 5 on Table 3.)

The other experiment to which we will refer is that of entry 2 (also on Table 3), the study of Cherenkov light in L or D material with chiral e^- (β-decay) [20]. The main results from this experiment are shown in Fig. 5. The authors claim an effect of perhaps 10^{-2} magnitude. This is, in our belief, far too large an effect, but a future study of Cherenkov radiation from L or D materials with polarized e^- beams could be promising.

In Fig. 3, we show the current limits or observations of asymmetry. The e^+ measurement of Gidley *et al.* [18] is a very nice experiment. Our conclusion is that no present experiment has reached the level of sensitivity needed to observe an effect. Therefore, it is premature to count out the weak force as a determining factor in the origin of homochirality in life.

3.   The Conditions for the Small Electroweak Asymmetries in the Prebiotic Medium to Be Amplified – Simulations

If it is possible for the WNC interaction to provide additional parity-selection energy difference in an organic biosystem, the handednesss of the organic molecules might appear. Such additional parity-selection energy difference could be thought of as a small bias in this transition process. For simulation, this small bias is converted into a non-dimensional parameter, $g(\Delta E/k_B T)$, to characterize the relative difference in the time to effect the phase transition. The work of Kondepudi and Nelson is seminal in this regard [1]. At UCLA, a series of simulations with constant g has been done that shows that an amplification process can occur [8,24].

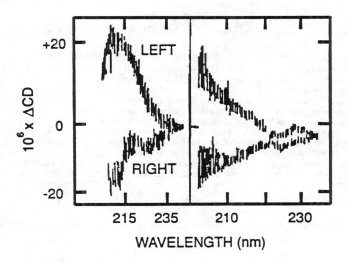

Figure 4. Effects of polarized light on racemic organic molecules.

271

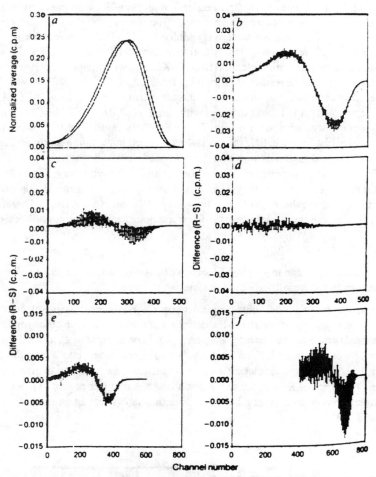

FIG. 1. *a*. Cerenkov pulse-height spectrum of ^{32}P dissolved in R- or S-PBA. Differential spectra (*b-f*) are generated by subtraction of the count rate in each channel of S from those of R. Differential spectra of *b*, R- and S-PBA shown in *a*; *c*, external ^{32}P and *d*, Compton electrons from ^{137}Cs. Scintillation differential spectra for *e*, ^{35}S and *f*, ^{32}P. Scintillation solutes were PPA and POPOP, see ref. 12 for method. A smaller amplitude for these processes may reflect the multiple achiral energy transfer processes of the liquid scintillation measurement. The first part of the ^{32}P spectra was removed to minimize the Cerenkov effect recorded in low-energy channels and concomitant irradiation from β-particles from unavoidable ^{33}P contamination. The latter may contribute to the high standard deviation on the ^{32}P differential scintillation spectrum. Note the different scales on the ordinate of the differential spectra. All spectra represent the average of four PBA samples measured in random order; the instrument was calibrated before each measurement. Saturation of photomultiplier and overspill between channels were avoided by assuring low count rates; samples of <200,000 total c.p.m. were used. All spectra were taken in five-channel increments and normalized to the total count rate. Error bars represent the standard deviation for each point. This set of experiments was repeated eight times with different batches of enantiomers and ^{32}P, a total of 32 spectra. All measurements yielded consistent results.

Figure 5. Possible evidence for different intensities of Cherenkov light from chiral electrons on dissolved PBA (liquid) samples [20].

In contrast to the gradual buildup of symmetry breaking [1,8], as proposed by Kondepudi, we also studied a more general condition that also allows for an impulse of chiral symmetry breaking [24,25]. For example, a natural fission reaction, where a large amount of U^{235} may have accumulated by accident, or a nearby supernova explosion could provide such a chiral impulse. Initially this might seem to be an unimportant effect that has been ignored in the past literature on this subject but, with increasing interest currently, more attention should be paid to this issue. Here we present some results based on our previous work [8] [($\alpha = (X_L X_D)/(X_L + X_D)$, the concentration of L and D isomers]:

$$\frac{d\alpha}{dt} = -A\alpha^3 + B(\lambda - \lambda_c)\alpha + \epsilon^{1/2}f(t) + Cg \qquad (1)$$

becomes

$$\frac{d\alpha}{dt} = -A\alpha^3 + B(\lambda - \lambda_c)\alpha + \epsilon^{1/2}f(t) + Cg + d\delta_{\gamma\gamma'} \quad , \qquad (2)$$

where α is the amplitude of the symmetry-breaking solution, λ is the control parameter, λ_c is the symmetry-breaking transition (critical) point, $\epsilon^{1/2}$ is the rms value of fluctuation (noise), $f(t)$ is the normalized fluctuation (noise), g is the interaction or bias symmetry-breaking selector, d is the amplitude of a quasi δ-pulse, and $\delta_{\gamma\gamma'}$ is the δ-function, which gives $\delta_{\gamma\gamma'} = 1$ when $\lambda = \lambda'$ and $\delta_{\gamma\gamma'} = 0$ when $\lambda \neq \lambda'$. Therefore,

$$\lambda = \lambda_c + \gamma t \quad , \qquad (3)$$

where λ_c is the initial value of λ, γ is the evolution rate, and t is the evolution time.

To solve the first-order stochastic equation numerically, we designated the initial symmetry-breaking amplitude, α, as zero at time $t = 0$, but the $f(t)$ is a random number generated by the computer. To obtain the trace of α at a different time for each trial, we just sampled it at different times. (Our studies confirm the results of Kondepudi and Nelson [1].) One simulation assumes that the impulse takes place near and after the critical point ($\lambda_c = 1.0$) at $\lambda = 1.5$, and $\delta = d/g = 100$ is shown in Fig. 6. In this case, the impulse plays a dominant role to push the process toward a favored state! We have studied over 40,000 trials and find a trend that the early chiral impulse produces a significant bias towards the preferred state. Furthermore, we found an 88% probability of the favored state being selected [8,24]. This illustrates the effect of a small symmetry-breaking process on a slowly evolving system. In essence, the small bias is magnified by the process of signal averaging, where the noise effect essentially cancels out. This process is sometimes called the Kondepudi effect [1].

The addition of a chiral impulse to the gradual bifurcation process of Kondepudi and Nelson could provide arguments in favor of a rapid buildup of the homochiral prebiotic medium, either in the ISM before the solar system was formed, and/or during the evolution of the medium in the water-based system on Earth. We have also studied the system near the critical point in our simulations and do not observe the effect discussed in Refs. [3, 13, and 14].

Recently, we have developed an electronic circuit to study the bifurcation process, as shown in Fig. 7(A) [25]. We have paid special attention to the noise generator, which is crucial if a noise-averaged process is to be studied (as in Ref. [1]). Figure 7(B) shows the scope trace for the bifurcation studies, showing clear evidence that a small bias can shift a state even in the presence of large noise! We are continuing parameter studies with this electronic circuit to further investigate this noise-averaged bifurcation process.

273

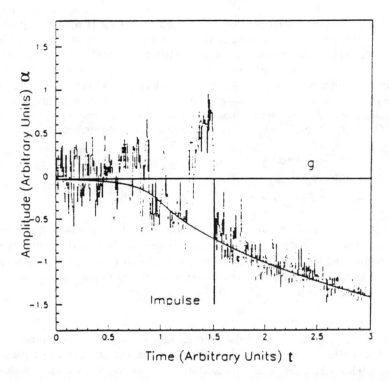

Figure 6. Single trial with an impulse at $\lambda = 1.5$, where $\delta = d/g = 100$. The impulse strongly affects the phase transition.

(A)   (B)

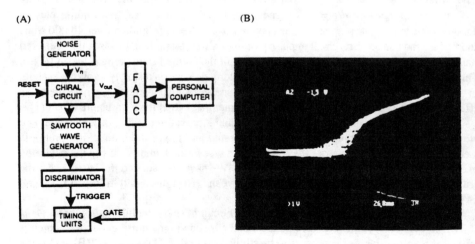

Figure 7. (A) Schematic of the UCLA chiral symmetry-breaking electronic circuit for bifurcation processes. (B) Scope trace of the output of the electronic circuit of (A), showing that a small bias in the presence of large noise can still switch the system into a preferred state.

274

4. The Santa Monica Symposium

For many years, there have been several issues associated with the homochiral structure of biomolecules, as first observed by L. Pasteur in 1848:

(A) Is a homochiral structure necessary for life as we know it?

(B) Did homochirality precede the formation of life (homochiral prebiotic medium hypothesis)?

(C) Is there any reasonable physical mechanism that could have produced the large chiral symmetry breaking in the prebiotic medium or in the observed homochiral structure?

(D) Is the homochiral structure an accident that occurred in the biological systems and was later amplified?

(E) Can the homochirality be used as a signature for existing or previous living systems in the solar system or other parts of our galaxy?

(F) Are there any experiments that can be carried out now to clarify the origin of homochirality?

These and many other questions were the basis for organizing the first symposium on this subject in Santa Monica, California. Some of the leaders in this field attended the meeting, made presentations, and joined the discussions. The following is a brief review of this meeting.

The Santa Monica meeting started with an overview of the importance of homochirality in biomolecular life by W. Bonner (Stanford) and V. I. Goldanskii (Moscow). Their conclusions were that homochirality is almost certainly required for the formation of the complex, self-reproducing biomolecular structures (*i.e.*, DNL) required for life. They both stated that the prebiotic medium should have been homochiral, although they differed on how and even where this may have happened. A. K. Mann (U. Penn.) and D. B. Cline (UCLA) discussed the possible physical processes that could have helped produce chiral symmetry breaking. Radioactive β-particle interactions and the influence of WNC all "violate parity" but are also very small effects. There was a brief discussion of the role that nearby supernovas could have played in the ISM before the solar system formed. It was agreed that some powerful amplification mechanism must have intervened to achieve any large effect.

On February 16, the first session was "Biomolecular Aspects of Homochirality." There were four very interesting presentations: Goldanskii (Moscow), G. Gilat (Technion), V. A. Avetisov (Moscow), and S. Miller (UCSD). Goldanskii discussed the Frank process (invented by and named after the British chemist), which can be used to amplify physical or spontaneous chiral symmetry breaking in so far as no definite chemical reaction has been proposed as an example of the Frank process. Goldanskii showed that formaldehyde could be assembled in a Frank-type process. Gilat indicated some other types of chiral symmetry-breaking processes that should be studied in the laboratory. Avetisov discussed the very interesting question of assembling a complex structure like DNA out of homochiral monomers. He indicated that even with a homochiral prebiotic medium, a "complexity threshold" would be reached, the passing through of which is hard to understand. Part of the final talk in this session by Miller dealt with the famous Miller–Urey experiment, which showed many years ago how amino acids can be formed using an electrical discharge in a methane–water, *etc.* gas medium. In his talk, Miller indicated that homochirality is not needed to give birth to biomolecules, however, it is a very great aid to life. He also pointed out that the organic materials produced in the famous Miller–Urey experiment were tested and found to be racemic!

The next part of the meeting, "Physical Origin of Homochirality–Symmetry Breaking," took the better part of February 16 and the first part of the morning of February 17, and caused the most controversy. There were three themes:

1. The weak interaction, which is the only universal chiral symmetry-breaking system -- A. MacDermott (Oxford Univ.), R. A. Hegstrom (Wake Forest Univ.);
2. Small effects might be amplified by autocatalytic Frank-type processes -- D. D. Kondepudi (Wake Forest Univ.), Goldanskii, and MacDermott;
3. There may be more useful mechanisms for breaking the symmetry (*e.g.*, a biochemical process or if the organic molecules are formed in the ISM and subjected to polarized light from neutron stars) --A. Eschenmoser (Zurich), P. E. Nielsen (Copenhagen), and Bonner.

Let us elaborate a little on some of the newer results presented at the meeting. MacDermott has carried out some very interesting calculations of the energy difference of several chirality biomolecular configurations resulting from the WNC and has found that some systems could have 10^{-14} kT energy difference, which is orders of magnitude larger than all previous estimates. Kondepudi once again emphasized the importance of the Frank/autocatalytic process and has even tested one of the ideas by studying salt crystals where the symmetry is broken by stirring the liquid! Eschenmoser showed how p-RNA could have symmetry breaking effects. Finally, Bonner elaborated on the "abiotic" possibilities of the homochirality and indicated that the time available on the early Earth for such a process to take place was less than 300 million years and could be even less than that. He indicated that a more viable possibility could be the formation of homochiral molecules in the ISM, and he proposed a specific mechanism caused by the absorption of polarized UV light from one or many neutron stars in the ISM. These ideas and others led to a healthy debate at the meeting. One issue was associated with so-called "false" chirality caused by electric and magnetic fields recently proposed by L. Barron (Univ. of Glasgow), whose talk was presented by MacDermott; similar ideas were discussed by Gilat.

The third session was devoted to "Astrophysical and Planetary Aspects of Homochirality and Origin of Life on Earth." The presentations by J. Bada (Scripps/UCSD) and J. M. Greenberg (Leiden) were very interesting. Bada first showed that biomolecular systems (*i.e.*, human teeth) progressively racemize as they age and indicated that the age of the Swiss Iceman is being determined this way. Then Bada pointed out that there is no evidence for non-racemized organic molecules in micrometeorites in the polar ice or at the K-T edge in sediments. This is a serious constraint on the theories that the organic molecules were brought in by comets, *etc.* Of course, it may not be directly relevant to what happened four billion years ago. The highlights of Greenberg's talk, "Photochemical Production of Non-Chiral and Chiral Organics in the Interstellar Dust: Laboratory, Observation, and Theory," were: Organic materials make up ~10^{-4} of the mass in the galaxy (as dust); Halley's comet carries 10% of the biomass on earth; and Organisms in dust undergo complex chemical reactions with the help of UV light (this has been studied in the laboratory where amino acids and other biochemicals are produced). Greenberg's main theme was that homochiral biomolecules could be produced from the polarized light from neutron stars over a long evolutionary period of the large gas clouds in the galaxy (this concept was first discussed by

Bonner, see Ref. [12] for example). Greenberg went on to describe an experiment carried out with polarized UV light where selective destruction of L or D molecules was observed, thus confirming the Bonner hypothesis that comets could have brought in both homochiral organisms and water to the early Earth approximately 4 billion years ago. (There was also a discussion that supernova in the ISM could provide a chiral impulse by producing the antineutrino and $A\ell^{26}$.)

The final section of the meeting was "Future Perspectives and Experiments." It is clear that some experimental results are needed if this is to be a viable field of study. I. Khriplovich (Novosibirsk) discussed a possible investigation of high-frequency radiation on homochiral substances, a careful study of which could yield the energy differences resulting from the WNC. An open discussion on future experiments was held with many participants providing suggestions. One interesting possibility is to use the new high-intensity, polarized electron guns for linear colliders in order to study the effects of polarized electrons on homochiral materials. This suggestion and other proposed experiments will be summarized in the Proceedings.

5. Organic Molecules in Space and the Possible Role of Nearby Supernovae and Neutron Stars

One of the main themes of the Santa Monica meeting was the likelihood that most of the early organic material on Earth was brought in by comets and asteroids. References [26-29] give a nice introduction, from different points of view, to this concept. There are some interesting "large numbers" to consider in this regard:

1. The estimated amount of dust matter in the galaxy is $\sim 10^{-4}$ m_G, or $\leq 10^7$ solar masses, largely in the form of dust grains. A fraction of that material is in the form of organic materials [26].
2. It has not been possible to measure the amount of interstellar dust that has accumulated on the Earth (bringing organic material) [27,28].
3. In a molecular cloud of density 10^4 M/cm^3 and 1 parsec radius, there could be complex organic matter equal to 100 solar masses.
4. The Earth revolves around the galaxy with a period of ~ 250 million years, and it likely encounters several dense molecular clouds in this trajectory.
5. It is likely that large quantities of organic material were deposited in the Earth in the first one billion years.

The above information is obtained by the infrared scattering from the dust in the galaxy and by modeling various UV-driven processes here on Earth [26]. Ultraviolet photo processing plays an important role in the organic chemistry of the dust particles [27]; see also Ref. [29], from which Fig. 8, showing the nature of the dust gram with prebiotic molecules inside, is taken.

We discuss here two scenarios where chiral interactions in the ISM could have led to a preponderance of homochiral molecules. These two concepts are outlined in Table 4, with Fig. 9 giving a fairly complete description of the hypothesis. The second hypothesis is illustrated in Fig. 10, where the possibility of the relative survival of the $e^{\pm}$ polarization deep within the cloud is also illustrated. Of course, during this time the effect of the WNC can be driving the system towards a homochiral state, as shown previously by Eqs. (1) and (2).

277

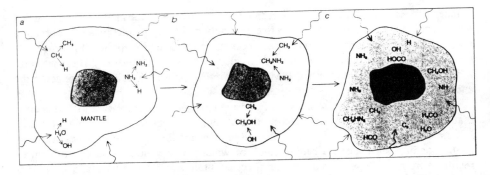

STRUCTURE OF GRAINS when they initially accrete is inferred from laboratory simulations in which mixtures of water, methane, ammonia and other simple molecules are subjected to ultraviolet irradiation at a temperature of 10 degrees K. Each grain begins as a silicate core that condensed in the atmosphere of a cool giant star. Around this core a mantle of ices forms. Ultraviolet radiation breaks some of the mantle molecules into radicals, or reactive molecular fragments (a). The radicals can then recombine in new ways (b). Over a longer period the continued ultraviolet irradiation of the grain can give rise to ever more complex mixtures of molecules and radicals (c).

Figure 8. From [29] and used in the talk of M. Greenberg at the Homochirality Symp., Santa Monica, CA, 1995.

TABLE 4. Possible sources of CPL (UV/keV) radiation in dense molecular clouds

<hr>

Primordial Soup → Molecular Clouds (ISM)

<hr>

1. Synchrotron radiation from neutron stars ([12,26]
   - CPL helicity depends on position (*i.e.*, above or below star);
   - In principle, this mechanism works. However, on the 250-Myr orbit around the galaxy, this effect is expected to average out.

2. Radiation from weak interaction processes [30] – injection into molecular cloud
   - Processes: supernova II ν_e interaction, $A\ell^{26}$ from nearby supernovas, *etc.*
   - Because of grain structure, dE/dx will be very different from that of solids, gases, or liquids;
   - Always gives the same chiral symmetry breaking;
   - Can act as a chiral impulse along with WNC

Let us consider the rate of these three effects:

1. For $\overline{\nu}_e$ absorption and a supernova 1 parsec away (or inside a 1-parsec dense cloud), the number of interactions will be $\sim 10^{-3}$/ kg of material for 100 $M_\odot$ of organic material (which would be 10^{12} grams of organic matter that is *active*), therefore the positrons from the $\overline{\nu}_e$ interactions would lose energy at a rate of 10^{-19} MeV/cm and thus travel over a parsec.

2. For the coherent $\nu_x + N \to \nu_x + N$, and for the carbon in the hydrocarbons, we would have $\sim 10^2$ more or $\sim 10^{14}$ grams of *active* material. This effect could be very important in light of the small energy difference that separates L and D molecules and the possibility of large coherent effects.

3. For the $A\ell^{26}$ over the half-life, there would be $\sim 10^{50}$ decays producing $\sim 10^{50}$ positrons that lose energy at the rate of 10^{-19} MeV/cm; for MeV positrons, the range of which would be of order of a parsec (ignoring possible magnetic-field effects).

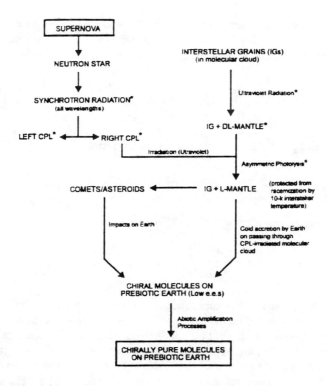

Figure 9. Possible extraterrestrial origin of terrestrial homochirality.

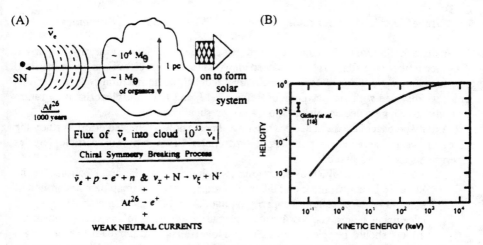

Figure 10. (A) Schematic of the effect of a supernova II on the organic molecules in a dense molecular cloud. (B) Estimated polarization of the β^+ decay particles as the energy decreases by ionization in H_2O and in the dense molecular cloud. One observation of the polarization of low energy e^+ by the Michigan group.

Consider the example where 0.001 $M_\odot$ of $A\ell^{26}$ is produced and assume, for simplicity, that the energy of the e^+ is 1 MeV and is contained in the gas cloud. Assume that the cloud has a density of 10^4 atoms/cm^3 and that 10^{-3} of the atoms are organic, then the stopping power for e^+ will be

$$\frac{dE}{dx} \sim \text{MeV/g/cm}^3 \quad , \tag{4}$$

and for a density of $\rho = 10^4$ atoms/cm^3 $\sim 10^{-17}$ g/cm^3, we find

$$dx \sim \frac{dE}{\rho}(\text{MeV}) \sim 10^{19}\,\text{cm}(\sim 3\,\text{parsec}) \quad , \tag{5}$$

and for an average energy exchange of 10 eV, we have

$$\frac{10^5_e \text{ collisions}}{A\ell^{26} \text{ decay}} . \tag{6}$$

For 0.001 $M_\odot$ of $A\ell^{26}$ and 10^{-3} organic fraction, we obtain a total of $\sim 10^{50}$ collisions of polarized positrons, with organic materials in the cloud assuming all of the e^+ stop in the cloud. (We assume that only one of the collisions can result in spin exchange.) There will also be the same order of polarized photons from the $e^+e^- \to \gamma\gamma$ annihilation. It is estimated in Refs. [17 and 24] that the asymmetry due to the weak interaction would be of order 10^{-11} to 10^{-6}, depending on the positron energy {it scales like $\alpha^2[\alpha/(v/c)]^2$}. Thus, it takes $N \sim 10^{22}$ interactions for the asymmetric to become statistically important. In this example there are far more interactions.

6. Future Tests in the Laboratory, Alternative Viewpoints, and Summary

In this brief review, we have only covered part of the many ideas for the physical origin of homochirality (see Ref. [31] for another viewpoint). However, we conclude that it is premature to rule out the weak force as the main effect behind homochirality in life. In this sense, we disagree with the viewpoint of Refs. [12, 13, and 15]. It is clear that only future experiments can help decide this question -- not personal opinions.

As a by-product of the Santa Monica meeting, there is now a renewed effort to study the effect of weak interactions on biological materials. Two types of particle physics experiments are being studied:

1. With the powerful RF electron guns, which provide polarized electrons for the Stanford Linear Collider and other laboratories, one can imagine a new round of experiments to search for parity violation in biological materials. In this experiment, it would be possible to decelerate the electrons to low energy where the effects should be relatively large. A Novosibirsk-UCLA team is preparing a proposal for these experiments. One initial experiment would be to study the Cherenkov light from L material (PBA) while reversing the polarization.

2. The possible use of reactor antineutrinos to simulate the effects of weak interactions. including coherent scattering from the nucleus, is being considered. (This could simulate the effect of supernova $\overline{\nu}_e$, *e.g.*, by taking very long time exposures to obtain an equivalent flux.) It is essential to get new calculations of the scattering asymmetry [24,32].

Other terrestrial experiments will include searches for homochiral materials in meteorites, and experiments with crossed electrical and magnetic fields and purely chemical methods of inducing homochirality.

Another class of extraterrestrial experiments will involve the search for homochiral materials in comets, asteriods and, perhaps, Mars. It is obvious that, if homochiral materials are found in these systems and all display the same homochirality as that of living systems on Earth, the case for a common physical origin is made much stronger. One intriguing possible experiment would use a future Rossetta Mission to a comet. The suggested detector, SETH (search for extraterrestrial homochirality), which would be the size of a cigar, would measure the optical activity of the samples in situ. (A. MacDermott is leading this effort and reported on it at this symposium.)

We may be entering into a new era where the very foundation of life and the physical forces of nature are seen to have a strong connection.

I wish to thank the attendees of the Santa Monica Symposium, and L. Keszthely (at this meeting), for many helpful comments.

7. References

1. Kondepudi, D. D. and Nelson. G. W.: *Nature* **314** (1985), 438-441.
2. Zel'dovich, Y. B. and Mikhailov, A. S.: *Sov Phys Usp* **30** (1987), 977-992.
3. Avetisov. V. A., Goldanskii, E. I., and Kuz'min, V. V.: *Physics Today*, July 1991, pp. 33-41.
4. Salam, A.: *J Mol Evol* **33** (1991), 105.
5. Avetisov, V. A., Goldanskii, V. I., and Kuz'min, V. V.: *Physics Today*, April 1992, p. 199.
6. Hegstrom. R. Q., Rein, D. W., and Sandars, P. G. H.: *J Chem Phys* **73** (1980), 2329-2341.
7. Mason, S. F. and Tranter, G. E.: *Proc R Soc London* **A397** (1985), 45.
8. Cline, D. B., Liu, Y., and Wang, H.: in *Proc of Int Symp on 30 Years of WNC*, AIP Press, March 1994.
9. Ulbricht, T.: *Quart Rev* **13** (1959) 48.
10. Vester, F. *et al.*: *Naturweissenschaften* **40** (1958), 68.
11. Yamagata, Y.: *J Theor Biol* **11** (1966), 495.
12. Bonner, W.: *Origins of Life Evol Biosphere* **21** (1991), 59-111.
13. Goldanskii, V. I. and Kuz'min, V. V.: *Sov Phys Usp* **32** (1989), 1-29.
14. Keszthely, L.: "Origin of the Homochirality of Biomolecules," *Quarterly Rev Biophys* (1995), to be published.
15. D. Cline (ed.): *Proc of the 1st Symp on the Physical Origins of Homochirality of Life*, Santa Monica, CA, Feb. 1995, AIP Press, to be published.
16. Hegstrom. R. A.: *Nature* **315** (1985), 749-750.
17. Meiring, W. R. *Nature* **329** (1987), 712-714.
18. Gidley, D.: *Nature* **297** (1982), 639-643.
19. Norden, B.: *Nature* **266** (1976), 567-568.
20. Garay, A.: *Nature* **346** (1990), 451-453.
21. Akabosh, N.: *Origins of Life* **20** (1990), 111.
22. Conte, G.: *Nuovo Cimento* **44** (1985), 641-647.
23. Campbell, D. and Farago, P.: *Nature* **318** (1985), 52-53.
24. Cline, D. B., Liu, Y. and Wang, H.: *Origins of Life* **25** (1995), 201.
25. Park, J. *et al.*: "The Origins of Homochirality: A Simulation by Electrical Current," *Photoelectronic Detectors, Cameras, and Systems, SPIE* **2551** (1995), 40-52.

26. Greenberg, J. M. and Mendoza-Gomez, C. X.: *The Chemistry of Life's Origins*, Kluwer Acad. Pub., 1993, pp. 1-32.
27. Khare, B. U. and Sagan, C.: "Experimental Interstellar Organic Chemistry, preliminary findings," in *Molecules in the Galactic Environment*, Wiley, 1973.
28. Chyba, C. and Sagan, C.: *Nature* 355 (1992), 125-131.
29. A very useful reference on the origin of biomaterials in the ISM is Whittet, D. and Chiar, J.: "Cosmic Evolution of the Biorganic Elements and Compounds," *Astro Rev* 5 (1993), 1.
30. Cline, D.: "Homochiral Prebiotic Molecule Formation in Dense Molecular Clouds," UCLA preprint No. PPH0072-10/95, *Nature*, submitted.
31. Chela-Flores, J.: *Chirality* 6 (1994), 165.
32. Pospelov, M.: "On the Scattering of Polarized Particles by Chiral Molecules," *Proc of the 1st Symp on the Physical Origins of Homochirality of Life*, Santa Monica, CA, Feb. 1995, D. Cline (ed.), AIP Press (to be published), submitted.

Author Index

AIP Conference Proceedings

Title	L.C. Number	ISBN
No. 307 Gamma-Ray Bursts Second Workshop (Huntsville, AL 1993)	94-71317	1-56396-336-1
No. 308 The Evolution of X-Ray Binaries (College Park, MD 1993)	94-76853	1-56396-329-9
No. 309 High-Pressure Science and Technology—1993 (Colorado Springs, CO 1993)	93-72821	1-56396-219-5 (Set)
No. 310 Analysis of Interplanetary Dust (Houston, TX 1993)	94-71292	1-56396-341-8
No. 311 Physics of High Energy Particles in Toroidal Systems (Irvine, CA 1993)	94-72098	1-56396-364-7
No. 312 Molecules and Grains in Space (Mont Sainte-Odile, France 1993)	94-72615	1-56396-355-8
No. 313 The Soft X-Ray Cosmos ROSAT Science Symposium (College Park, MD 1993)	94-72499	1-56396-327-2
No. 314 Advances in Plasma Physics Thomas H. Stix Symposium (Princeton, NJ 1992)	94-72721	1-56396-372-8
No. 315 Orbit Correction and Analysis in Circular Accelerators (Upton, NY 1993)	94-72257	1-56396-373-6
No. 316 Thirteenth International Conference on Thermoelectrics (Kansas City, Missouri 1994)	95-75634	1-56396-444-9
No. 317 Fifth Mexican School of Particles and Fields (Guanajuato, Mexico 1992)	94-72720	1-56396-378-7
No. 318 Laser Interaction and Related Plasma Phenomena 11th International Workshop (Monterey, CA 1993)	94-78097	1-56396-324-8
No. 319 Beam Instrumentation Workshop (Santa Fe, NM 1993)	94-78279	1-56396-389-2
No. 320 Basic Space Science (Lagos, Nigeria 1993)	94-79350	1-56396-328-0
No. 321 The First NREL Conference on Thermophotovoltaic Generation of Electricity (Copper Mountain, CO 1994)	94-72792	1-56396-353-1
No. 322 Atomic Processes in Plasmas Ninth APS Topical Conference (San Antonio, TX)	94-72923	1-56396-411-2

	Title	L.C. Number	ISBN
No. 323	Atomic Physics 14 Fourteenth International Conference on Atomic Physics (Boulder, CO 1994)	94-73219	1-56396-348-5
No. 324	Twelfth Symposium on Space Nuclear Power and Propulsion (Albuquerque, NM 1995)	94-73603	1-56396-427-9
No. 325	Conference on NASA Centers for Commercial Development of Space (Albuquerque, NM 1995)	94-73604	1-56396-431-7
No. 326	Accelerator Physics at the Superconducting Super Collider (Dallas, TX 1992-1993)	94-73609	1-56396-354-X
No. 327	Nuclei in the Cosmos III Third International Symposium on Nuclear Astrophysics (Assergi, Italy 1994)	95-75492	1-56396-436-8
No. 328	Spectral Line Shapes, Volume 8 12th ICSLS (Toronto, Canada 1994)	94-74309	1-56396-326-4
No. 329	Resonance Ionization Spectroscopy 1994 Seventh International Symposium (Bernkastel-Kues, Germany 1994)	95-75077	1-56396-437-6
No. 330	E.C.C.C. 1 Computational Chemistry F.E.C.S. Conference (Nancy, France 1994)	95-75843	1-56396-457-0
No. 331	Non-Neutral Plasma Physics II (Berkeley, CA 1994)	95-79630	1-56396-441-4
No. 332	X-Ray Lasers 1994 Fourth International Colloquium (Williamsburg, VA 1994)	95-76067	1-56396-375-2
No. 333	Beam Instrumentation Workshop (Vancouver, B. C., Canada 1994)	95-79635	1-56396-352-3
No. 334	Few-Body Problems in Physics (Williamsburg, VA 1994)	95-76481	1-56396-325-6
No. 335	Advanced Accelerator Concepts (Fontana, WI 1994)	95-78225	1-56396-476-7 (Set) 1-56396-474-0 (Book) 1-56396-475-9 (CD-Rom)
No. 336	Dark Matter (College Park, MD 1994)	95-76538	1-56396-438-4
No. 337	Pulsed RF Sources for Linear Colliders (Montauk, NY 1994)	95-76814	1-56396-408-2

	Title	L.C. Number	ISBN
No. 338	Intersections Between Particle and Nuclear Physics 5th Conference (St. Petersburg, FL 1994)	95-77076	1-56396-335-3
No. 339	Polarization Phenomena in Nuclear Physics Eighth International Symposium (Bloomington, IN 1994)	95-77216	1-56396-482-1
No. 340	Strangeness in Hadronic Matter (Tucson, AZ 1995)	95-77477	1-56396-489-9
No. 341	Volatiles in the Earth and Solar System (Pasadena, CA 1994)	95-77911	1-56396-409-0
No. 342	CAM -94 Physics Meeting (Cacun, Mexico 1994)	95-77851	1-56396-491-0
No. 343	High Energy Spin Physics Eleventh International Symposium (Bloomington, IN 1994)	95-78431	1-56396-374-4
No. 344	Nonlinear Dynamics in Particle Accelerators: Theory and Experiments (Arcidosso, Italy 1994)	95-78135	1-56396-446-5
No. 345	International Conference on Plasma Physics ICPP 1994 (Foz do Iguaçu, Brazil 1994)	95-78438	1-56396-496-1
No. 346	International Conference on Accelerator-Driven Transmutation Technologies and Applications (Las Vegas, NV 1994)	95-78691	1-56396-505-4
No. 347	Atomic Collisions: A Symposium in Honor of Christopher Bottcher (1945-1993) (Oak Ridge, TN 1994)	95-78689	1-56396-322-1
No. 348	Unveiling the Cosmic Infrared Background (College Park, MD, 1995)	95-83477	1-56396-508-9
No. 349	Workshop on the Tau/Charm Factory (Argonne, IL, 1995)	95-81467	1-56396-523-2
No. 350	International Symposium on Vector Boson Self-Interactions (Los Angeles, CA 1995)	95-79865	1-56396-520-8
No. 351	The Physics of Beams Andrew Sessler Symposium (Los Angeles, CA 1993)	95-80479	1-56396-376-0
No. 352	Physics Potential and Development of $\mu^+ \mu^-$ Colliders: Second Workshop (Sausalito, CA 1994)	95-81413	1-56396-506-2
No. 353	13th NREL Photovoltaic Program Review (Lakewood, CO 1995)	95-80662	1-56396-510-0
No. 354	Organic Coatings (Paris, France, 1995)	96-83019	1-56396-535-6

	Title	L.C. Number	ISBN
No. 355	Eleventh Topical Conference on Radio Frequency Power in Plasmas (Palm Springs, CA 1995)	95-80867	1-56396-536-4
No. 356	The Future of Accelerator Physics (Austin, TX 1994)	96-83292	1-56396-541-0
No. 357	10th Topical Workshop on Proton-Antiproton Collider Physics (Batavia, IL 1995)	95-83078	1-56396-543-7
No. 358	The Second NREL Conference on Thermophotovoltaic Generation of Electricity	95-83335	1-56396-509-7
No. 359	Workshops and Particles and Fields and Phenomenology of Fundamental Interactions (Puebla, Mexico 1995)	96-85996	1-56396-548-8
No. 360	The Physics of Electronic and Atomic Collisions XIX International Conference (Whistler, Canada, 1995)	95-83671	1-56396-440-6
No. 361	Space Technology and Applications International Forum (Albuquerque, NM 1996)	95-83440	1-56396-568-2
No. 362	Two-Center Effects in Ion-Atom Collisions (Lincoln, NE 1994)	96-83379	1-56396-342-6
No. 363	Phenomena in Ionized Gases XXII ICPIG (Hoboken, NJ, 1995)	96-83294	1-56396-550-X
No. 364	Fast Elementary Processes in Chemical and Biological Systems (Villeneuve d'Ascq, France, 1995)	96-83624	1-56396-564-X
No. 365	Latin-American School of Physics XXX ELAF Group Theory and Its Applications (México City, México, 1995)	96-83489	1-56396-567-4
No. 366	High Velocity Neutron Stars and Gamma-Ray Bursts (La Jolla, CA 1995)	96-84067	1-56396-593-3
No. 367	Micro Bunches Workshop (Upton, NY, 1995)	96-83482	1-56396-555-0
No. 368	Acoustic Particle Velocity Sensors: Design, Performance and Applications (Mystic, CT, 1995)	96-83548	1-56396-549-6
No. 369	Laser Interaction and Related Plasma Phenomena (Osaka, Japan 1995)	96-85009	1-56396-445-7
No. 370	Shock Compression of Condensed Matter-1995 (Seattle, WA 1995)	96-84595	1-56396-566-6